S0-DFE-277

SOCIETY FOR NEUROSCIENCE SYMPOSIA

Volume IV

SOCIETY FOR NEUROSCIENCE

SOCIETY FOR NEUROSCIENCE SYMPOSIA

Volume IV

ASPECTS OF DEVELOPMENTAL NEUROBIOLOGY

James A. Ferrendelli, M.D., Editor
Department of Pharmacology and Neurology
Washington University School of Medicine

Gerry Gurvitch, Assistant Editor
Society for Neuroscience

Published by the Society for Neuroscience
Bethesda, Maryland

Library of Congress Catalog Card Number 75-27110

International Standard Book Number 0-916110-09-5

Papers presented at five symposia during the
Eighth Annual Meeting of the
Society for Neuroscience, held in St. Louis, Missouri,
November 1978.

9650 Rockville Pike, Bethesda, MD 20014

Printed in the United States of America

PREFACE

Interest in the neurosciences continues to grow at a rapid rate. With this increasing interest the population of neuroscientists has increased markedly, research has expanded, and the body of scientific literature on the nervous system has become so large that it is impossible for any person to peruse even a small fraction. Now, more than ever, it is necessary to have available reviews on various aspects of neurobiology so that one can, without too much difficulty, venture out from his or her own small area of interest or research. We think that the *Society for Neuroscience Symposia* volumes meet this need to some extent, and their enthusiastic reception confirms our belief.

The present volume, *Aspects of Developmental Neurobiology*, is the fourth in a series that began in 1976. Like its predecessors, Volume IV consists of selected symposia presented at the preceding Society for Neuroscience Annual Meeting. Also, as in previous years, only a few of the several excellent symposia presented at the November 1978 meeting in St. Louis could be published because of limitations of time and space. The five symposia selected cover subjects not considered in previous volumes, and with only slight stretching, all fall under the rubric of "Developmental Neurobiology."

Obviously, only a few aspects of the several areas that make up developmental neurobiology are considered. However, all of the subjects covered here are areas of research where important advances have been made recently; these, in turn, have provided new insights into nervous system structure and function. The topics range from detailed molecular mechanisms to organization of complex nervous systems and include cell differentiation and cellular interactions, as well as several other subjects. The inclusion of such a wide variety of topics demonstrated relationships that I, for one, did not previously appreciate.

The excellent quality of the content of this book is a direct result of the efforts of Drs. Xandra O. Breakefield, Ray D. Lund, Paul H. Patterson, Jack Diamond, and Peter S. Spencer, who organized and chaired the five symposia. The Communication Committee wishes to express its gratitude to these organizers and to the individual contributors of each symposium. Our thanks also go to Mrs. Marjorie Wilson, Executive Secretary of the Society, and to Ms. Gerry Gurvitch, Assistant Editor of the symposia volumes—the two individuals who deserve most of the credit for the organization and production of each volume. Finally, we thank Ms. Dorothy A. Kinscherf for help with editing.

James A. Ferrendelli
St. Louis, April 1979

CONTENTS

PARTICIPANTS

Albert J. Aguayo
Division of Neurology
Montreal General Hospital
1650 Cedar Avenue
Montreal, Quebec H3G 1A4, Canada

Ira B. Black
Laboratory of Developmental Neurology
Cornell University Medical College
515 East 71st Street
New York, New York 10021

Mark M. Black
Department of Neuroscience
Children's Hospital Medical Center
300 Longwood Avenue
Boston, Massachusetts 02115

Garth M. Bray
Division of Neurology
Montreal General Hospital
1650 Cedar Avenue
Montreal, Quebec H3G 1A4, Canada

Xandra O. Breakefield
Department of Human Genetics
Yale University School of Medicine
333 Cedar Street
New Haven, Connecticut 06510

Marina Brunetti
Istituto di Chimica Biologica
Universita di Perugia
06100 Perugia, Italy

David E. Burstein
Department of Neuroscience
Children's Hospital Medical Center
300 Longwood Avenue
Boston, Massachusetts 02115

Richard M. Cawthon
Department of Human Genetics
Yale University School of Medicine
333 Cedar Street
New Haven, Connecticut 06510

Philippe Cochard
Laboratory of Developmental Neurology
Cornell University Medical College
515 East 71st Street
New York, New York 10021

Maria R. Castro Costa
Department of Human Genetics
Yale University School of Medicine
333 Cedar Street
New Haven, Connecticut 06510

Michael D. Coughlin
Laboratory of Developmental Neurology
Cornell University Medical College
515 East 71st Street
New York, New York 10021

David P. Cross
Department of Neurosciences
McMaster University
1200 Main Street West
Hamilton, Ontario L8S 4J9, Canada

Max S. Cynader
Department of Psychology
Dalhousie University
Halifax, Nova Scotia B3H 4J1, Canada

Luigi Di Giamberardino
Département de Biologie
Commissariat à l'Energie Atomique
C.E.N. de Saclay
B.P. 2, 91190 Gif sur Yvette, France

Bernard Droz
Département de Biologie
Commissariat à l'Energie Atomique
C.E.N. de Saclay
B.P. 2, 91190 Gif sur Yvette, France

Ian D. Duncan
Division of Neurology
Montreal General Hospital
1650 Cedar Avenue
Montreal, Quebec H3G 1A4, Canada

Susan B. Edelstein
Department of Human Genetics
Yale University School of Medicine
333 Cedar Street
New Haven, Connecticut 06510

Patricia M. Frair
Department of Neurosciences
McMaster University
1200 Main Street West
Hamilton, Ontario L8S 4J9, Canada

Jill C. Gardner
Department of Psychology
Dalhousie University
Halifax, Nova Scotia B3H 4J1, Canada

Lloyd A. Greene
Department of Neuroscience
Children's Hospital Medical Center
300 Longwood Avenue
Boston, Massachusetts 02115

Ralph J. Greenspan
Department of Biology
Brandeis University
Waltham, Massachusetts 02154

Jeffrey C. Hall
Department of Biology
Brandeis University
Waltham, Massachusetts 02154

William A. Harris
Department of Neurobiology
Harvard Medical School
25 Shattuck Street
Boston, Massachusetts 02115

Morris Hawkins, Jr.
Department of Microbiology
Howard University College of Medicine
Washington, D.C. 20059

Douglas R. Kankel
Department of Biology
Yale University
New Haven, Connecticut 06520

Herbert L. Koenig
Laboratoire de Cytologie
Université Pierre et Marie Curie
12 rue Cuvier
75005 Paris, France

Simon LeVay
Department of Neurobiology
Harvard Medical School
25 Shattuck Street
Boston, Massachusetts 02115

Raymond D. Lund
Department of Biological Structure
University of Washington School of Medicine
Seattle, Washington 98195

Jeffrey C. McGuire
Chemotherapy Fermentation Laboratory
Frederick Cancer Center
Frederick, Maryland 21701

Laurence J. McIntyre
Section on Myelin and Brain Development
Developmental and Metabolic Neurology Branch
National Institute of Neurological and Communicative Disorders and Stroke
National Institutes of Health
Bethesda, Maryland 20205

Donald E. Mitchell
National Vision Research Institute of Australia
386 Cardigan Street
Carlton, Victoria 3053, Australia

Michael J. Mustari
Department of Ophthalmology
University of Washington School of Medicine
Seattle, Washington 98195

John Palka
Department of Zoology
University of Washington
Seattle, Washington 98195

Paul H. Patterson
Department of Neurobiology
Harvard Medical School
25 Shattuck Street
Boston, Massachusetts 02115

C. Suzanne Perkins
Division of Neurology
Montreal General Hospital
1650 Cedar Avenue
Montreal, Quebec H3G 1A4, Canada

Alan C. Peterson
Department of Neurosciences
McMaster University
1200 Main Street West
Hamilton, Ontario L8S 4J9, Canada

Giuseppe Porcellati
Istituto di Chimica Biologica
Universita di Perugia
06100 Perugia, Italy

Richard H. Quarles
Section on Myelin and Brain Development
Developmental and Metabolic Neurology Branch
National Institute of Neurological and Communicative Disorders and Stroke
National Institutes of Health
Bethesda, Maryland 20205

Helen R. Rayburn
Department of Neurosciences
McMaster University
1200 Main Street West
Hamilton, Ontario L8S 4J9, Canada

Carla J. Shatz
Department of Neurobiology
Stanford University School of Medicine
Stanford, California 94305

Peter S. Spencer
Department of Neuroscience and Pathology
Albert Einstein College of Medicine
1410 Pelham Parkway South
Bronx, New York 10461

Nancy H. Sternberger
Cellular Neuropathology Section
Laboratory of Neuropathology and Neuroanatomical Sciences
National Institute of Neurological and Communicative Disorders and Stroke
National Institutes of Health
Bethesda, Maryland 20205

Michael P. Stryker
Department of Physiology
University of California School of Medicine
San Francisco, California 94143

Richard M. Weinshilboum
Department of Pharmacology
Mayo Foundation
Rochester, Minnesota 55901

GENETIC CONTROL OF NEUROTRANSMITTER METABOLISM

NEURAL DEFECTS INDUCED BY GENETIC MANIPULATION OF ACETYLCHOLINE METABOLISM IN *DROSOPHILA*

Jeffrey C. Hall, Ralph J. Greenspan, and Douglas R. Kankel

Brandeis University, Waltham, Massachusetts, and Yale University, New Haven, Connecticut

INTRODUCTION

Genetic perturbations have potential as precise and powerful probes in studies of the nervous system, particularly in cases where the mutant gene product has been identified. For a known product, specific questions can be asked about the consequences of its elimination. We have set out to find mutant genes for enzymes of neurotransmitter metabolism in *Drosophila melanogaster* for the express purpose of applying genetic lesions in these essential functions and monitoring their impact on several aspects of the animal's nervous system. In particular, our intention was to probe the requirement of normal transmitter metabolism and function for the development and maintenance of neuroanatomical structures and to use highly localized lesions in transmitter metabolism to perturb small and defined parts of the brain in order to test for their involvement in a variety of functions.

We have focused in particular on the enzymes of acetylcholine (ACh) metabolism because ACh appears to be a major neurotransmitter present and functional in the central nervous system (CNS) of insects and other invertebrates (Pitman, 1971; Gerschenfeld, 1973; Callec, 1974; Kehoe and Marder, 1976). Evidence in *Drosophila melanogaster* suggesting that ACh could be a neurotransmitter comes from (1) descriptions of the ontogeny of the synthesizing enzyme choline acetyltransferase (CAT) (Dewhurst, McCaman, and Kaplan, 1970; Dewhurst and

Seecof, 1975) and the purification of the CAT protein (Driskell, Weber, and Roberts, 1978); (2) studies of the degradative enzyme acetylcholinesterase (AChE) at different stages of the life cycle and in specific cell and tissue types (Dewhurst et al., 1970; Dewhurst and Seecof, 1975; Hall and Kankel, 1976; Best-Belpommé and Courgeon, 1977; Cherbas, Cherbas, and Williams, 1977; Deak, 1977) and (3) in vivo and in vitro investigations of receptors that bind cholinergic ligands, either nicotinic ones (Hall and Teng, 1975; Dudai, 1977*a*, 1978; Dudai and Amsterdam, 1977; Schmidt-Nielsen, Gepner, Teng, and Hall, 1977; Rudloff, 1978) or those that are muscarinic (Dudai and Ben-Barak, 1977; Haim, Nahum, and Dudai, 1979).

As a first step toward the goal of using lesions in the ACh system to probe neural phenomena, we isolated and are analyzing genetic variants affecting AChE and CAT. We have found that there is one gene in *D. melanogaster* for each enzyme and have isolated mutations in each of the two genes. Some of the induced mutations are conditional, thus permitting manipulation of the enzyme levels at various stages of ontogeny. We have further examined ACh metabolism mediated by these enzymes through the use of special genetic techniques that eliminate AChE or CAT activity from limited portions of the CNS. The results of these experiments show that AChE and CAT activities are essential to the animal's survival and proper functioning. Also, studies of viable flies missing AChE or CAT in only part of the CNS have revealed informative and sometimes unexpected deviations from normal processes. For example, an AChE-minus genotype and phenotype can lead to marked structural damage to the *Drosophila* nervous system. In using these genetic tools as probes of the functional anatomy of the fly brain, we have found that a local block in ACh metabolism can produce anomalies in courtship and visual optomotor behavior not predictable from previous methods of study.

MUTATIONS IN THE AChE AND CAT GENES

The genes coding for the two key enzymes in ACh metabolism were identified by techniques specifically designed to focus on the gene loci. Most genes coding for known enzymes in *Drosophila* have been localized through mapping of electrophoretic (allozyme) variants (reviewed by O'Brien and MacIntyre, 1978). For AChE, no electrophoretic variants were ever detected, in spite of extensive searches (Kojima, Gillespie, and Tobari, 1970; Kankel and DeNiro, unpublished). Yet, an enzyme gene can be putatively identified in this organism, without any initially detectable allelic variation, by a technique called "segmental aneuploidy" (Lindsley, Sandler, Baker, Carpenter, Denell, Hall, Jacobs, Miklos, Davis, Gethmann, Hardy, Hessler, Miller, Nozawa, Parry, and Gould-Somero, 1972).

The general strategy of enzyme gene identification using aneuploids is reviewed by O'Brien and MacIntyre (1978) and is diagrammed schematically in Figure 1*A*. Kankel and Hall (1976) used these techniques in *D. melanogaster* to find one segment on chromosome *3* (one of the three autosomes) for which AChE activity levels responded to changes in gene dosage. In spectrophotometrically measured AChE activity, this segment showed about 40% increase or decrease for duplication or deletion, respectively (Figure 1*B*); no other part of the genome showed a dosage effect. Subsequent assays of smaller deletions partitioned the segment initially discovered and narrowed the AChE-coding region down to 3–5 polytene chromosome "bands" (i.e., 87E1-87E5 in Figure 1*B*). The *Cat* gene was localized by Greenspan (1979) using the same aneuploid strategy and a radiometric assay for the transferase activity (Dewhurst and Seecof, 1975; Fonnum, 1975). The segment coding for this ACh-synthesizing enzyme was finally pinned down to a 15-band region, also on chromosome *3*, but at least 230 bands away from the AChE-coding segment (Figure 1*C*).

Mutations in the genes coding for AChE and CAT were isolated through techniques that depended on the initial cytogenetic mapping. It was assumed that a mutation eliminating either of these enzymes would be lethal when homozygous, and that such a lethal mutation would have subnormal activity when heterozygous with a normal allele, as does a deletion heterozygote. Indeed, Hall and Kankel (1976) found four lethal mutations that mapped within the 3–5-band region for AChE on chromosome *3*; each mutation reduced the enzyme activity to 50–60% of normal in homogenates of *lethal*/+ flies. These four allelic mutations thus defined the *Ace* locus. Subsequently, 22 additional *Ace* mutations have been induced by screening chemically mutagenized third chromosomes for lethality when heterozygous with either a previously identified *Ace*-lethal or an *Ace*-deletion (Greenspan, Finn, and Hall, 1979).

Greenspan (1979) has induced mutations in the *Cat* gene by using a chemical mutagen. Of the many lethal mutations that were detected initially by their lethality when heterozygous with a deletion of the CAT-controlling region of the third chromosome (Figure 1*C*), only four

A. Gene localization by aneuploidy

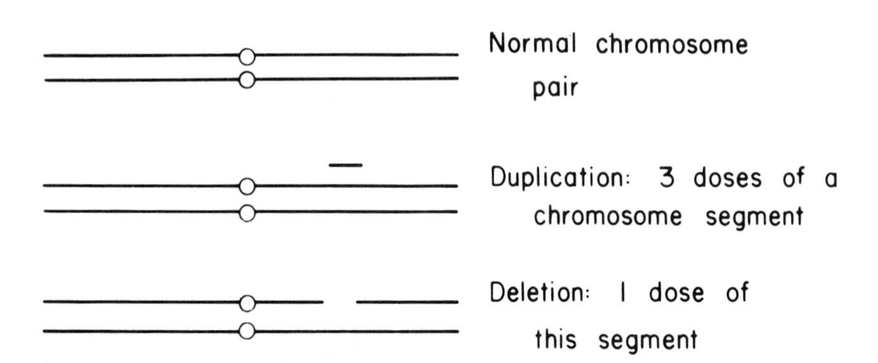

B. AChE gene localization

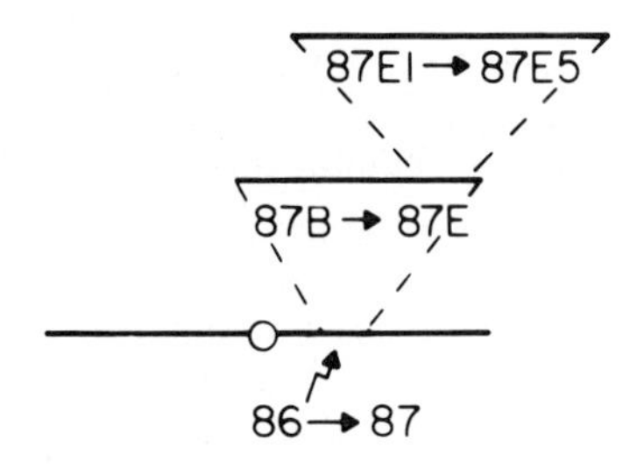

C. CAT gene localization

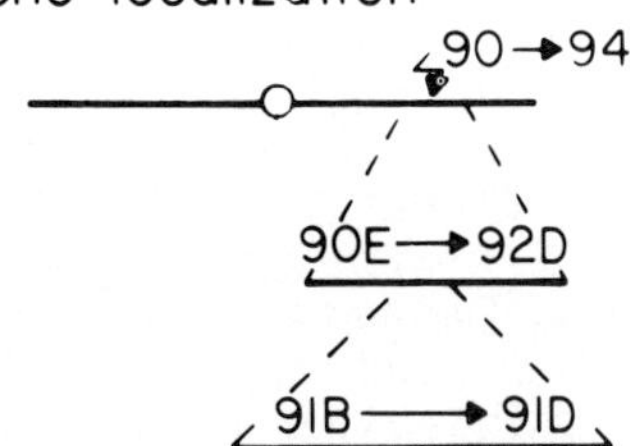

FIGURE 1. Scheme for localization of genes for AChE and CAT by aneuploidy. *A*: Doses of chromosome segments in duplicated or deleted flies. A series of duplications and deletions are constructed for nearly all the different subsegments of the four chromosomes of *D. melanogaster*. A schematic example of a duplication and a deletion of one such subsegment is shown. Duplication of a segment gives ca. 40% increase in activity of an enzyme whose structural locus is in the segment; deletion of a segment gives about a 40% decrease. *B*: Localization by dosage sensitivity of structural gene for AChE to successively smaller segments of the right

had half-normal CAT activity in heterozygous *lethal*/+ flies. These mutations, which are allelic, define the *Cat* locus.

It should be noted parenthetically that other enzyme genes in *Drosophila* have been identified using the segmental aneuploid technique (reviewed by O'Brien and MacIntyre, 1978). Of potential interest for neurochemical work not involving ACh is the identification and subsequent isolation of mutants in the dopa decarboxylase gene (Wright, Bewley, and Sherald, 1976; Wright, Hodgetts, and Sherald, 1976). Also, L. M. Hall and colleagues (unpublished) have searched the genome for chromosome regions that affect levels of quantitatively detectable α-bungarotoxin binding. No such regions were found, though other methods are under way to identify genes coding for or influencing the putative ACh receptors in *Drosophila* (Hall, von Borstel, Osmond, Hoeltzli, and Hudson, 1978).

Conditional mutations are valuable because they permit gene functions to be "turned off" or "turned on" in a way that allows tests of such functions during particular periods of the ontogeny. Thus, we screened for *Ace* mutations and *Cat* mutations at an elevated temperature (29° C), hoping that some of the lethals could be viable at a lower temperature (e.g., 18–22° C). Indeed, two of the four *Cat* mutations met this criterion, opening the possibility of directly manipulating transmitter synthesis (cf. manipulations of a wide variety of temperature-sensitive mutants in *Drosophila*, reviewed by Suzuki, Kaufman, and Falk, 1976). For the *Ace* locus, only one new mutation was temperature-sensitive by itself. This has a cold-sensitive lethal phenotype: it is viable if development occurs at 29° C, but dies in the embryonic stage when developing at 18° C. The cold-sensitive mutation did not prove to have a cold-labile AChE molecule in vitro, nor could behavioral defects be induced by cold treatment of adults that had matured at the permissive (high) temperature. This mutation may have a cold-sensitive defect in the assembly of AChE subunits or a defect in putative packaging or sequestering of the active protein into membranous components of neurons (Dudai, 1977*b*). These arguments are made by analogy to cold-sensitive assembly processes (vs. catalytic ones) known from

arm of chromosome *3*. Numbers and letters labeling each segment refer to the standard map of *Drosophila* chromosome bands drawn from the polytene chromosomes in the larval salivary glands (Lindsley and Grell, 1968). *C*: Localization by dosage sensitivity of structural gene for CAT to successively smaller segments of the right arm of chromosome *3*. Segments labeled as in *B*.

protein work in other organisms (Olmsted and Borisy, 1973; Nomura, Morgan, and Jaskunas, 1977).

The molecular properties of the somewhat unorthodox cold-sensitive *Ace* mutant have not been investigated, because it was most desirable to obtain mutant *Ace* genes whose products could be turned off by in vivo or in vitro temperature treatments. Yet no such mutants were initially obtained. We remembered, though, that AChE is a multimeric protein in many organisms (Leuzinger, Goldberg, and Cauvin, 1969; reviewed by Silver, 1974) and that this could be the case in *Drosophila* as well, since the molecular weight of the smallest detectable active molecule is 200,000–300,000 (Dudai, 1977*b*; Kankel, unpublished). These facts led to the suggestion that there might be cases of intragenic complementation for some of the heteroallelic *Ace* combinations, of which we could make several hundred, given the many extant *Ace* alleles. That is, differentially mutant subunits might cooperate to form an active multimeric molecule. Some cases of such complementation might be temperature-sensitive, by analogy to similar biochemical genetic results from other organisms (Schlesinger and Levinthal, 1965; Ullmann and Perrin, 1970). Indeed, six of the relevant *Ace* heterozygotes were viable at low temperature but completely lethal at high temperature (Greenspan et al., 1979). Two of these genotypes had 70–80% of normal viability when grown at 18° C but were totally lethal when grown at 29° C. These mutant combinations both involve a particular *Ace* allele, j40. Thus, Ace^{j40}/Ace^{j19} and Ace^{j40}/Ace^{j50} flies survive at low but not high temperature.

BASIC PHENOTYPE OF MUTANTS WITH ALTERED ACETYLCHOLINE METABOLISM

The lethal stage for presumptive AChE-minus and CAT-minus mutants is at the end of the embryonic period (which lasts 1 day at 25° C). This means that the mutant embryos do not hatch into the larval stage, when coordinated movement is first required. The early lethality was assessed in mutants carrying an *Ace*-lethal or a *Cat*-lethal in heterozygous condition with a deletion of the relevant locus, or another independently derived lethal mutation at the locus. This method of testing the lethals was desirable because these chromosomes carrying the enzyme mutations have been mutagenized, and it is very likely that at least one additional lethal mutation was induced elsewhere on each treated chromosome. This means that an *Ace*-lethal homozygote, for example, would be gratuitously homozygous for another deleterious mutation, which could ruin the interpretation of abnormalities in these genotypes (e.g., lethal phase, defects in nerve structure or function). Heterozygotes involving two *Ace* variants, though, will express recessive factors only at the *Ace* locus.

At the lethal stage in an *Ace* mutant or a *Cat* mutant, there is no detectable activity for the respective enzyme. For AChE, sections from late embryos or early larvae that are wild-type have easily demonstrable AChE activity (Figure 2*A*), yet no AChE can be seen in a lethal mutant (Figure 2*B*). These sections were stained for AChE activity (Hall and Kankel, 1976). More sensitive quantitative assays on one particular *Ace*-lethal/deletion combination showed no detectable enzyme activity as compared to control assays without embryos. This kind of assay (Dewhurst and Seecof, 1975) had been used earlier to show that AChE activity first makes its appearance at about 30% of the way through embryogenesis and increases steadily until the end of the embryonic period (Poulson and Boell, 1946; Dewhurst et al., 1970; Dewhurst and Seecof, 1975). The sections from mutants show that the animal has been able to assemble a grossly normal CNS in the absence of AChE throughout the 16 hours or so when it is usually present, since the volume and basic form of the brain and ventral ganglion do not appear to be abnormal (Figure 2*B*).

Ace-lethal embryos, carrying a temperature-sensitive genotype and grown to the end of embryogenesis at a permissive temperature (18° C), can have all noticeable AChE activity turned off by heat treatment (45° C) of the sections before staining (Greenspan et al., 1979), while similarly treated wild-type embryos still have easily detectable activity after such treatment. The AChE molecule also appears to be thermolabile in homogenates made from adults—carrying temperature-sensitive alleles and grown permissively—and heated in vitro before quantitative assays (Figure 3). We have also shown that five *Ace*-lethal mutations, when heterozygous with the *wild-type allele*, have thermolabile in vitro enzyme activity (40–50° C) that is again much more sensitive than normal enzyme. This is found even for three of the *Ace*-lethals that are not temperature-sensitive for viability under any circumstances, i.e., when heterozygous with any other *Ace* mutants. A fly that is heterozygous for a deletion of the *Ace* locus and for the wild-type gene does *not* have a thermolabile AChE. These findings reinforce the

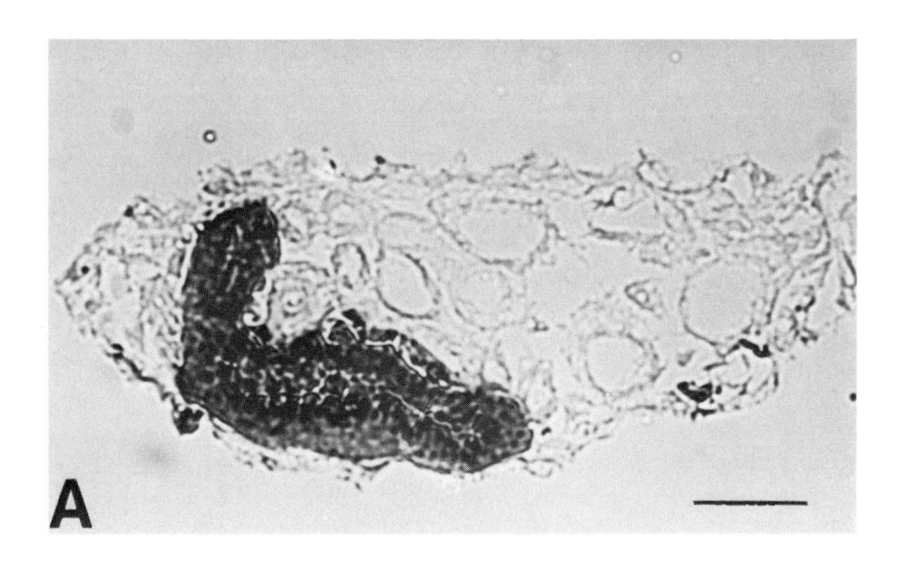

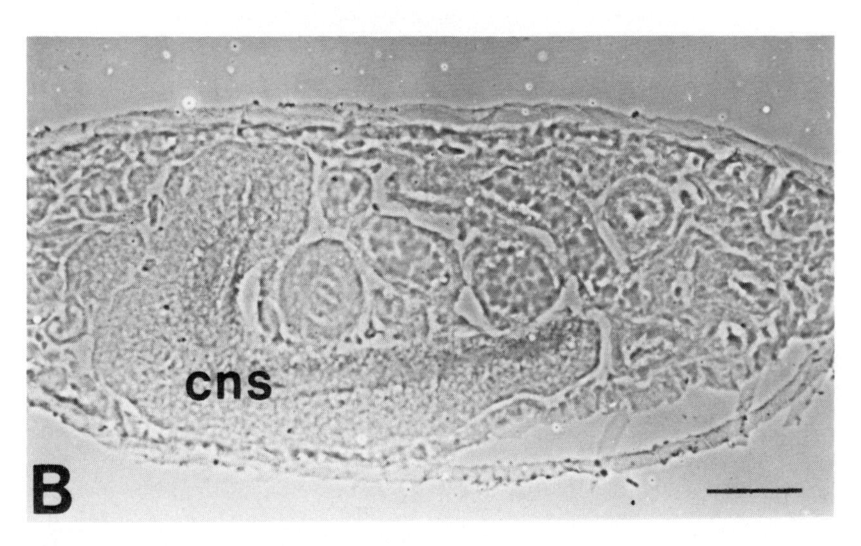

FIGURE 2. AChE activity in early development. *A*: Wild-type early first instar larva stained for AChE activity. *B*: *Ace*^{j40}/*Ace*^{j19} late embryo grown at nonpermissive temperature (29° C) and stained for AChE activity. The CNS in the mutant embryo is present but does not stain for AChE. Bars = 50 μm.

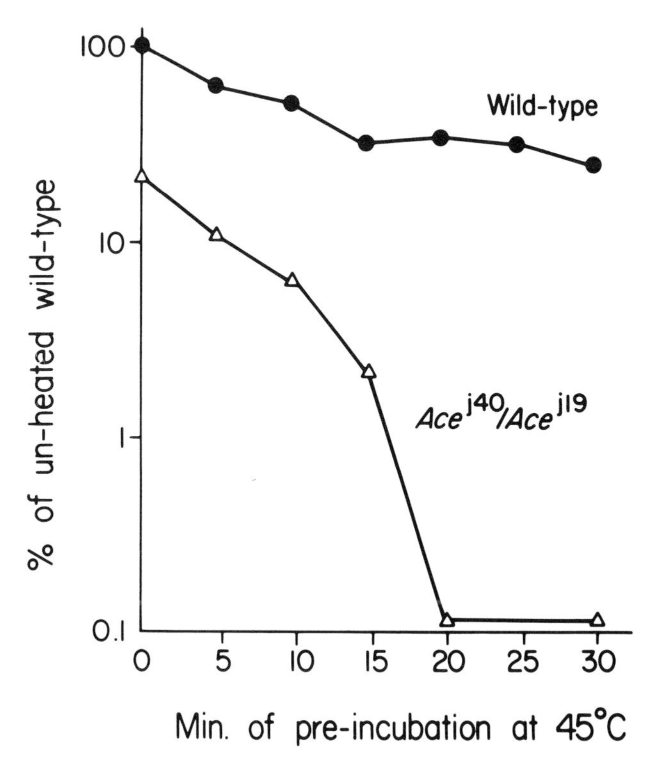

FIGURE 3. Thermal inactivation of AChE activity in temperature-sensitive *Ace* mutants and wild-types. Mutants of the temperature-sensitive combination Ace^{j40}/ Ace^{j19} were raised at 18° C, and a homogenate made of their heads was incubated at 45° C. Aliquots were taken from this homogenate at the times indicated and assayed for AChE activity at room temperature. Similar assays were performed on a homogenate of wild-type heads. AChE activity was measured as the change in OD_{412}/min using the assay of Ellman, Courtney, Andres, and Featherstone (1961) with acetylthiocholine as substrate. Enzyme activities are plotted as the log % wild-type activity with no heat incubation. For details see Greenspan et al. (1979).

idea that AChE is a multimer. Any molecule from an *Ace*-lethal/+ fly that has a mixture of mutant and wild-type subunits may thus be abnormal under appropriate conditions.

In studies of the transferase mutants, *Cat*-lethal embryos also have no detectable activity for this enzyme. This was shown by making a stock of uniform genotype, i.e., homozygous for a temperature-sensitive *Cat*-lethal. This is not possible for *Ace*-lethals, since their temperature-sensitive genotypes are heterozygotes, and the relevant stocks (carrying two different *Ace*-lethals) must be kept heterozygous with "balancer" chromosomes (marked, inverted chromosomes carrying the normal enzyme gene; see Figure 7). For the *Cat* mutants, however, no cases of intragenic complementation were found; this is as expected, since the enzyme is a monomer (Driskell et al., 1978). The *Cat*-temperature-sensitive animals were grown to the end of the embryonic period under nonpermissive conditions (30° C); then the embryos were assayed quantitatively (Fonnum, 1975). They had no detectable CAT activity compared with the readily demonstrated enzyme activity from wild-type embryos (Greenspan, 1979). This assay had been used earlier to show that CAT is first detectable about 4 hours after the first embryonic appearance of AChE (Dewhurst and Seecof, 1975).

When *Cat*-temperature-sensitive animals are grown to adulthood under permissive conditions (18° C), CAT activity is thermolabile in vitro. The mutant activities fall rather faster than do wild-type activities after homogenates are heated at 38° C (Figure 4).

The *Cat*-mutant embryos, grown to the end of that stage without any transferase activity, have a CNS with ostensibly normal morphology, as is also found for *Ace*-lethal mutants. Yet we realized that only a small part of the eventual adult nervous system is assembled during embryogenesis (White and Kankel, 1978), and that we have asked for the normal morphology to be maintained for only a short period of time in the embryonic lethals. Thus it was imperative to assess postembryonic effects of these genes on various properties of the CNS, including possible abnormalities of structure. First, we determined temperature-sensitive lethal periods (TSPs) at 29° C in *Ace*-temperature-sensitive combinations. These were determined by temperature-shift experiments. The TSP for the two relevant genotypes is at the end of embryogenesis (Figure 5*A*,*B*), the same stage as the non-conditional lethal phase. The TSPs are perhaps unexpected, because they mean, for instance, that all postembryonic shifts from 18° C to 29° C allow survival. Thus there is an absolute requirement for a low temperature only during the embryonic period, when AChE is first being synthesized, not during larval, pupal, and adult stages, despite the fact that the mutant enzyme in adults is itself thermolabile (see above). A postembryonic defect of an *Ace*-mutant genotype is, however, shown in the temperature-shift experiment involving one of the temperature-sensitive combinations, Ace^{j40}/Ace^{j19}. Here, development throughout at

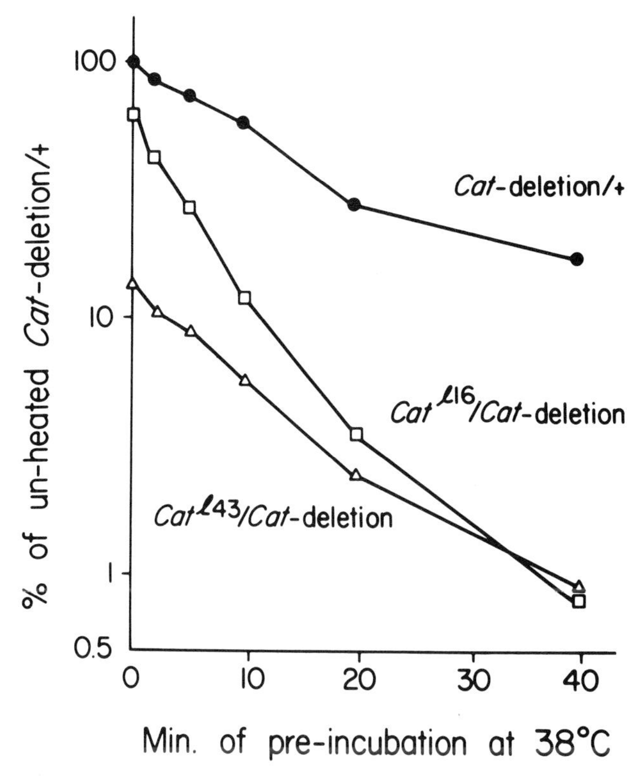

FIGURE 4. Thermal inactivation of CAT activity in temperature-sensitive *Cat* mutants heterozygous for the *Cat*-deletion, and in wild-type flies heterozygous for the same deletion. Flies of the genotypes *Cat*^{ℓ16}/*Cat*-deletion, *Cat*^{ℓ43}/*Cat*-deletion and *Cat*-deletion/+ were grown at 18° C and homogenates were made of their heads. Aliquots of each homogenate were incubated at 38° C in the assay mixture, minus [^{3}H]AcCoA substrate, for the indicated times. Then the substrate was added and the assays performed at 30° C for 15 min. Enzyme activities are plotted as the log % *Cat*-deletion/+ activity with no heat incubation. For details see Greenspan (1979).

18° C or development with very early shifts from 29° C to 18° C allows good survival (ca. 0.8 normal). Yet postembryonic shifts from 18° C to 29° C, while allowing non-zero survival, lead to quantitatively poor viability (Figure 5*A*). Moreover, the emerging adults are often sluggish in their movements and may die within a week or less after eclosion. These postembryonic effects of high temperature are not seen in the other *Ace* combination (Figure 5*B*).

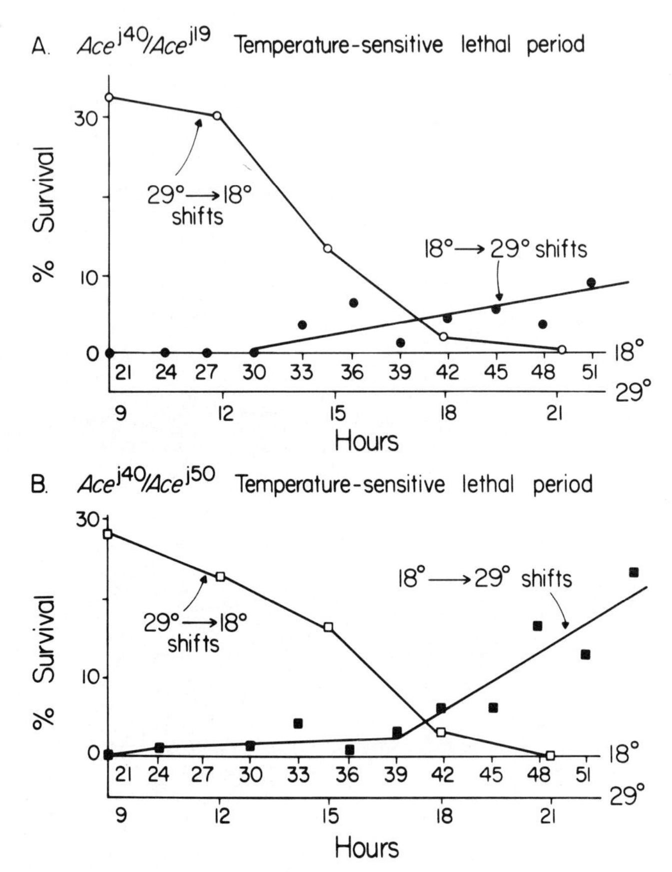

FIGURE 5. Temperature-sensitive lethal period of *Ace* mutants. *Ace*-temperature-sensitive mutant combinations were generated from crosses of (*A*) *Balancer/Ace*^{j40} to *Balancer/Ace*^{j19} or (*B*) *Balancer/Ace*^{j40} to *Balancer/Ace*^{j50}. This means that the maximum possible survival of *Ace*^{j40}/*Ace*$^{j19 \text{ or } 50}$ is 33% for any cross. Progeny from these crosses were collected, and one set of cultures was started at 18° C and subjected to a series of shifts to 29° C at 3-h intervals from the start to the end of embryogenesis (i.e., 0–48 h at 18° C), and thereafter was shifted at daily intervals through to adulthood. The other set of cultures was started at 29° C and shifted to 18° C at intervals of 3 h during embryogenesis (0–21 h at 29° C) and at daily intervals thereafter until adulthood. The percent of *Ace* mutant flies surviving to adulthood for each shift is plotted on the ordinate. The time scale for 18° C is shown above that for 29° C in each case. ● and ■: 18–29° C shifts; ○ and □: 29–18° C shifts. The time scale on the abscissa does not show the whole experiment: all points to the left of the origin at earlier times give zero survival for 18–29° C shifts, and give 25–35% survival for 29–18° C shifts, for either *A* or *B*.

Flies that develop to adulthood carrying the temperature-sensitive *Ace* alleles, but that are grown under permissive conditions, are debilitated by temperatures higher than those used to test TSPs. They pass out and die within 1–3 hours of treatment at 33° C (Figure 6). The wild-type is much less severely affected, since 80% are still viable after 10 hours of treatment. If an *Ace*-lethal fly that has stopped moving at 33° C is returned to 18° C immediately, it will not recover. The relatively long time required for heat-induced paralysis and death is in contrast to the much more rapid temperature-sensitive paralysis seen in other mutants of *Drosophila* (Suzuki, Grigliatti, and Williamson, 1971; Siddiqi and Benzer, 1976; Wu, Ganetzky, Jan, Jan, and Benzer, 1978). Also, whereas mutants in these other genes may lead to rapid passing out at 29° C, temperatures below 30–31° C have no noticeable effect on the *Ace* mutants. This is in distinct contrast to the total lethality caused by 29° C treatments of the *Ace* embryos, but is in keeping with the absence of blatant effects of 29° C on larvae and pupae.

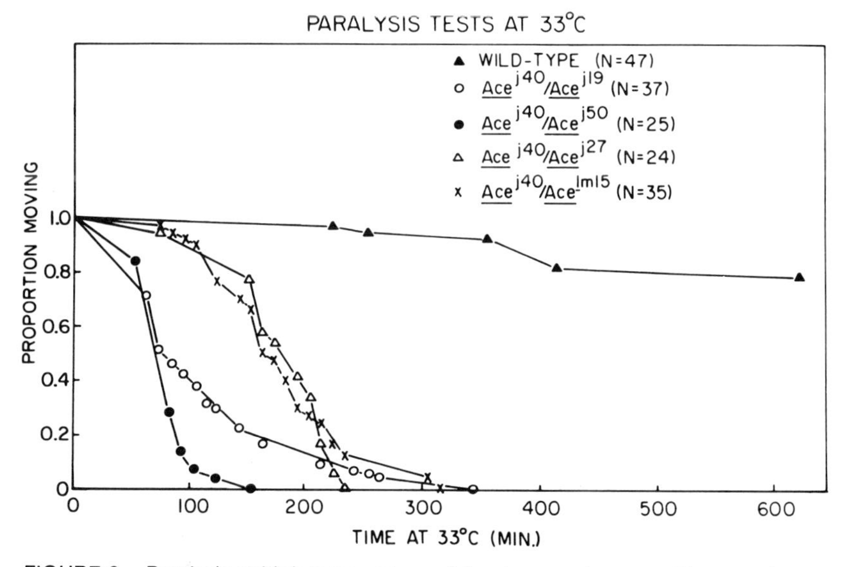

FIGURE 6. Paralysis at high temperature of *Ace*-temperature-sensitive genotypes. *Ace* mutant and wild-type flies were grown to adulthood at 18° C; each fly was placed in an individual test chamber; these small chambers were then heated by placing them in a larger chamber surrounded by circulating water at 33° C. The proportion of flies still moving was recorded at the indicated time intervals.

Heterozygotes involving an *Ace*-lethal (not necessarily one from a temperature-sensitive combination) and the wild-type gene are also heat-sensitive for paralysis, yet the amount of time required at 33° C to show these dominant effects of the mutants is about 4–5 hours. An *Ace*-deletion/+ heterozygote does not show paralysis any earlier than normal. Again, these dominant effects are comprehensible in light of the putative multimeric structure of AChE.

The two temperature-sensitive *Cat* mutants are also heat-sensitive paralytics. In this case, the embryonically lethal temperature of 29° C is sufficient to cause permissively grown adults to pass out. This requires 2–3 days of high-temperature treatment; the same treatment administered to wild-type adults has no effect. Moreover, the *Cat* mutants, having been returned relatively quickly to 18° C after loss of consciousness, will recover within 1 day. A 3–4-day 29° C treatment of such adults, though, is lethal to these conditional mutants.

In summary, we have learned that the *Ace* locus and the *Cat* locus each define essential genes, and that the function of these genes is required at a variety of stages of the life cycle. These loci are probably the only genes among the four chromosomes of *D. melanogaster* that code for AChE and CAT, respectively. This conclusion comes from the demonstration that virtually all activity of the relevant enzyme can be eliminated after appropriate assay of sections or homogenates. Also, these findings greatly facilitate the further analysis of acetylcholine metabolism, to be described below, since the appropriate genotypes for assessing the importance of these genes in different portions and various functions of the nervous system can be readily created.

ABNORMALITIES IN FLIES MOSAIC FOR ACETYLCHOLINE-METABOLIZING ENZYMES

Production of Mosaics for Acetylcholinesterase and Choline Acetyltransferase Activities

To examine postembryonic effects of altered ACh metabolism in detail, with respect to particular neural tissues and the behavior they control, we examined genetic mosaics missing AChE or CAT in only part of their adult nervous system. Although a fly completely devoid of the ACh-synthesizing or -degrading enzyme is dead, an individual missing either enzyme in a fraction of the nervous system may survive and have informative abnormalities of structure or function. Such

AChE-minus or CAT-minus mosaics were constructed by a scheme shown in Figure 7. In these crosses, somatic loss of an X-chromosome from an initially diplo-X zygote creates an XO cell line or clone in the mosaic. The chromosome loss is induced by the maternal effects of the mitotic loss (*mit*) mutation of Gelbart (1974); the resulting XO clones comprise about 10% of each mosaic, implying early loss of an X-chromosome, perhaps from one nucleus at the eight-nucleus stage (reviewed by Hall, Gelbart, and Kankel, 1976). The haplo-X parts of the mosaics, which are male in genotype, are also yellow in color externally and are missing AChE or CAT internally. The en-

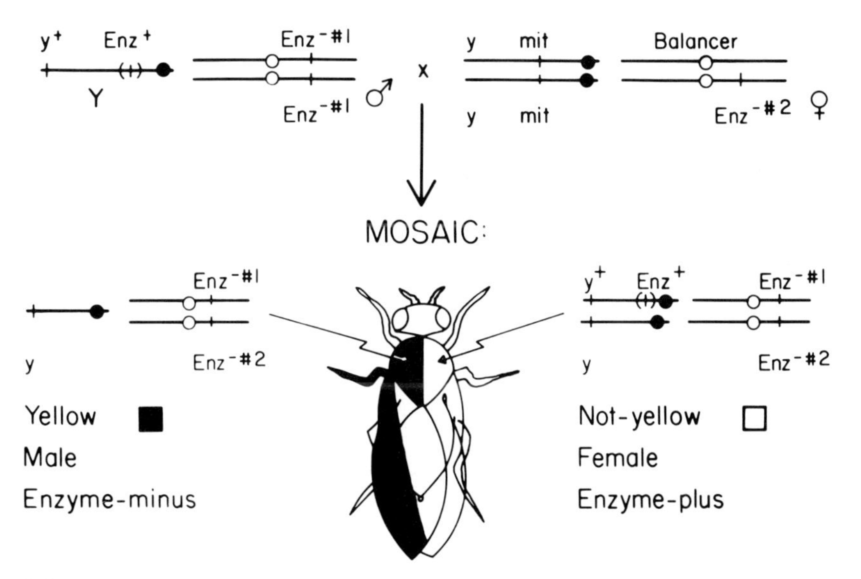

FIGURE 7. Scheme for generating genetic mosaics for enzyme mutations. The male parent donates an X-chromosome (centromere = solid circle) carrying an insertion of the wild-type enzyme locus (*Enz*$^{+}$) and the wild-type allele of the cuticle marker *y*$^{+}$ (not-yellow); and an autosome (centromere = open circle) carrying a mutant allele of the enzyme locus (*Enz*$^{-\#1}$). The female parent donates an X-chromosome carrying the mutant allele of the cuticle marker *y* (yellow) and an autosome carrying another mutant allele of the same enzyme locus (*Enz*$^{-\#2}$). Since the mother is homozygous for the mutation *mit* (mitotic-loss inducer) on her X-chromosomes, her progeny will have mitotically unstable chromosomes. Thus, a zygote that starts out as diplo-X can undergo loss of one of its X-chromosomes. If the father's *Enz*$^{+}$-bearing X is lost, it will be detectable by the fact that the recessive cuticle marker *y* will be uncovered. Patches of yellow cuticle on adults will be diagnostic of mosaicism in that animal, and some of these mosaics will have lost the X with *Enz*$^{+}$ in their nervous systems as well.

zyme-minus feature of the XO clones depends on the fact that the normal allele of Ace^{+} is inserted into an X-chromosome by a translocation (Hall and Kankel, 1976); the same is true for another X-chromosome that bears Cat^{+} (Oliver, 1937). Somatic loss of such an X-chromosome, from developing flies that also carry the appropriate third chromosomal mutants, is necessary to create these enzyme mosaics. The reason is that, while loss of an X-chromosome creates a quite viable mosaic (or gynandromorph), somatic loss, early in development, of a large autosome such as chromosome *3* is invariably lethal, irrespective of any lethal mutations uncovered on the remaining chromosome *3* (Wright, 1970). It should be reiterated that the mutant genotype that is "uncovered" in tissues that have lost the normal enzyme gene is comprised of two independently isolated *Ace* mutations or *Cat* mutations (i.e., $Enz^{-\#1}/Enz^{-\#2}$ in Figure 7). Thus, any neurological defects seen in the mosaics will be caused by recessive factors only in the ACh-metabolizing gene under study.

Morphological Abnormalities and Mapping of Lethal Defects

In the initial work in *Ace* mosaics, we isolated 155 gynandromorphs with patches of yellow, male tissue externally, of which 80 were shown to be mosaic for AChE when sectioned and stained for the enzyme. For these mosaics, it was quickly apparent that many parts of the CNS can tolerate an absence of enzyme activity from a portion of the cellular cortex of a ganglion. The enzyme-null clones tend to be relatively small, given the specific mosaic technique that was used and the possibility that most extensively null clones would be lethal or extremely abnormal (see below). Yet, in all 80 mosaics that have been serially sectioned and subjected to the usual enzyme stain (cf. Figure 2), all separate portions of the cortices in the brain and ventral ganglia have been found to contain AChE-minus clones. Since AChE is limited almost exclusively to the nervous system and is present throughout that system (Hall and Kankel, 1976; Deak, 1977), analysis of internal mosaicism has been restricted to neural tissues.

It was perhaps surprising that no individual portion of a ganglion must be AChE-plus in order that the fly survive to adulthood. Possibly then, the function of the enzyme is not as interesting and important as would have been expected. This notion was immediately undermined by the striking finding of morphological abnormalities in the ganglionic neuropil associated with essentially all patches of AChE-minus cortex (i.e., the rind of nerve cell bodies that surrounds the synaptic neuropil). An example from the brain is shown in Figure 8. Here, the anterior cortex on the right side is AChE-minus; the left side of the brain is entirely AChE-plus. The neuropil just beneath (posterior to) the null clone, which happens to be in the vicinity of the antennal nerve's input to the anterior brain, shows subjectively detected abnormalities: compacted, denser appearance; darker staining of the enzyme; and possibly disorganized geometry of gross axonal associations. Although the latter impression is quite subjective, it appears to be true no matter where in the CNS the mutant clone is found. The possibility of disarranged "wiring" in the neuropil is currently being examined in mosaics whose axons are silver-stained (Meyerowitz and Kankel, 1978). A more objective metric related to the condensed, compacted appearance of neuropil adjacent to null cortex came from the readily measured reduction in volume of a ganglion associated with cortical nullness, compared with the volume measured for the genetically normal ganglion in the opposite side of the bilaterally symmetrical CNS. These volume reductions (ranging from 12–18%) were measured in three mosaics selected for the analysis because they had fairly extensive amounts of null tissue unilaterally.

Another AChE mosaic in the CNS is shown in Figure 9, in which essentially the entire cortex on one side of the thoracic-abdominal ganglionic complex has the null phenotype; again, the morphology of the neuropil is clearly abnormal in comparison with the normal ganglia on the left side. It is apparent that the neuropil on the mutant side is heavily stained, implying the presence of many AChE-plus axons. Neuropil adjacent to null cortex is, in fact, usually stained in the *Ace* mosaics. This is at first confusing, since the nerve cell bodies adjacent to the right-hand neuropil are mutant, but probably occurs because most areas of neuropil in *Drosophila* and other Dipterans receive axonal input from several different cortical regions (Power, 1943; Strausfeld, 1976). Thus, in Figure 9, there are many normal Ace^{+} axons projecting from the left side to the mutant side.

It should be noted that the morphological abnormalities seen in association with the *Ace*-lethal genotype in mosaics were routinely observed with two different *Ace* mutations. Moreover, several *Ace* mutants were studied in another kind of genetic mosaic that has much smaller clones. This mosaic technique involved x-ray-induced mitotic recombination (reviewed by Becker, 1976) in developing animals that were heterozygous for a given *Ace* mutation. The resulting mutant

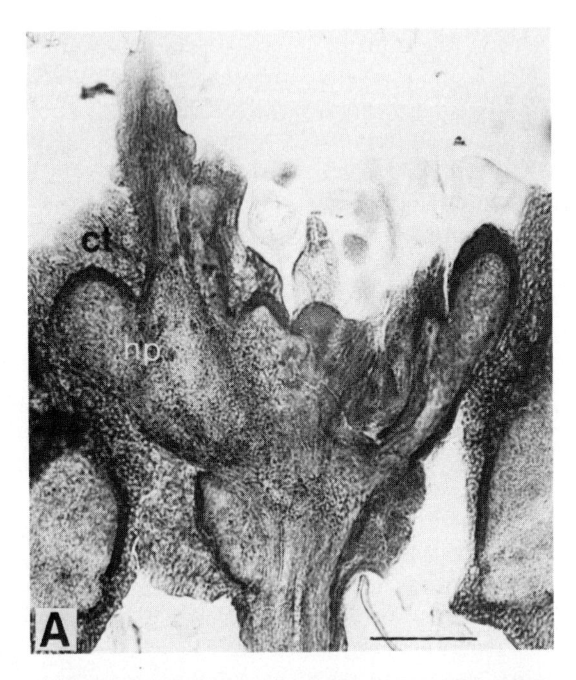
ct
np
A

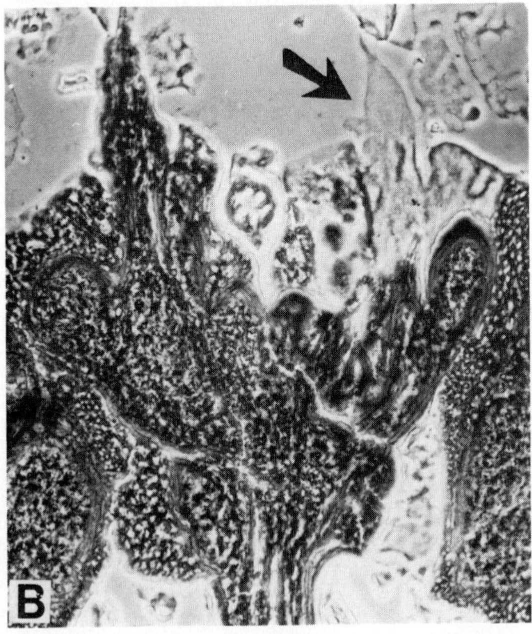
B

patches were induced during larval development, much later than the stage of *mit*-induced mosaicism; thus the patches of nonstaining cortex were small, but adjacent neuropil of abnormal appearance could still be seen (Ferrus and Kankel, unpublished).

The only ganglia in the adult's CNS that are not quite so abnormal in structure when AChE-minus are the optic lobes, four pair of ganglia flanking the brain throughout most of its dorsoventral distribution in the head capsule. Some mosaics have been detected with extensively null phenotypes in this portion of the CNS, in some cases including large areas of unstained neuropil (see below and Figure 11). These large null clones allowed survival, since visual functioning is not necessary for overt survival in *Drosophila* (Pak, 1975; Meyerowitz and Kankel, 1978). However, the extensive null areas were not grossly deformed. One remarkable AChE mosaic was found with a very large mutant clone in its brain: virtually the entire dorsal brain on one side was AChE-minus in the cortex and the neuropil (Figure 10*A*). In the ventral brain, the cortex was still extensively null, but there was appreciable staining in several synaptic areas in the neuropil. This mosaic was severely impaired in its overall behavior (see below). Its survival (for 6 days before sectioning and staining) is even more remarkable given the extreme morphological defects seen in the nonstaining neuropil. There are many holes in the brain (Figure 10*B*), suggesting degeneration even of the wild-type axons projecting into the mutant areas from the normal brain on the right side of this mosaic. A clone of this extent, with such widespread absence of AChE in the neuropil, was observed only once, as might be expected, since very poor survival is likely to be associated with such a widespread AChE deficit in the brain itself. Perhaps for this reason, neuropil phenotype of such an extreme degree was also seen only in this mosaic.

The rare mosaic with its large clone mutant for AChE probably represents the tail end of a distribution of mosaics with large patches, most of which failed to survive. For smaller AChE-minus clones, mo-

---

FIGURE 8. Ventral brain of an AChE mosaic adult shown in horizontal section and stained for AChE activity. *A*: Brightfield optics showing normal staining on the left side and a large mutant patch on the right anterior side, showing the characteristic morphology of the neuropil in AChE mosaics (see text). Cortex (ct) and neuropil (np) are labeled on the right side, and the darkly staining border between them can be seen all around the brain except in the AChE-null patch. *B*: Phase contrast optics of the same section indicating the presence of brain tissue on the right (AChE-null) side. Arrow points to mutant antennal nerve. Bar = 50 μm.

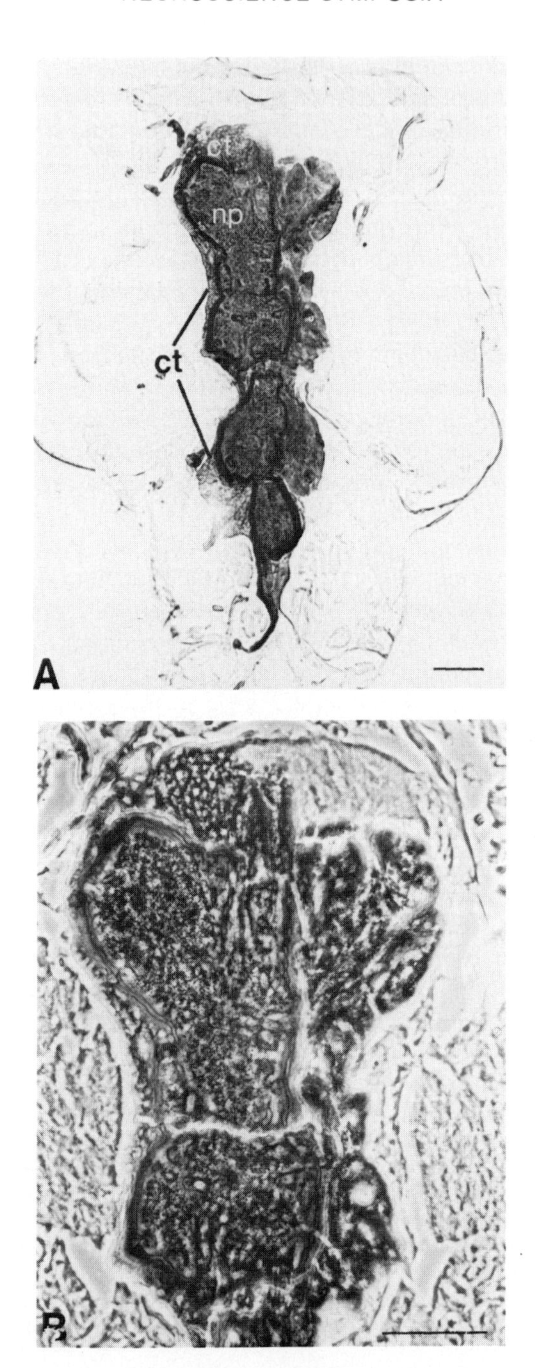
ct
np
ct
A
B

saics may initially have been generated with equal probabilities of all the progenitors of different parts of the CNS being mutant. However, there could have been highly nonrandom survival of different mosaics, depending on the particular part of the nervous system that was attempting to develop in the face of the defects of ACh metabolism. Furthermore, the morphological defects associated with mutant nerves could arise either early in neural development, or later, as a consequence of long-term deprivation of AChE during the continual cell differentiation and pattern formation occurring throughout the development of the nervous system in *Drosophila* (White and Kankel, 1978). Such defects, induced early or late, could kill particular classes of mosaics preferentially. AChE is distributed throughout the CNS in *Drosophila* and other insects (reviewed by Klemm, 1976), but ACh itself is not (Colhoun, 1959), so the *requirement* for AChE function may not be uniformly distributed in the CNS.

Against this background of facts and assumptions, lethality induced by *Ace* mutations was mapped within the nervous system. Simply put, the probabilities of different portions of the many ganglia in the head and thorax being AChE-minus were calculated for the surviving mosaics. The mean probability that any region of the CNS is unilaterally mutant is 0.06 ± 0.04. In contrast, the average mutant value for *external* structures in these same mosaics was higher (0.12 ± 0.03), as expected, since clones of yellow, *Ace*-minus male cuticular tissue are not lethal. The general pattern of surviving mosaics with internal AChE mosaicism is that null clones are much less frequent in some regions of the CNS than in others. First, and more generally, there is a large standard deviation in the average mutant value. This variation is not a consequence of the technique used to produce mosaics. In gynandromorphs marked with a nonlethal acid phosphatase internal marker (Kankel and Hall, 1976; Hall, 1977, 1979; Schilcher and Hall, 1979), different parts of the CNS had very little variation in mutant probabilities, i.e., 0.29 ± 0.02, almost the same as the mean value for their external tissues. (These mosaics have a higher mean probability of mutant tissue because their

FIGURE 9. Thoracic-abdominal nervous system of an AChE mosaic in horizontal section, stained for AChE activity. *A*: Brightfield optics showing normal AChE activity on the left side in both the cortex (ct) and neuropil (np), and AChE-null clone encompassing most of the cortex on the right side. *B*: Same section under phase contrast optics at higher magnification. Mutant tissue in the cortex on the right side can be clearly seen, as well as altered neuropil morphology. Bars = 50 μm.

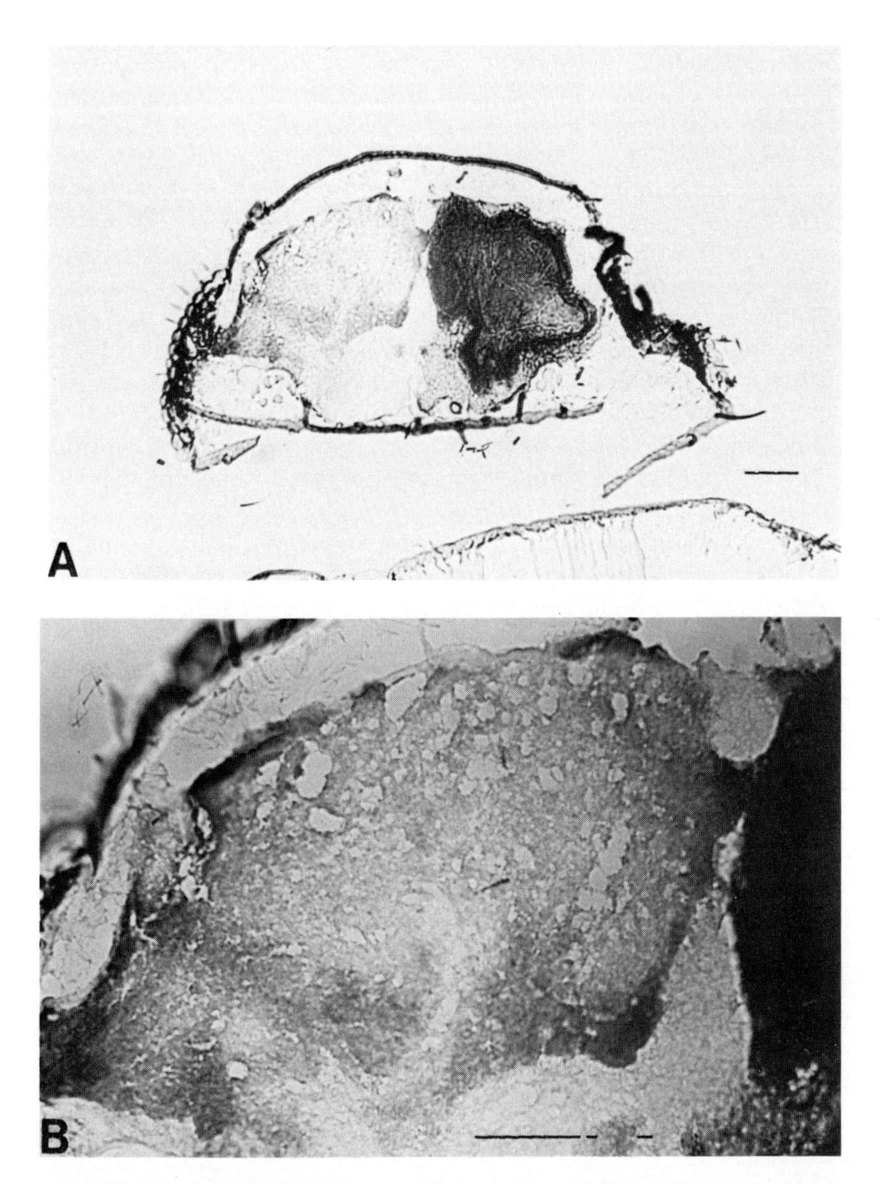

FIGURE 10. Dorsal brain of AChE mosaic with largest mutant clone. *A*: Horizontal section stained for AChE activity under low power, showing absence of enzyme activity on the left side of both cortex and neuropil. *B*: Mutant side of the same section under higher power and printed darker in order to show the degenerate appearance of the neuropil. Bars = 50 μm.

X-chromosome loss occurred relatively early, induced by a technique different from that used in *Ace* mosaics.)

A more specific feature of the nonrandom array of AChE-minus clones concerns the portion of the posterior area of the midbrain, near the region where the esophagus traverses the brain. This brain region had the lowest probability of being mutant, even on only one side of the brain. Other nearby brain regions that were more anterior, more dorsal, or more ventral had 3- 10-fold higher probabilities of being mutant. The areas with the highest mutant values among the mosaics were in the dorsal brain and in several parts of the thoracic-abdominal ganglion. Many parts of the brain (including the dorsal brain and the optic lobes) had a few *bilateral* clones. No bilateral null clones in the mid-posterior brain ever survived. Bilateral clones are generally less frequent even in acid phosphatase gynandromorphs (simply because the mosaic technique gives mutant clones that encompass only 10–30% of the embryo [Hall et al., 1976]), though they are readily obtainable with this nonlethal marker. The bilateral patches of AChE-mutant tissue, though, are found in an especially subnormal frequency in the mosaics, even for small bilateral clones (e.g., they were never found in the ventral brain). It is as if malfunction in both the left and homologous right parts of many areas of the brain is poorly tolerated.

In summary, AChE functioning in development or physiology is not of uniform importance in different parts of the CNS in *Drosophila*. Thus, the *Ace* mosaics provide more information than ubiquitous distribution of AChE throughout the CNS of wild-type flies. As for the parts of the brain that seem least able to tolerate the AChE enzyme deficiency, results of some classical ablation experiments done on the nervous systems of other insects may be relevant. It is sometimes found that ablations to dorsal parts of the brain may lead to defects in complex, higher-order fixed action patterns, such as those involved in courtship and mating (Bullock and Horridge, 1965; Huber, 1965, 1967; Elsner, 1973), but these animals are still able to carry out basic functions and movements. This kind of result also tends to be found for lesions in one side or the other of the ventral nervous system (Bullock and Horridge, 1965). In contrast, when ablations are made in the midbrain of some insects—that is, in the vicinity of the near-lethal focus for the *Ace* mutants—the animal tends to be very generally debilitated and unable to carry out any kind of general locomotion (Bullock and Horridge, 1965). Thus, the midposterior brain of the fly, which seems to have an especially large number of synapses (Strausfeld, 1976), many of which may be cholinergic (L. M. Hall, personal communication), may have been revealed as an important center of neural functioning through the neurochemical lesions we have created by genetic dissection.

Behavioral and Physiological Defects

The lack of normal ACh metabolism and the structural defects in the CNS induced by mutant clones might be expected to have behavioral consequences. Many of the AChE mosaics were ostensibly normal in their ability to maintain basic posture and carry out general movement. An exception was the grossly mosaic adult (Figure 10), which was overtly sluggish and in fact fell over frequently. For the majority of mosaics, however, more systematic behavioral tests were required to reveal important abnormalities of neural functioning. Optomotor tests were carried out on several putative internal AChE mosaics (initially recognized only on the basis of external mosaicism for the yellow marker, Figure 7). A normal fly placed in the center of an illuminated, rotating cylinder of black and white vertical stripes will readily run in the direction of rotation (i.e., 12 rpm for the whole cylinder, which has each pair of stripes subtending an angle of 19°). Thus, controls expressing the external markers used in the mosaic scheme but heterozygous for an *Ace*-lethal and Ace^{+} have a strong tendency to move in a small circle corresponding to the direction of stripe movement. To monitor this behavior, individual flies were placed within the cylinder on a small platform divided into four quadrants and covered with a watch glass. Control individuals ($n = 7$, tested for 5 min in a clockwise rotation test and 5 min in a separate, counterclockwise test), crossed 80–115 quadrant lines in the direction of stripe movement, and only about 20 in the "wrong" direction. Of the *Ace* mosaics, 35 flies gave normal or nearly normal optomotor responses. Several of these mosaics ($n = 21$) did have internal mosaicism for AChE, but almost no mutant clones in the optic lobes or the dorsal brain. Several mosaics were extremely abnormal in their optomotor response, crossing 10 lines or fewer in the direction of stripe rotation and about the same number of lines in the wrong direction. These eight nonresponding cases gave similar results in the clockwise tests and the counterclockwise tests. All of them proved to have mosaicism in the optic lobes and/or the dorsal brain, including cases of anterior or posterior mosaicism. Thus, there was a highly nonrandom distribution of internal mosaicism for the well-behaving vs. poorly behaving mosaics. If the mutant clones were indeed functioning poorly, the results suggest that altered input to the visual system or defective processing of the input in certain more central parts of the CNS can derange a visually triggered behavioral response.

It is important to compare the mosaics with poor optomotor response with flies expressing other types of visual defects and to obtain information on electrical activity in the visual system of *Ace* mosaics. Previous work has shown that insects with one eye covered still show a near-normal optomotor response (reviewed by Bullock and Horridge, 1965). Our analogous controls involved mosaics without defects in the optic lobes or brain, but with one eye blinded by a "no-receptor-potential" (*norpA*) mutation (reviewed by Pak, 1975). Mutants of this kind have defects at the level of retinal cells only (Hotta and Benzer, 1970) and have no light-induced electroretinogram (ERG), based on extracellular recordings from the retina (Pak, 1975). Each of seven *norpA* mosaics with one blind and one normal eye was quite vigorous in its optomotor response (ca. 100 correct responses per 5 min vs. ca. 50 incorrect ones), and the responses were equivalent whether the direction of stripe rotation was toward their blind eye or away from it, though these bilaterally split mosaics often held their bodies at an anomalous angle while running. Thus, CNS mosaicism for AChE is not simply equivalent to peripheral and unilateral blindness. Since the optomotor response in the *Ace* mosaics was defective in *either* test direction, the optic lobe-dorsal brain "focus" can be termed "domineering" (Hotta and Benzer, 1972). In other words, internal mosaicism in the optic lobe and/or the brain needs to be present in only *one* side of the animal in order to cause a defective behavioral response in *both* directional tests.

When ERGs were recorded from *Ace* mosaics, some of which had been tested behaviorally, five eyes showed a dramatic and unique type of defect. The "off-transient" of the ERG, a corneal-negative component triggered by turning off the light, was absent or grossly reduced. Two of the flies showing this defect had been tested for optomotor response and found to be highly abnormal. Each eye that gave a defective ERG proved to have an AChE-null clone, at least in the optic lobes proximal to the eye (Figure 11). Two mosaics had the clone only in the lamina (the first-order lobe just beneath the retina), and two with mixed laminas produced either greatly reduced or no off-transients, depending on the recording site (with the sites giving a partial defect roughly corresponding to the wild-type portion of the lamina beneath, e.g., Figure 11*C*). Many controls had normal ERGs; that is, eyes that proved to have no corresponding mosaicism in the optic lobes yielded ERGs with full on-transients (corneal-positive deflections) and off-transients. These controls include eyes on the nonmosaic side of flies showing the missing off-transient from the other eye. Other cases of

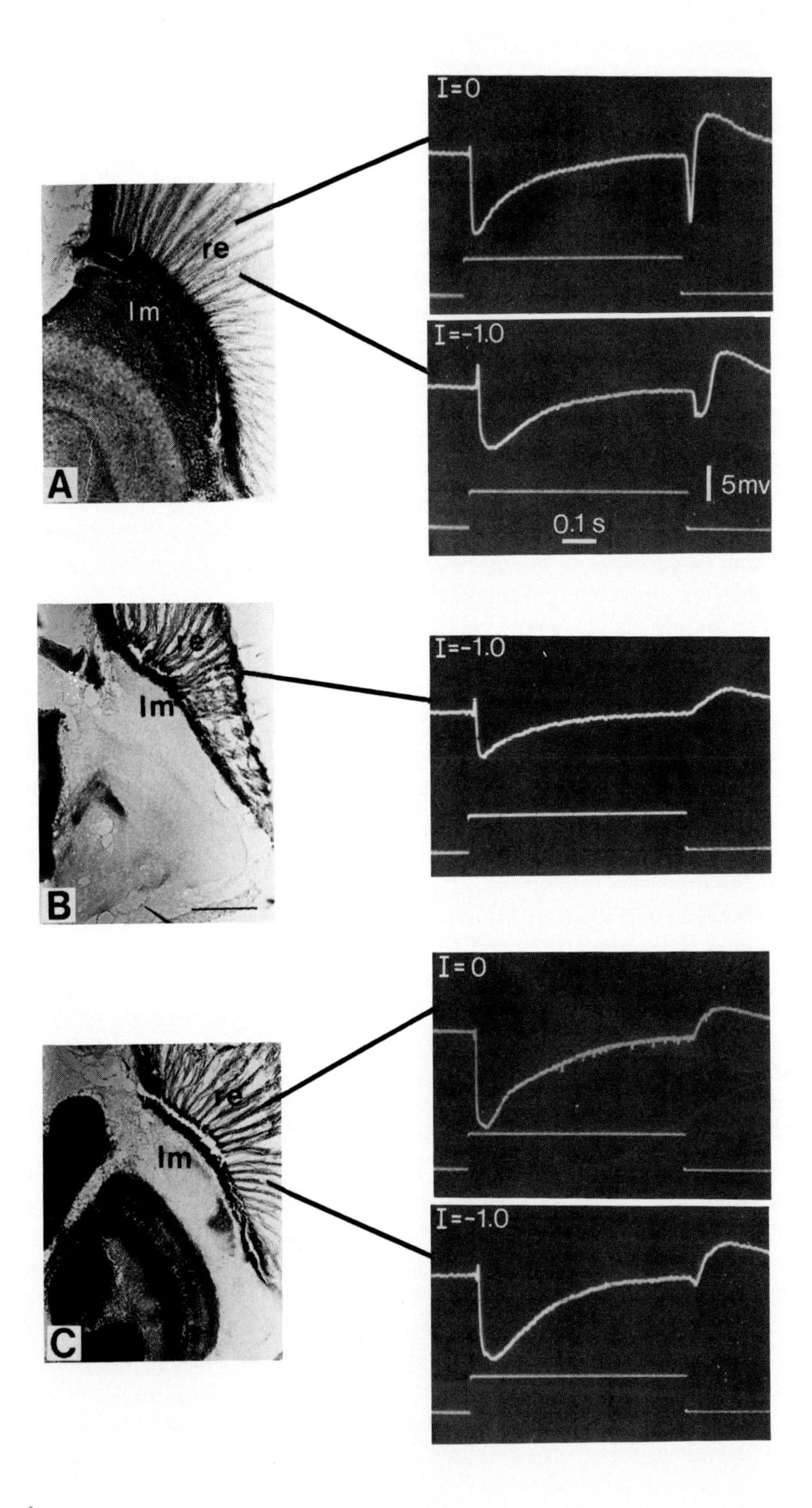

re
lm
A
I=0
I=-1.0
5mv
0.1 s
re
lm
B
I=-1.0
re
lm
C
I=0
I=-1.0

normal ERGs came from mosaics (n = 4) that had abnormal optomotor responses but *Ace*$^+$ optic lobes.

The transients of the ERG in flies are believed to originate from synaptic activity in the optic lobes, in particular, the lamina (Autrum, Zettler, and Jarvilehto, 1970; Pak, 1975; Zimmerman, 1978). The on-transient and the off-transient are eliminated in certain mutants in *Drosophila* that are apparently blind (Pak, 1975) but do have a normal photoreceptor potential, the major corneal-negative component of the ERG. The on-transient in all *Ace* mosaics was normal, even for eyes that yielded no off-transient (Figure 11*B,C*). Thus, there is probably some synaptic input to the optic system on the mutant side in these mosaics. This selective elimination of one component of the synaptic response in the second-order visual cells may be analogous to a phenomenon observed in the second-order cells of the dragonfly ocellus. There, Klingman and Chappell (1978) have shown that the "off" potential recorded intracellularly can be eliminated pharmacologically, leaving the "on" potential and the rest of the response intact.

Defective ERGs in *Ace* mosaics were compared with recordings from flies expressing a temperature-sensitive *Cat* mutation. For these tests, *Cat*$^{\ell 43}$/deletion flies were grown to adulthood at 18° C, then placed at 31° C for 2 days, during which time ACh synthesis might be turned off in the visual system as well as in the rest of the CNS. Indeed, at

FIGURE 11. Electroretinograms from normal and AChE mosaic flies. Photographs on the left side show horizontal sections through the compound eye and optic lobes of each fly tested. The sections were stained for AChE activity. The retina (re) in each eye shows the regular array of ommatidia containing the eye screening pigments. The optic lobes in *A*, including the first-order optic lobe, the lamina (lm), stain normally for AChE. The mosaic shown in *B* had no detectable AChE activity in its lamina and virtually none in the more central lobes. The mosaic shown in *C* had mutant tissue in most of the lamina, but not in the more central optic lobes. A small patch of AChE staining can be seen in the posterior lamina of this fly. The scale bar in *B* = 50 μm and applies to *A* and *C* as well.

Electroretinogram traces for each fly are shown in the photographs on the right. The top trace in each photo is the voltage difference recorded between an electrode placed just under the cornea at the level of the retina, and a reference electrode placed in the hemolymph of the abdomen, displayed on an AC coupled oscilloscope. The bottom trace in each photo indicates the onset, duration, and end of the white light stimulus. I values refer to log attenuation by neutral density filters of the full intensity (i.e., I = 0) where full intensity corresponds to about 4×10^{-3} W/cm^2 measured at 470 nm. The voltage and time scales, indicated in the second trace of *A*, are the same for all traces. For details see Greenspan et al. (1979).

the end of this period of heat treatment, these *Cat* mutants started to become paralyzed. ERGs recorded before paralysis set in showed a defect similar to that from the *Ace* mosaics with mutant optic lobes, i.e., an apparently normal on-transient and receptor potential, but no off-transient (Figure 12). It should be noted that the waveforms of the oscilloscope traces, corresponding to light-off, were not identical in the off-transient-minus ERGs from the *Ace* mosaic and the temperature-sensitive *Cat* mutant. Two kinds of controls yielded normal ERGs: wild-type flies placed at 31° C for 5 days after rearing at 18° C (Figure 12) and $Cat^{\ell 43}$/deletion flies raised and stored at 18° C before the recordings (not shown). A tentative conclusion from these experiments on visual physiology is that an absence of hydrolysis of ACh may have the same eventual physiological effect as an absence of synthesis of this putative neurotransmitter.

Because courtship and mating in *Drosophila* would be expected to involve higher orders of CNS functioning than the "reflexes" triggered by simple visual stimuli, sex behavior was analyzed in flies expressing mutants of ACh metabolism. When a male is in the presence of a female, fixed action patterns involving a variety of stereotyped male and female behaviors are triggered (Spieth, 1974). The importance of genetically male brain tissue for the control of male courtship actions in *Drosophila* has been probed by Hall (1977, 1979). Sex mosaics expressing a non-lethal acid phosphatase marker in the CNS were observed in the presence of normal females. Performance of any male behavior, such as following of females and male wing display at the beginning of the courtship sequence, was always associated with male neurons in the dorsal posterior brain in the left and/or right side of the head. These results were interpreted in regard to possible central processing of olfactory information from the female by the "mushroom body" axons in the dorsal brain (Hall, 1978). The interpretation of putative female pheromones by the mushroom bodies, which indeed are frequently related

FIGURE 12. Electroretinograms in wild-type, Cat^{ts}, and *Ace* mosaic flies. Wild-type and temperature-sensitive $Cat^{\ell 43}$ flies were raised at 18° C and shifted to 31° C when they emerged as adults. The wild-type fly was kept at 31° C for 5 days before testing; the $Cat^{\ell 43}$ fly was kept at the high temperature for 2 days before testing. This temperature treatment was sufficient to eliminate the off-transient in the mutant, while the wild-type remained normal. The bottom photograph shows a trace from the same *Ace* mosaic shown in Figure 11*C*, also missing the off-transient. See legend to Figure 11 for description of recordings and procedure. Light intensity corresponds to log I = 0 from Figure 11.

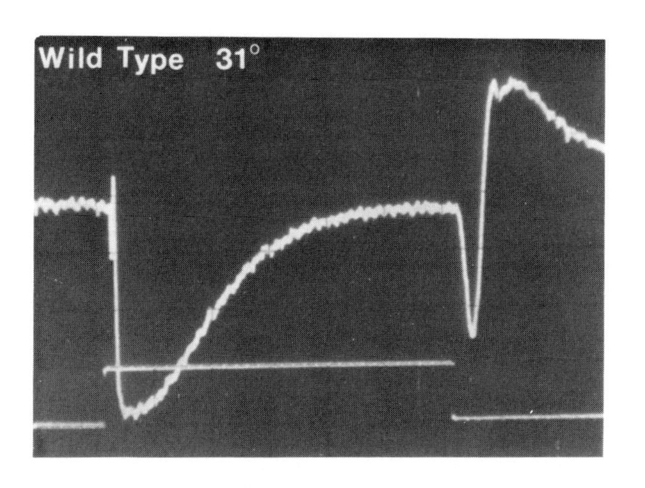
Wild Type 31°

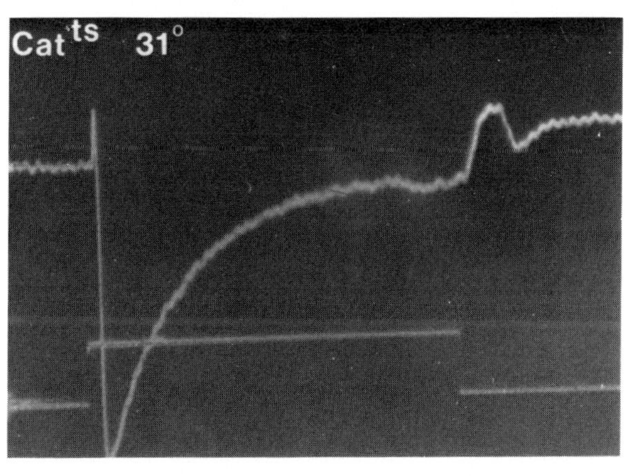
Cat ts 31°

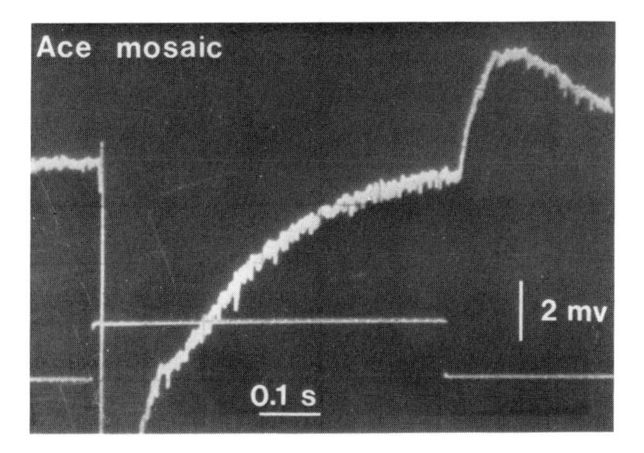
Ace mosaic
2 mv
0.1 s

to olfaction (Howse, 1975), could depend on neuronal connectivity or activity controlled specifically by male genotype.

An alternative way to view the mating behavior controlled by mixed male-female brains came from courtship tests of *Ace* gynandromorphs. Of 32 *Ace* mosaics tested, 16 courted virgin females. This is perhaps surprising, since gynandromorphs require male brain cells in order to behave as males, yet the only male tissues in these mosaics is AChE-minus and morphologically abnormal as well. All *Ace* mosaics that did court proved to have null tissue in the dorsal posterior brain regions, as expected. These courting cases showed a threefold decrement in the fraction of time spent courting, compared with controls heterozygous for *Ace* and Ace^{+} on chromosome *3* or in comparison with the acid phosphatase mosaics of Hall (1979). In addition, three mosaics with relatively extensive clones of AChE-minus tissue in the dorsal posterior brain, including the enfeebled mosaic with its entire dorsal male brain on one side (Figure 10), did not court a female at all. Other noncourting *Ace* mosaics (n = 13), which were not noticeably feeble in general, proved to have no male brain tissue in the critical region.

These results seem to imply that the AChE-minus male tissue in the CNS is able to function to allow some of the mosaics to court. This could mean that an absence of AChE in the relevant portions of the brain does not necessarily have severe consequences. Alternatively, it is possible that there is quasi-normal ACh degradation in the null clones due to nonautonomy of AChE activity (i.e., contribution of the normal enzyme from nearby Ace^{+} neurons to parts of the null clones). An additional explanation, which would not exclude those just mentioned, is that other transmitters besides ACh undoubtedly function in many parts of the brain, and may be relevant to the neurotransmission involved in the control of courtship behavior. In this context, the three dorsal brain mosaics that failed to court may have been the only ones with mutant tissue in the crucial cholinergic cells. Finally, we would like to suggest a possible interpretation of the courting *Ace* mosaics that still views the mutant clones in these gynandromorphs as severely impaired with respect to ACh metabolism and brain structure. There may be no particularly *male*-specific functioning of neuronal activity and pathways in the dorsal posterior brain. Instead, the key feature of male tissues in the brain of a mosaic is perhaps that they are nonfemale, or nonfunctioning female tissues. This condition is satisfied if one side of the brain is partially male, *whether it is functional* (i.e., Ace^{+}) *or not*. Then, the genetically female tissues on the other side of the brain (and in the rest of the CNS) are able to mediate male-like actions. This hypothesis views the female fly brain as inhibiting the possibility of male-specific courtship behavior; male tissue in one or both sides of the brain would then lead to a "release" of his inhibition. These ideas may be related to other findings on mating behavior in insects, such as the release of male copulatory movements that is triggered by damage to the ventral brain of the mantis (reviewed by Roeder, 1967), or a genetic variant in *Drosophila* (Cook, 1975) that causes females to behave like males and court other females. This mutant could conceivably have non-sex-specific defects in the functioning of the female brain.

A study of the CNS in flies expressing CAT activity in only part of the nervous system has been made possible by the ability to recognize an internally mosaic gynandromorph behaviorally. *Cat* mosaics were generated using the same general strategy employed in the production of *Ace* mosaics (Figure 7). Such gynandromorphs could not have their internal patterns of mosaicism scored directly, because no well-developed histochemical techniques now exist for this ACh-synthesizing enzyme (reviewed by Rossier, 1977). One could ask the *Cat* mosaics to perform some task in a feeble fashion (e.g., optomotor response) and then look for structural defects in the CNS that might be induced by the *Cat*-minus genotype in certain clones. It seems more desirable, though, to have a behavioral method that *positively* selects for mosaicism in the CNS; the performance of male actions (instead of the absence of a behavior) provides just such a technique. One *Cat* mosaic did court a female, and so must have had a male, *Cat*-minus clone at least in the dorsal posterior brain. This mosaic carried a *Cat*-temperature-sensitive allele on its third chromosomes; the gynandromorph had been grown at low temperature and was tested with females over the first few days of adult life, when it continued to be kept at 18° C (Figure 13). When shifted to 29° C, this mosaic improved in its courtship performance for 1 more day (Figure 13), but on subsequent days of testing, its ability to court was dramatically "turned off."

A normal gynandromorph in *Drosophila*, not expressing a *Cat*-minus mutation, will continue to court at its usual level in repeated tests (Hall, 1979). After the shift to high temperature, the *Cat* mosaic was not particularly debilitated in its general behavior, only in its quantitative and qualitative courtship performance; even after 6 days at 29° C, it had not become paralyzed. A comparative courtship test of a heat-treated male, uniformly expressing a *Cat*-temperature-sensitive mutation,

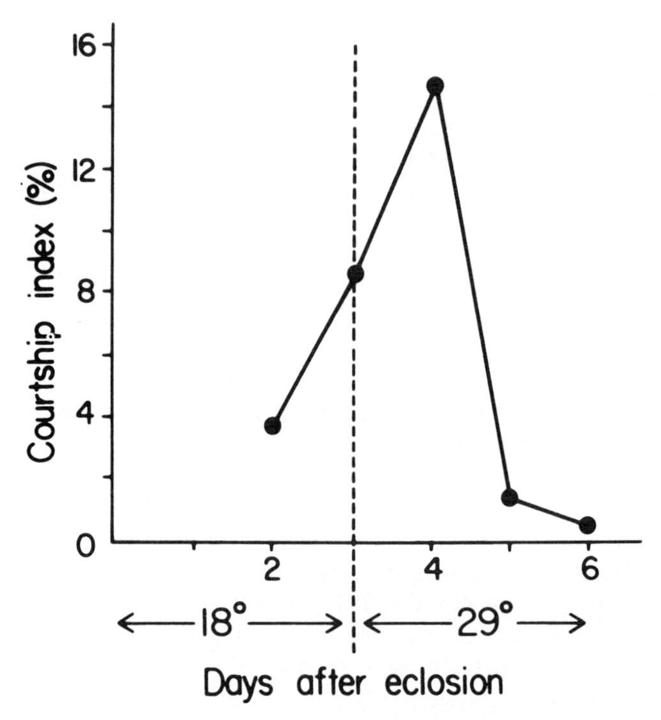

FIGURE 13. Courtship behavior of a *Cat*-temperature-sensitive mosaic. A gynandromorph raised at 18° C and mosaic for the temperature-sensitive allele $Cat^{\ell 16}$ was tested for its ability to exhibit any male courtship behavior toward females in 20-min observation periods on each day indicated. It was stored at 18° C for the first 3 days after eclosion, and then after its courtship test on day 3 it was shifted to 29° C. The courtship index plotted on the ordinate is the percentage of time spent performing male courtship behaviors during the observation period.

would be meaningless because the fly would become generally paralyzed over the same period of high-temperature treatment during which the mosaic ceased courting. With the mosaic, we were able to see if the postdevelopmental turn-off of *Cat* activity in part of the brain could lead to noticeable morphological abnormalities. No obvious ones were seen in the dorsal posterior regions of silver-stained brain sections. However, this general approach now allows us to ask if a *Cat* mosaic expressing the mutant phenotype throughout all or part of preadult stages, and subsequently showing some male behavior, would have morphological abnormalities in the CNS, possibly similar to those found in *Ace* mosaics.

CONCLUSIONS AND PROSPECTS

These initial findings on genetically disrupted ACh metabolism in *Drosophila*, including the many neurobiological defects associated with AChE-minus and CAT-minus genotypes, suggest several interesting questions. First, what are the reasons for morphological defects in the CNS of *Ace*-null clones? Are they due to defects in assembly processes in embryonic or postembryonic development of the nervous system? Or are they strictly degenerative changes—following hypothetically normal development—due to higher than normal levels of ACh? The latter possibility could be correlated with apparent degeneration induced in cholinergic muscle tissues after postdevelopmental treatment of the input plus target cells with anticholinesterases in vertebrates (Laskowski, Olson, and Dettbarn, 1975). Degeneration in target neurons appears to be induced by treatment of postsynaptic cells with abnormally high concentrations of a natural transmitter (glutamate) that impinges on such synapses in the mammalian brain (Olney and Gubareff, 1978), or by treatment with a powerful agonist of the natural transmitter (Coyle and Schwarcz, 1976; Herndon and Coyle, 1977; Schwarcz and Coyle, 1977). Such treatments could lead to effects formally analogous to abnormally high ACh concentrations in the fly's brain due to reduced or absent AChE activity. That altered ACh levels and effects of the neurotransmitter on target cells may be due to changes *during development* should also be considered. Examples of such developmental lesions in other organisms include cases of altered target cell differentiation (Black and Geen, 1974) or abnormal morphology (Drachman, 1974; Freeman, Engel, and Drachman, 1976; Landauer, 1976; Hervonen, 1977) seen after manipulation of ACh metabolism during the ontogeny of vertebrates or after treatments of vertebrate cells in culture.

If the mutant CNS morphology in our mosaics is the result solely of developmental perturbations, the defects in the fly may be analogous to the results from higher forms, and they may be caused by direct effects on pattern formation or by secondary effects of an initial defect in the differentiation of target cells that results from abnormal ACh metabolism long before adulthood. We should be able to distinguish between these two possibilities by temperature-shift experiments on *Ace* mosaics that carry a temperature-sensitive mutant combination on their third chromosome: At which stage of the life cycle is a heat-induced turn-off of AChE correlated with morphological defects in the neuropil? In addition to these mosaic tests with the temperature-sensitive mutants, it will be valuable to carry out heat treatments of uniformly mutant *Ace*-temperature-sensitive adults that have grown up at low temperature. In this case, the chronic postdevelopmental treatments would be done at a temperature below 33° C, since that high temperature makes the mutants pass out and die before putative long-term deficits in AChE could be shown to cause degenerative changes in the CNS.

Another question relating to abnormalities in CNS morphology induced by *Ace* mutations concerns functional defects: Are they caused directly by an absence of ACh breakdown in AChE-minus parts of the nervous system, or do the behavioral abnormalities, for instance, result from structural "damage" seen in the neuropils of the various CNS ganglia? Experiments aimed at answering this kind of question might involve mosaics expressing a temperature-sensitive *Ace* genotype in tissues that had lost the Ace^+ allele (see above). High-temperature treatment of such mosaics, after development at permissive temperatures, may lead to severe defects in the optomotor response or some other behavior in the absence of long-term turn-off of enzyme activity and without any apparent morphological abnormalities in the brains of these mosaics. The null clones in these mosaics will be scorable, because AChE activity in such portions of the CNS can be turned off by heat treatment of the sections prior to staining. This experimental approach points out the potential use of AChE as a simple marker in mosaic work aimed at determining parts of the CNS affected by behavioral and neurological mutants in general (reviewed by Hall, 1978). For these experiments, the expression of a temperature-sensitive *Ace* genotype would be induced only after development under permissive conditions. The *Ace*-minus clone whose appearance was induced by heat treatment would then mark neural tissues expressing the mutant under study, such as hyperkinetic (Ikeda and Kaplan, 1974), paralytic-temperature-sensitive (Suzuki et al., 1971), or dunce (Dudai, Jan, Byers, Quinn, and Benzer, 1976).

If some of the behavioral problems observed in *Ace* mosaics are due to physiological defects, then it is still not clear whether the higher than normal levels of ACh that might be present in mutant clones lead to potentiation of postsynaptic responses or desensitization and blockade of such responses. The latter possibility is plausible given the desensitization that has been observed in vertebrates (Katz and Thesleff,

1957) and in invertebrates (Callec, 1974) under the influence of artificially high levels of ACh at synapses. Some of the experiments showing such desensitization have used anticholinesterases (Kuba, Albuquerque, Daly, and Barnard, 1974) whose effects may be similar to mutationally induced AChE deficits. The mutants, though, may lead to a more specific and limited defect in AChE per se. In addition, the genetic variants allow one to ask more extensive and detailed questions about the effects of faulty ACh hydrolysis on the ability of the fly to perform complex behavioral tasks, and about physiological defects during development that may directly cause the structural damage we have observed. For instance, it is possible to turn off CAT activity throughout development or during limited developmental periods to see if an absence of ACh synthesis in various parts of the CNS has the same morphological consequences as an absence of breakdown of the presumptive neurotransmitter. If this turns out to be true, it may buttress the notion that the disrupted anatomy and function in *Ace* mutants are due to a failure of postsynaptic cells to respond after exposure to an oversupply of ACh, which hypothetically would result in desensitization of the target neurons.

A few more specific questions are raised by the current state of the genetic and neurobiological evidence on ACh metabolism in the fly. Why is a relatively low-temperature treatment (29° C) of *Ace*-temperature-sensitive mutants completely lethal to embryos, while having little if any effect on larvae and adults? It may be that these particular mutations affect the synthesis as well as the functioning of the AChE molecules. AChE is first being synthesized and accumulated in reasonable amounts at about the same time as the embryonic temperature-sensitive lethal period. Thus, synthesis of the molecule, which is hypothetically blocked by 29° C treatments during embryogenesis, may include assembly of monomers into the complete, functional protein and packaging of this presumed multimer into parts of neuronal membranes. Such events, in particular the assembly of mutant molecules, may be more heat-sensitive than the catalytic functions of the mutant enzyme in the postembryonic stages.

Alternatively, the embryonic TSPs may indicate that cholinergic portions of the nervous system that are essential throughout the life cycle are being made in the embryonic period. In this regard, it is interesting to note that some of the cells persisting in the adult that appear to have been made during embryogenesis occupy medial sites in the supraesophageal ganglion (White and Kankel, 1978). This may correlate with the lethal focus for *Ace* mutant tissue in the posterior midbrain, a part of the brain that is apparently crucial for survival.

Defects in optomotor performance have raised the question of the surprising nonequivalence of unilateral blindness, brought about by the *norpA* mutation, and unilateral AChE mutant tissue. For those mosaics that had mutant tissue in the dorsal brain, the response may have been blocked at some step involving integration of visual input and motor output, a function consistent with the inferred role of the dorsal brain in many insects (Howse, 1975; Strausfeld, 1976). However, flies with mutant tissue on one side in the optic lobes were also optomotor-defective, and in one case this was virtually the only mutant brain tissue.

If AChE-null optic lobes are responding to the input of light but are failing to register either the contrast between black and white stripes or their motion, then perhaps the brain receives contradictory and overriding signals from the mutant vs. the normal sides. In this context, it may be relevant to note that lateral inhibition, a function that is likely to be necessary for pattern and/or movement detection, has been demonstrated in the lamina cells of larger flies (Arnett, 1972; Zettler and Jarvilehto, 1972; Mimura, 1976). If such mechanisms are cholinergic in *Drosophila*, then they could be selectively blocked in *Ace* mosaics. It is not certain that the lamina is directly involved in optomotor processing, but Heisenberg and Buchner (1977) have found that the visual pathways that synapse in the lamina must be intact for normal optomotor behavior.

Another unexpected aspect of the behavior of *Ace* mosaics was their ability to exhibit male courtship. That is, male tissue in the brain is necessary for male courtship (Hall, 1977), but the only male brain tissue in the current mosaics lacks AChE and is morphologically abnormal. This had led us to speculate on the possible role of normal female dorsal brain tissue in *inhibiting* programmed male-specific behavior. This is readily testable, because it is possible to generate AChE mosaics in the brains of flies that are *entirely female* (i.e., not gynandromorphs). Such a method would involve the somatic loss of a small chromosome fragment carrying the Ace^+ gene from a diplo-X zygote that also carries *Ace* mutations on its third chromosomes. Some of the resulting mosaics would have *Ace*-minus cells in one side of the dorsal brain, and if the clone is not too large, its defective nature may release male behavior in a totally diplo-X fly.

Although some of the suggestions we have made concerning our observations on mutants of ACh metabolism are not immediately testable, many others are uniquely approachable with the genetic technology at hand. The temperature-sensitive alleles and techniques for making *Ace* and *Cat* mosaics allow the usual lethal stage for these mutations to be bypassed, and make possible developmental manipulations of ACh metabolism at any stage from the onset of embryogenesis through to the adult period, by pulses or chronic treatments, and in a localized manner not possible with standard pharmacological tools. Similarly, the lesions caused by these mutations provide a probe for behavioral and physiological manipulations that is more selective than ablation and more controlled than pharmacological agents. After disrupting ACh metabolism centrally, the effects on very complex behaviors can be tested without otherwise disturbing or encumbering the animal with recording or iontophoretic electrodes. The neurogenetic lesions allow meaningful questions to be asked about how particular physiological and neurochemical disruptions are related to behavioral events ranging from the simplest reflexes to the most sophisticated and complex higher functions.

ACKNOWLEDGMENTS

This work was supported by a grant from the United States Public Health Service (NS-12346) and in part by USPHS Biomedical Institutional Research Support Grant (FR 07044). J.C.H. was also supported by a Research Career Development Award from the USPHS (GM 00297). R.J.G. was supported by a USPHS Training Grant (GM 07122) and in part by a Brandeis University Goldwyn Fellowship. We thank James Finn, Susan Wayne, and Kristin White for assisting in data collection, and we appreciate comments on the manuscript from Laurie Tompkins, Kalpana White, and Eve Marder.

REFERENCES

Arnett, D. W. (1972). Spatial and temporal integration properties of units in first optic ganglion of Dipterans, *J. Neurophysiol.* **35**:429–444.

Autrum, H., F. Zettler, and M. Jarvilehto (1970). Postsynaptic potentials from a single monopolar neuron of the ganglion opticum I of the blowfly *Calliphora, Z. Vgl. Physiol.* **70**:414–424.

Becker, H. J. (1976). Mitotic recombination, pp. 1020–1089 in *The Genetics and Biology of Drosophila*, Vol. lc, Ashburner, M. and E. Novitski, eds. Academic Press, New York.

Best-Belpommé, M. and A.-M. Courgeon (1977). Ecdysterone and acetylcholinesterase activity in cultured *Drosophila* cells, *FEBS Lett.* **82**:345–347.

Black, I. B. and S. C. Geen (1974). Inhibition of the biochemical and morphological maturation of adrenergic neurons by nicotinic receptor blockade, *J. Neurochem.* **22:**301–306.

Bullock, T. H. and G. A. Horridge (1965). *Structure and Function in the Nervous Systems of Invertebrates*. W. H. Freeman, San Francisco.

Callec, J.-J. (1974). Synaptic transmission in the central nervous system of insects, pp. 119–185 in *Insect Neurobiology*, Treherne, J. E., ed. Elsevier/North-Holland, New York.

Cherbas, P., L. Cherbas, and C. M. Williams (1977). Induction of acetylcholinesterase activity by β-ecdysone in a *Drosophila* cell line, *Science* **197:** 275–277.

Colhoun, E. (1959). Physiological events in organophosphorus poisoning, *Can. J. Biochem. Physiol.* **37:**1127–1134.

Cook, R. (1975). "Lesbian" phenotype of *Drosophila melanogaster? Nature (London)* **254:**241–242.

Coyle, J. T. and R. Schwarcz (1976). Lesion of striatal neurones with kainic acid provides a model for Huntington's chorea, *Nature (London)* **263:**244–246.

Deak, I. I. (1977). A histochemical study of the muscles of *Drosophila melanogaster*, *J. Morphol.* **153:**307–316.

Dewhurst, S. A., R. E. McCaman, and W. D. Kaplan (1970). The time course of development of acetylcholinesterase and choline acetyltransferase in *Drosophila melanogaster*, *Biochem. Genet.* **4:**499–508.

Dewhurst, S. A. and R. L. Seecof (1975). Development of acetylcholine metabolizing enzymes in Drosophila embryos and in cultures of embryonic Drosophila cells, *Comp. Biochem. Physiol.* **50C:**53–58.

Drachman, D. B. (1974). The role of acetylcholine as a neurotrophic transmitter, *Ann. N. Y. Acad. Sci.* **228:**160–175.

Driskell, W. J., B. H. Weber, and E. Roberts (1978). Purification of choline acetyltransferase from *Drosophila melanogaster*, *J. Neurochem.* **30:**1135–1141.

Dudai, Y. (1977*a*). Demonstration of an α-bungarotoxin-binding nicotinic receptor in flies, *FEBS Lett.* **76:**211–213.

Dudai, Y. (1977*b*). Molecular states of acetylcholinesterase from Drosophila melanogaster, Drosophila Information Service **52:**65–66.

Dudai, Y. (1978). Properties of an α-bungarotoxin-binding cholinergic nicotinic receptor from *Drosophila melanogaster*, *Biochim. Biophys. Acta* **539:**505–517.

Dudai, Y. and A. Amsterdam (1977). Nicotinic receptors in the brain of *Drosophila melanogaster* demonstrated by autoradiography with [^{125}I]α-bungarotoxin, *Brain Res.* **130:**551–555.

Dudai, Y. and J. Ben-Barak (1977). Muscarinic receptor in *Drosophila melanogaster* demonstrated by binding of [^{3}H] quinuclidinyl benzilate, *FEBS Lett.* **81:**134–136.

Dudai, Y., Y. Jan, D. Byers, W. Quinn, and S. Benzer (1976). *dunce*, a mutant of *Drosophila* deficient in learning, *Proc. Natl. Acad. Sci. USA* **73:**1684–1688.

Ellman, G. L., K. D. Courtney, V. Andres, Jr., and R. M. Featherstone (1961). A new and rapid colorimetric determination of acetylcholinesterase activity, *Biochem. Pharmacol.* **7:**88–95.

Elsner, N. (1973). The central nervous control of courtship behaviors in the grasshopper *Gomphocerippus rufus* L. (Orthoptera: Acrididae), pp. 261–287 in *Mechanisms of Rhythm Regulation*, Salanki, J., ed. Akademiai Kiado, Budapest.

Fonnum, F. (1975). A rapid radiochemical method for the determination of choline acetyltransferase, *J. Neurochem.* **24**:407–409.

Freeman, S. S., A. G. Engel, and D. B. Drachman (1976). Experimental acetylcholine blockade of the neuromuscular junction. Effects on endplate and muscle fiber ultrastructure, *Ann. N. Y. Acad. Sci.* **274**:46–59.

Gelbart, W. M. (1974). A new mutant controlling mitotic chromosome disjunction in *Drosophila melanogaster*, *Genetics* **76**:51–63.

Gerschenfeld, H. M. (1973). Chemical transmission in invertebrate central nervous systems and neuromuscular junctions, *Physiol. Rev.* **53**:1–119.

Greenspan, R. J. (1979). Genetics and neurobiology of choline acetyltransferase mutants in *Drosophila*, in preparation.

Greenspan, R. J., J. A. Finn, Jr., and J. C. Hall (1979). Acetylcholinesterase mutants in *Drosophila* and their effects on the structure and function of the central nervous system, submitted for publication.

Haim, N., S. Nahum, and Y. Dudai (1979). Properties of a putative muscarinic cholinergic receptor from *Drosophila melanogaster, J. Neurochem.* **32**: 543–552.

Hall, J. C. (1977). Portions of the central nervous system controlling reproductive behavior in *Drosophila melanogaster*, *Behav. Genet.* **7**:291–312.

Hall, J. C. (1978). Behavioral analysis in Drosophila mosaics, pp. 259–305 in *Genetic Mosaics and Cell Differentiation*, W. J. Gehring, ed. Springer-Verlag, New York.

Hall, J. C. (1979). Control of male reproductive behavior by the central nervous system of Drosophila: dissection of a courtship pathway by genetic mosaics, *Genetics*, in press.

Hall, J. C., W. M. Gelbart, and D. R. Kankel (1976). Mosaic systems, pp. 265–314 in *The Genetics and Biology of Drosophila*, Vol. 1a, Ashburner, M. and E. Novitski, eds. Academic Press, New York.

Hall, J. C. and D. R. Kankel (1976). Genetics of acetylcholinesterase in *Drosophila melanogaster*, *Genetics* **83**:517–535.

Hall, L. M. and N. N. H. Teng (1975). Localization of acetylcholine receptors in *Drosophila melanogaster*, pp. 282–289 in *Developmental Biology—Pattern Formation—Gene Regulation*, McMahon, D. and C. F. Fox, eds. W. A. Benjamin, Menlo Park, Calif.

Hall, L. M., R. W. von Borstel, B. C. Osmond, S. D. Hoeltzli, and T. H. Hudson (1978). Genetic variants in an acetylcholine receptor from *Drosophila melanogaster*, *FEBS Lett.* **95**:243–246.

Heisenberg, M. and E. Buchner (1977). The role of retinula cell types in visual behavior of *Drosophila melanogaster*, *J. Comp. Physiol.* **117**:127–162.

Herndon, R. M. and J. T. Coyle (1977). Selective destruction of neurons by a transmitter agonist, *Science* **198**:71–72.

Hervonen, H. (1977). Effect of cholinesterase inhibitors on differentiation of cultured sympatheticoblasts, *Experientia* **33**:1215–1217.

Hotta, Y. and S. Benzer (1970). Genetic dissection of the *Drosophila* nervous system by means of mosaics, *Proc. Natl. Acad. Sci. USA* **67:**1156–1163.

Hotta, Y. and S. Benzer (1972). Mapping of behaviour in *Drosophila* mosaics, *Nature (London)* **240:**527–535.

Howse, D. E. (1975). Brain structure and behavior in insects, *Annu. Rev. Entomol.* **20:**359–379.

Huber, F. (1965). Brain controlled behaviour in Orthopterans, pp. 233–246 in *The Physiology of the Insect Central Nervous System*, Treherne, J. E. and J. W. L. Beament, eds. Academic Press, New York.

Huber, F. (1967). Central control of movements and behavior in invertebrates, pp. 333–351 in *Invertebrate Nervous Systems: Their Significance for Mammalian Neurophysiology*, Wiersma, C. A. G., ed. Univ. of Chicago Press, Chicago.

Ikeda, K. and W. D. Kaplan (1974). Neurophysiological genetics in *Drosophila melanogaster*, *Am. Zool.* **14:**1055–1066.

Kankel, D. R. and J. C. Hall (1976). Fate mapping of nervous system and other internal tissues in genetic mosaics of *Drosophila melanogaster*, *Dev. Biol.* **48:**1–24.

Katz, B. and S. Thesleff (1957). A study of the "desensitization" produced by acetylcholine at the motor end-plate, *J. Physiol. (London)* **138:**63–80.

Kehoe, J. and E. Marder (1976). Identification and effects of neural transmitters in invertebrates, *Annu. Rev. Pharmacol. Toxicol.* **16:**245–268.

Klemm, N. (1976). Histochemistry of putative neurotransmitter substances in the insect brain, *Prog. Neurobiol. (Oxford)* **7:**99–169.

Klingman, A. and R. L. Chappell (1978). Feedback synaptic interaction in the dragonfly ocellar retina, *J. Gen. Physiol.* **71:**157–176.

Kojima, K., J. Gillespie, and Y. N. Tobari (1970). A profile of *Drosophila* species enzymes assayed by electrophoresis. I. Number of alleles, heterozygosities, and linkage disequilibrium in glucose-metabolizing systems and some other enzymes, *Biochem. Genet.* **4:**627–637.

Kuba, K., E. X. Albuquerque, J. Daly, and E. A. Barnard (1974). A study of the irreversible cholinesterase inhibitor di-isopropyl phosphorofluoridate on the time course of end-plate currents in the frog sartorius muscle, *J. Pharmacol. Exp. Ther.* **189:**499–512.

Landauer, W. (1976). Cholinomimetic teratogens. V. The effect of oximes and related cholinesterase reactivators, *Teratology* **15:**33–42.

Laskowski, M. B., W. H. Olson, and W.-D. Dettbarn (1975). Ultrastructural changes at the motor end-plate produced by an irreversible cholinesterase inhibitor, *Exp. Neurol.* **47:**290–306.

Leuzinger, W., M. Goldberg, and E. Cauvin (1969). Molecular properties of acetylcholinesterase, *J. Mol. Biol.* **40:**217–225.

Lindsley, D. L. and E. H. Grell (1968). *Genetic variations of Drosophila melanogaster*, Carnegie Institution of Washington Publication No. 627. Washington, D.C.

Lindsley, D. L., L. Sandler, B. S. Baker, A. T. C. Carpenter, R. E. Denell, J. C. Hall, P. A. Jacobs, G. L. G. Miklos, B. K. Davis, R. C. Gethmann, R. W. Hardy, A. Hessler, S. M. Miller, H. Nozawa, D. M. Parry, and

M. Gould-Somero (1972). Segmental aneuploidy and the genetic gross structure of the Drosophila genome, *Genetics* **71**:157–184.

Meyerowitz, E. M. and D. R. Kankel (1978). A genetic analysis of visual system development in *Drosophila melanogaster*, *Dev. Biol.* **62**:112–142.

Mimura, K. (1976). Some spatial properties in the first optic ganglion of the fly, *J. Comp. Physiol.* **105**:65–82.

Nomura, M., E. A. Morgan, and S. R. Jaskunas (1977). Genetics of bacterial ribosomes, *Annu. Rev. Genet.* **11**:297–348.

O'Brien, S. J. and R. J. MacIntyre (1978). Genetics and biochemistry of enzymes and specific proteins of *Drosophila*, pp. 396–551 in *The Genetics and Biology of Drosophila*, Vol. 2a, Ashburner, M. and T. R. F. Wright, eds. Academic Press, New York.

Oliver, C. P. (1937). Evidence indicating that facet in Drosophila is due to a deficiency, *Am. Nat.* **71**:560–566.

Olmsted, J. B. and G. G. Borisy (1973). Characterization of microtubule assembly in porcine brain extracts by viscometry, *Biochemistry* **12**:4282–4289.

Olney, J. and T. Gubareff (1978). Glutamate neurotoxicity and Huntington's chorea, *Nature (London)* **271**:557–559.

Pak, W. L. (1975). Mutations affecting the vision of *Drosophila melanogaster*, pp. 703–733 in *Handbook of Genetics*, Vol. 3, King, R. C., ed. Plenum Publishing Corp., New York.

Pitman, R. M. (1971). Transmitter substances in insects: a review, *Comp. Gen. Pharmacol.* **2**:347–371.

Poulson, D. and E. Boell (1946). A comparative study of cholinesterase activity in normal and genetically deficient strains of *Drosophila melanogaster*, *Biol. Bull. (Woods Hole)* **91**:228.

Power, M. E. (1943). The brain of *Drosophila melanogaster*, *J. Morphol.* **72**:517–559.

Roeder, K. D. (1967). *Nerve Cells and Insect Behavior*. Harvard University Press, Cambridge, Mass.

Rossier, J. (1977). Choline acetyltransferase: a review with special reference to its cellular and subcellular localization, *Int. Rev. Neurosci.* **20**:284–337.

Rudloff, E. (1978). Acetylcholine receptors in the central nervous system of *Drosophila melanogaster*, *Exp. Cell Res.* **111**:185–190.

Schilcher, F. V. and J. C. Hall (1979). Neural topography of courtship song in sex mosaics of *Drosophila melanogaster*, *J. Comp. Physiol. A Sens. Neural Behav. Physiol.* **129**:85–95.

Schlesinger, M. J. and C. Levinthal (1965). Complementation at the molecular level of enzyme interaction, *Annu. Rev. Microbiol.* **19**:267–284.

Schmidt-Nielsen, B. K., J. I. Gepner, N. N. H. Teng, and L. M. Hall (1977). Characterization of an α-bungarotoxin binding component from *Drosophila melanogaster*, *J. Neurochem.* **29**:1013–1031.

Schwarcz, R. and J. T. Coyle (1977). Striatal lesions with kainic acid: neurochemical characteristics, *Brain Res.* **127**:235–249.

Siddiqi, O. and S. Benzer (1976). Neurophysiological defects in temperature-sensitive paralytic mutants of *Drosophila melanogaster*, *Proc. Natl. Acad. Sci. USA* **73**:3253–3257.

Silver, A. (1974). *The Biology of Cholinesterases.* Elsevier/North-Holland, New York.

Spieth, H. T. (1974). Courtship behavior in *Drosophila, Annu. Rev. Entomol.* **19**:385–406.

Strausfeld, N. H. (1976). *Atlas of an Insect Brain.* Springer-Verlag, Berlin.

Suzuki, D. T., T. Grigliatti, and R. Williamson (1971). Temperature-sensitive mutations in *Drosophila melanogaster.* VII. A mutation (*para*^{ts}) causing reversible adult paralysis, *Proc. Natl. Acad. Sci. USA* **68**:890–893.

Suzuki, D. T., T. Kaufman, and D. Falk (1976). Conditionally expressed mutations in *Drosophila melanogaster*, pp. 208–263 in *The Genetics and Biology of Drosophila*, Vol. 1a, Ashburner, M. and E. Novitski, eds. Academic Press, New York.

Ullmann, A. and D. Perrin (1970). Complementation in β-galactosidase, pp. 143–172 in *The Lactose Operon*, Beckwith, J. R. and D. Zipser, eds. Cold Spring Harbor Laboratory, Cold Spring Harbor, N.Y.

White, K. and D. R. Kankel (1978). Patterns of cell division and cell movement in the formation of the imaginal nervous system in *Drosophila melanogaster, Dev. Biol.* **65**:296–321.

Wright, T. R. F. (1970). The genetics of embryogenesis in *Drosophila, Adv. Genet.* **15**:261–395.

Wright, T. R. F., G. C. Bewley, and A. F. Sherald (1976). The genetics of dopa-decarboxylase in *Drosophila melanogaster.* II. Isolation and characterization of dopa-decarboxylase-deficient mutants and their relationship to the α-methyl-dopa-hypersensitive mutants, *Genetics* **84**:287–310.

Wright, T. R. F., R. B. Hodgetts, and A. F. Sherald (1976). The genetics of dopa-decarboxylase in *Drosophila melanogaster.* I. Isolation and characterization of deficiencies that delete the dopa-decarboxylase-dosage-sensitive region and the α-methyl-dopa-hypersensitive locus, *Genetics* **84:** 267–285.

Wu, C.-F., B. Ganetzky, L. Y. Jan, Y.-N. Jan, and S. Benzer (1978). A *Drosophila* mutant with a temperature-sensitive block in nerve conduction, *Proc. Natl. Acad. Sci. USA* **75**:4047–4051.

Zettler, F. and M. Jarvilehto (1972). Lateral inhibition in an insect eye, *Z. Vgl. Physiol.* **76**:233–244.

Zimmerman, R. P. (1978). Field potential analysis and the physiology of second-order neurons in the visual system of the fly, *J. Comp. Physiol.* **126**:297–316.

A GENETIC VIEW OF MONOAMINE OXIDASE

Xandra O. Breakefield, Richard M. Cawthon, Maria R. Castro Costa, Susan B. Edelstein, and Morris Hawkins, Jr.

Yale University School of Medicine, New Haven, Connecticut

BIOCHEMICAL AND GENETIC CHARACTERIZATION OF MONOAMINE OXIDASE

Introduction

Monoamine oxidase (MAO, monoamine:O_2 oxidoreductase, EC 1.4.3.4) is primarily responsible for the neuronal degradation of amine transmitters, and activity of the enzyme is thought to be able to modulate synaptic transmission (for reviews see Neff and Yang, 1974; Houslay, Tipton, and Youdim, 1976; Murphy, 1978). Although this enzyme has been studied extensively, many fundamental questions remain concerning its molecular nature. Because MAO is membrane-bound, its solubilization, purification, and characterization have proven difficult. Further, MAO has a heterogeneous expression of activity with regard to substrate specificity and drug sensitivity when compared in different tissues and species at various stages of development. This variability in properties is reminiscent of the isozymic forms of a number of other enzymes (e.g., lactate dehydrogenase and phosphoglucomutase; see Harris, 1977) in which variations result from differential expression of evolutionarily related gene loci. A genetic basis for variation in MAO activity has been alluded to (White and Glassman, 1977; Murphy, 1978), but the conceptual framework and implications of such a theory have not been presented in detail. An understanding of the genetic control of MAO activity is critical in determining the role of enzyme polymorphisms in mammalian physiology.

Function and Properties of the Enzyme

Monoamine oxidase is present in essentially all mammalian tissues. It deaminates amine neurotransmitters, such as dopamine, norepinephrine, and serotonin, as well as amines that can serve as "false" transmitters, such as tyramine, tryptamine, and phenylethylamine. In neurons that use catecholamines or indoleamines as transmitters, this enzyme serves as the primary means of degradation. Inhibition of MAO activity in experimental animals leads to increased concentrations of these transmitters in the nervous system (see reviews cited above), decreased spontaneous activity in norepinephrine cells in the locus coeruleus, and decreased numbers of β-adrenergic receptors in the brain (Campbell, Gallagher, Murphy, and Tallman, 1978). In humans, MAO inhibitors elevate mood, presumably through potentiation of catecholaminergic pathways in the brain (for review see Berger, 1977). Further, inhibition of MAO activity in the periphery results in a reduced rate of degradation of other amines, many of which can replace neurotransmitter stores at nerve endings and disrupt synaptic transmission. More extensive therapeutic use of MAO inhibitors will require an understanding of differences in individual responsiveness to these drugs.

Certain properties of MAO are expressed fairly consistently from tissue to tissue within the same species, and among the same tissues in different species. These properties include differences in affinities for various substrates and in sensitivities to certain drugs. For the most part such differences can be classified by assuming two types of activity, termed "A" and "B." Several recent reviews discuss the nature of these two types of activity, as well as variant properties of the enzyme that do not fit into this dual classification scheme (Neff and Yang, 1974; Houslay et al., 1976; Fowler, Callingham, Mantle, and Tipton, 1977; Murphy, 1978).

The A and B types of MAO activity are classified as follows. The A type preferentially deaminates serotonin and norepinephrine and is inhibited selectively by low concentrations of clorgyline (Johnston, 1968). The B type is more effective against phenylethylamine and benzylamine and more sensitive to inhibition by deprenyl (Knoll and Magyar, 1972). Several amines, e.g., tyramine and tryptamine, serve equally well as substrates for either form of the enzyme; some other inhibitors, e.g., pargyline, are not as selective for either form. The levels of A and B activity vary independently among different tissues, and apparently among different cell types within a tissue. Studies with isolated rat liver parenchymal cells (Tipton, Houslay, and Mantle, 1976) and cultured rat hepatoma cells (Hawkins, Costa, and Breakefield, 1978) suggest that both types of activity can be expressed together in the same cell. However, some cells express only one form of the enzyme, e.g., type A in most continuous rodent lines tested (Donnelly, Richelson, and Murphy, 1976; Murphy, Donnelly, and Richelson, 1976*b*; Hawkins and Breakefield, 1978) and type B in platelets (Edwards and Chang, 1975; Donnelly and Murphy, 1977) and several continuous human lines (Powell and Craig, 1978). In the developing brains of several species—mouse (Diez and Maderdrut, 1977), rat (Bourgoin, Artaud, Adrieu, Hery, Glowinski, and Hamon, 1977), and humans (Lewinsohn, Glover, and Sandler, unpublished; Suzuki and Yagi, 1977)—the A type of activity appears prior to the B type, suggesting separate, temporal control of gene expression.

Structure of the Enzyme

Knowledge of the molecular nature of MAO is necessary to understand genetic mechanisms of control. One would like to know how many polypeptides (subunits) form the enzyme, whether these polypeptides are identical or different, whether the polypeptides are chemically modified after translation, what other factors are involved in expression of activity, and whether there is any difference in the structure of the enzymes responsible for A and B types of activity. Such information would allow some estimate of the number of genes necessary for expression of activity. Unfortunately, few of these questions have been answered, since MAO has not been completely purified. The following information is available. The enzyme is tightly bound to the outer mitochondrial membrane (Sottocasa, Kuylenstierna, Ernster, and Bergstrand, 1967) and requires both a covalently bound flavin nucleotide (Walker, Kearney, Seng, and Singer, 1971; Chuang, Patek, and Hellerman, 1974) and iron (Youdim, 1976) for activity. Several laboratories have partially purified MAO using multistep procedures involving either treatment with nonionic detergents (Nara, Gomes, and Yasunobu, 1966; Erwin and Hellerman, 1967; Youdim and Sandler, 1968) or organic solvents (Hollunger and Oreland, 1970). Assuming a ratio of one flavin residue per enzyme molecule, the purified enzyme has been estimated to have a molecular weight of 100,000 to 150,000. However, it has been disturbing that the kinetic properties of the enzyme change during purification (Kandaswami, Diaz Borges, and D'Iorio, 1977) and, as a rule, only the B type of activity remains after purification. This seems to be due to the preferential inactivation of the A type of activity through removal of lipids, rather than conversion of A to B activity (Houslay and Tipton, 1973; Ekstedt and Oreland, 1976; McCauley, 1978).

Both purified enzyme and mitochondrial preparations have been labeled with radioactive inhibitors—[^{3}H]- and [^{14}C]pargyline, [^{14}C]-phenylhydrazine, and [^{14}C]deprenyl (Hellerman and Erwin, 1968; Erwin and Deitrich, 1971; Collins and Youdim, 1975; Youdim, 1976; Costa and Breakefield, 1979). These inhibitors are thought to bind tightly to the flavin residue of the enzyme (Chuang et al., 1974). Resolution by sodium dodecyl sulfate (SDS) polyacrylamide gel electrophoresis has revealed a single band with a MW of 60,000–70,000 (Collins and Youdim, 1975; McCauley, 1976; Costa and Breakefield, 1979). These results suggest that both types of activity reside in either one or several polypeptides of similar molecular weight.

The molecular basis of differences in A and B activity has been studied using antibodies against purified enzyme preparations. These studies have been inconclusive because only partially purified MAO has been used, and an inactive A form may copurify with the B form (McCauley, 1978). Both antigenic similarity (Dennick and Mayer, 1977) and dissimilarity (McCauley and Racker, 1973) between these forms have been reported. Immunologic analysis is further complicated by the fact that different antisera contain varying amounts of antibodies directed against both shared and unshared antigenic determinants.

Hypotheses on Molecular Basis of Differences in Activity

The differences between A and B types of activity could reside in the enzyme molecules themselves or in other properties specific to different cell types. Here we speculate on the mechanisms that could explain the existence of two distinct forms of enzyme activity. The means of generation of two forms could be extended to include additional forms. Three theories are illustrated in Figure 1. These mechanisms are not mutually exclusive, and all three could be operating simultaneously. Current evidence supports the idea that differences between A and B types of activity are conferred in part by the lipids associated with the enzyme in the outer mitochondrial membrane (Houslay and Tipton, 1973; Houslay, 1977). In Theory I, expression of A and B types of activity results from differences in the membrane microenvironment. In cells with both forms of activity, the lipid composition

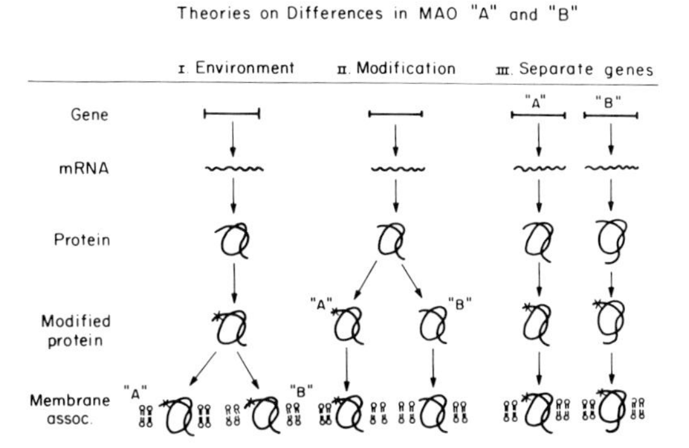

FIGURE 1. Three models are illustrated to explain the molecular differences between A and B types of MAO activity. Genes are those coding for the MAO protein itself. The final MAO molecule is associated (assoc.) with lipids in the outer mitochondrial membrane.

would have to be different among mitochondria within a cell or among domains within the mitochondrial membrane. In this theory the genetic basis of differences in activity would reside in genes controlling lipid metabolism and mitochondrial composition. In another theory (II), differences in activity might be conferred by posttranslational processing of the enzyme. The processed forms of the enzyme might associate inherently with different lipids. In this theory, genetic differences would reside in the structural genes and the regulated expression of enzymes necessary for processing of MAO. In the third theory (III), differences in activity might reside in the structure of the enzymes as determined by different structural gene loci. Differences in the primary amino acid sequence of the enzyme subunit(s) might determine in turn the mode of posttranslational modification and/or the type of lipid interactions. This theory would be compatible with an evolutionary divergence of duplicated gene loci accounting in part for varying expression of enzyme activity.

Inheritance of Activity

Studies on the inheritance of MAO activity may also shed light on the genetic control of this enzyme and the relationship between A and B activities. In the human population, MAO activity in platelets is inherited. Activity in any one individual is fairly constant over time (Murphy, Wright, Buchsbaum, Nichols, Costa, and Wyatt, 1976*c*), with only secondary changes resulting from age or physiologic state. Further, twin studies (Nies, Robinson, Lamborn, and Lampert, 1973; Murphy, Belmaker, and Wyatt, 1974) show a close correlation (0.88–0.95) in activities between monozygotic twins, who possess an identical genetic complement. Activities in dizygotic twins, who share approximately half their genes, have a correlation of about 0.50. As expected, activity in siblings and related individuals is more similar than in unrelated individuals (Nies et al., 1973; Murphy et al., 1974; Wyatt and Murphy, 1976). Although these findings do not elucidate the number of genes that control MAO activity, they do demonstrate that activity is primarily under genetic control.

Heritability is also important in evaluating the role of variations in MAO activity in human behavior and disease. If enzyme activity is primarily determined by inherited variations in the structure of the enzyme, one would expect a correlation of levels of activity throughout the tissues of an individual. On the other hand, if the activity is established by differentiated properties specific to certain cell types, there would be no reason to expect a correlation of activities among tissues from the same individual. In a comparison of MAO activity in inbred strains of mice, those with relatively lower activity in brain also had lower activity in liver (MacPike and Meier, 1976). Further studies are required to substantiate this observation. As discussed by Weinshilboum (1979), an understanding of the molecular basis of inherited levels of enzyme activity can establish the validity of sampling one tissue and inferring levels of activity in other tissues in vivo. If a correlation does not exist, it is essentially impossible to establish the association between levels of enzyme activity and other aspects of physiology.

DO THE SAME GENES CONTROL EXPRESSION OF A AND B TYPES OF ACTIVITY?

Genes that might be involved in expression of MAO include structural genes encoding the amino acid sequences of MAO itself and of other enzymes involved in posttranslational modification of MAO or metabolism of lipids critical to its function. Further, regulatory genes could modulate expression of the structural genes in several ways. If A and B types of MAO are encoded in different gene loci, their relative levels of activity could be controlled by varying the rates of transcription or translation of the genes, or the rates of posttranslational modification or degradation of these polypeptides. If both types of MAO are encoded in the same gene loci, other genes involved in posttranslational modification of MAO or control of the lipid environment may be important in determining the levels of each type of activity.

Molecular Studies in Rat Hepatoma Cells

In a number of continuous lines tested, only one rat hepatoma line, MH_1C_1 expressed both types of MAO activity (Hawkins and Breakefield, 1978; Hawkins et al., 1978). This was demonstrated in living cells by chromatographic analysis of metabolites formed from labeled amine substrates in the presence of varying concentrations of MAO inhibitors (Figure 2). Deamination of 5-hydroxytryptamine (A substrate) was preferentially blocked by low concentrations of clorgyline, while deamination of phenylethylamine (B substrate) was inhibited by low concentrations of deprenyl. Clorgyline inhibition of tryptamine (common substrate) deamination was biphasic, in agreement with the differential sensitivity of A and B types of activity.

These hepatoma cells have provided a homogeneous population of cells for analysis of possible differences in the structure of the protein(s) responsible for A and B types of MAO activity. The enzyme can be completely solubilized to remove associated lipids that could interfere with electrophoretic mobility. Since solubilization of the enzyme results in loss of activity, we have labeled the catalytic site by specific binding of [^{3}H]pargyline (Costa and Breakefield, 1979). This drug forms a stable adduct with the enzyme via the flavin cofactor (Chuang et al., 1974). Crude mitochondrial preparations of MH_1C_1 cells were incubated with 0.1 μM [^{3}H]pargyline for 1 hour at 37° C to allow maximal irreversible binding (Figure 3). Unbound label was removed by centrifugation, and proteins in the pellet were solubilized by boiling in SDS with β-mercaptoethanol. Electrophoresis was performed in polyacrylamide gels in the presence of SDS to separate proteins on the basis of molecular weight. At this concentration, [^{3}H]pargyline was bound to only one protein band with a MW of 57,000. Labeling of this band was inhibited by low concentrations of both clorgyline and deprenyl (Figure 4). Further, binding conditions resulted in loss of activity against both A and B substrates (Costa and Breakefield, 1979). We have concluded that the protein(s) responsible for the A and B types of activity in hepatoma cells have the same MW. These findings do not elucidate whether A and B types of MAO activity reside in the same or similar protein molecules.

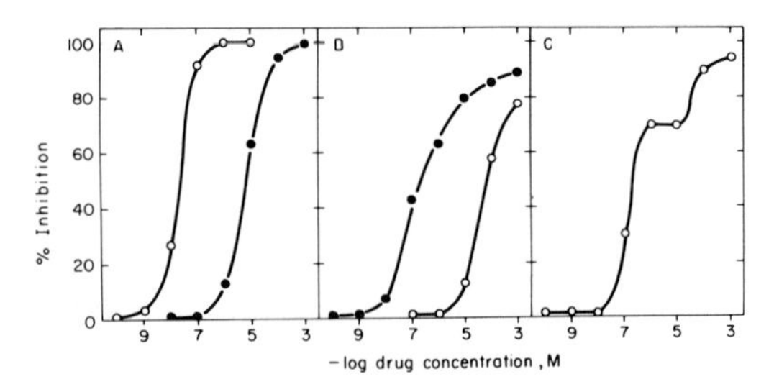

FIGURE 2. Inhibition of MAO activity in rat hepatoma cells. Cells were incubated with labeled amines and varying concentrations of clorgyline (○) and deprenyl (●) under conditions of culture. Metabolites in the cells and medium were identified and quantitated by thin layer chromatography. The formation of deaminated products was determined as the percent radioactivity recovered from the chromatogram comigrating with the appropriate standard. Cells were incubated with (*A*) 25 μM [^{3}H]-5-hydroxytryptamine, (*B*) 20 μM [^{14}C]phenylethylamine, or (*C*) 10 μM [^{3}H]-tryptamine. Details of the experimental protocols are given in the articles in which these figures were first published.

(Reprinted with permission from: Hawkins, M., Jr. and X. O. Breakefield [1978]. Monoamine oxidase A and B in cultured cells, *J. Neurochem.* **30:**1391–1397; and from Hawkins, M., Jr., M. R. C. Costa, and X. O. Breakefield [1978]. Distinct forms of monoamine oxidase expressed in hepatoma and HeLa cells in culture, *Biochem. Pharmacol.* **28:**525–528. Copyright both journals 1978 by Pergamon Press, Ltd.)

Activities in Human Platelets and Fibroblasts

Another way to examine the relationship between the A and B forms of MAO is to study the enzyme in cells that have only one type of activity. Nature has provided us with two human cell types in which one form of the enzyme predominates: peripheral blood platelets with type B activity (Donnelly and Murphy, 1977) and cultured skin fibroblasts with type A activity (Roth, Breakefield, and Castiglione, 1976; Groshong, Gibson, and Baldessarini, 1977).

Activity in platelets and fibroblasts can be distinguished by the biochemical characteristics of A and B types of activity. Thus platelets preferentially deaminate phenylethylamine and benzylamine, since B type activity has a higher affinity for these substrates (Donnelly and Murphy, 1977). Fibroblasts, on the other hand, preferentially deaminate the A substrate 5-hydroxytryptamine (Roth et al., 1976; Groshong et al., 1977). Common substrates, tryptamine and tyramine, are acted upon by MAO from both cell types with some difference in substrate affinity; thus in platelets the K_m for tryptamine is 12–24 μM (Edwards and Chang, 1975; Murphy, Donnelly, Miller, and Wyatt, 1976*a*; Berretini,

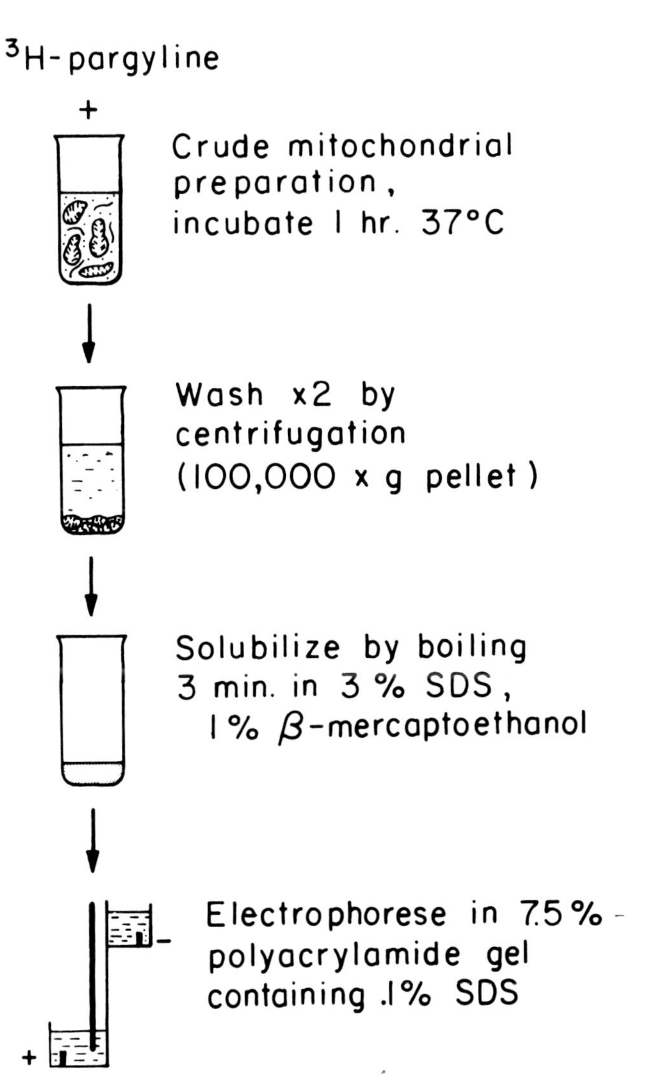

FIGURE 3. Procedure for labeling MAO from crude mitochondrial preparations with [^{3}H]pargyline. Details of this technique are given in the text and in Costa and Breakefield, 1979.

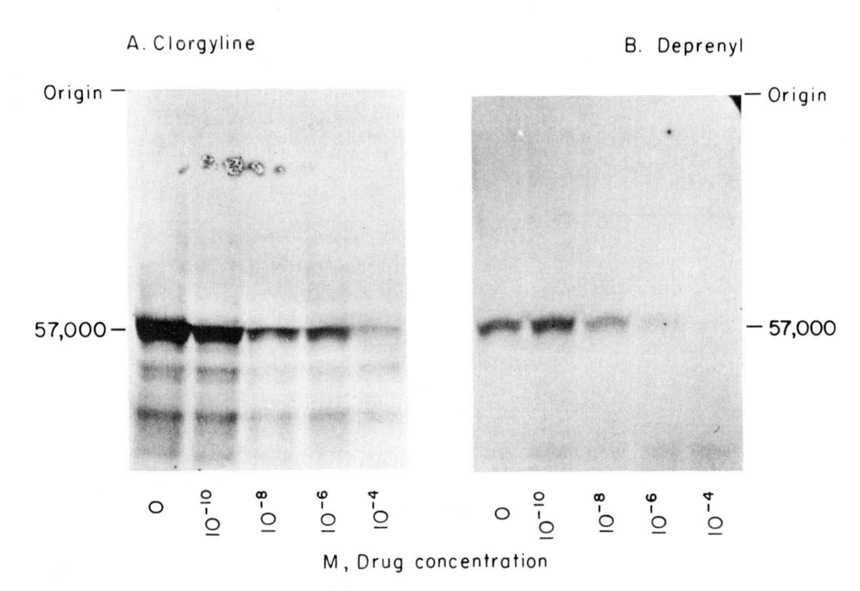

FIGURE 4. Inhibition of [^{3}H]pargyline binding to the 57,000-dalton protein by clorgyline and deprenyl. Following pre-incubation with varying concentrations of clorgyline or deprenyl for 30 min at 37° C, crude mitochondrial preparations were incubated with [^{3}H]pargyline. The samples were solubilized and electrophoresed on SDS polyacrylamide slab gels. Gels were exposed to pre-flashed X-Omatic R film for 12 days at −70° C. Autoradiograms of the dose-dependent inhibition of [^{3}H]-pargyline binding by clorgyline and deprenyl are shown in panels *A* and *B*, respectively. Results shown are from one of two similar experiments.

(Reprinted with permission from Costa, M. R. C. and X. O. Breakefield [1979]. Electrophoretic characterization of monoamine oxidase by ^{3}H-pargyline binding in rat hepatoma cells with A and B activity, *Mol. Pharmacol.*, in press. Copyright 1979 by Academic Press, Inc.)

Vogel, and Clouse, 1977), and in fibroblasts, 5–8 μM (Groshong et al., 1977; Costa, Edelstein, Castiglione, Chao, and Breakefield, 1979). As expected, MAO activity in platelets is especially sensitive to deprenyl (Donnelly and Murphy, 1977), and that in fibroblasts to clorgyline (Costa et al., 1979). These differences in substrate affinity and drug sensitivity could be explained by differences in the structure or conformation of A and B types of MAO, or by differences in associated lipids and other proteins.

Levels of MAO activity in platelets have been shown to be under genetic control (see above). Differences in activity among individuals may reflect allelic variations in the genes coding for the MAO molecule itself, or other regulatory proteins or enzymes responsible for MAO turnover and activity. A comparison of the range and distribution of activities in human fibroblasts and platelets could provide insight into their separate or common modes of inheritance. In platelets from 680 control individuals, Murphy et al. (1976c) observed a unimodal distribution of activity over a 10-fold range. This pattern of distribution provides no information as to the number of gene loci or of allelic variants at these loci that determine levels of activity. In a smaller study in cultured fibroblasts from 24 control individuals, we have observed an apparent bimodal distribution of activity over a 50-fold range (Figure 5). The different patterns of distribution of B and A activity in platelets and fibroblasts suggest they are under different genetic control.

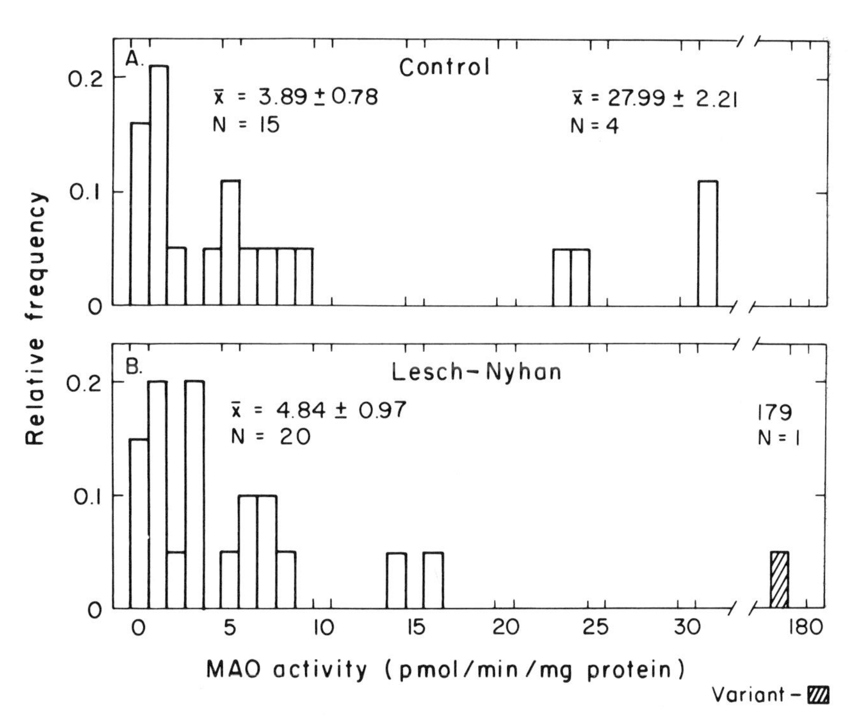

FIGURE 5. Distribution of MAO activity in fibroblasts from control individuals and patients with the Lesch-Nyhan syndrome. MAO activity was measured against saturating concentrations of tryptamine using homogenates prepared from cells grown under standard conditions of culture. Mean ± SEM values are presented; N = number of lines examined. This is a diagrammatic representation of data presented elsewhere (Costa et al., 1979).

The discrete distribution of activity in fibroblasts is consistent with levels being predominantly determined by two alleles at a single locus. The large number of individuals in the high and low categories, about 80% and 20% of the population, respectively, indicates this polymorphism is common. So far, in the small number of lines examined (Costa et al., 1979), we have observed no difference in the distribution of activity on the basis of sex. This argues against X-linkage of this locus and suggests an autosomal location.

If levels of A and B activity were under the same genetic control, we would anticipate a correlation between activity in platelets and fibroblasts from the same individual. No correlation has been observed in these two cell types from 16 individuals we (Giller, Breakefield, Young, and Cohen, in preparation) and others (Groshong et al., 1977) have studied. This contrasts with the finding of Bond and Cundall (1977) that levels of B activity in platelets and lymphocytes from the same individual are similar.

The biochemical and genetic relationship between A and B forms of MAO activity remains to be resolved. Purification of different forms of MAO will be necessary to elucidate the extent of their similarities. If A and B types of activity are encoded in different genes, these genes may be distinguished by their location on different chromosomes or by independent mutations at these loci.

DO ALLELIC VARIANTS OF MAO EXIST IN THE HUMAN POPULATION?

Monoamine oxidase is one of the few enzymes critical to neuronal transmitter metabolism that can be measured in tissues obtained from living individuals. If inherited variations in the functioning of this enzyme occur in the population, these variations may be expressed in cells throughout the body. Measurements of MAO in peripheral cell types then may allow us to assess the role of altered expression of this enzyme in vivo.

Polymorphisms of B Activity in Platelets

Monoamine oxidase activity has been measured in platelets from a large number of control individuals and patients suffering from neurologic and psychiatric disorders. Although conflicting evidence exists (e.g., see Groshong, Baldessarini, Gibson, Lipinski, Axelrod, and Pope, 1978), many studies have reported levels of B type activity in the low normal range in individuals suffering from chronic schizophrenia (Murphy et al., 1974), alcoholism (Sullivan, Stanfield, and Dackis, 1977), and bipolar depression (Murphy and Weiss, 1972). Further, control individuals with MAO activity in the low normal range tend to have a distinctive personality type and a higher incidence of psychopathologic behavior (Buchsbaum, Coursey, and Murphy, 1976; Buchsbaum, Harir, and Murphy, 1977; Murphy, Belmaker, Buchsbaum, Martin, Ciaranello, and Wyatt, 1977). These studies suggest that individuals who inherit low levels of enzyme activity may be more susceptible to certain environmental stresses and may have a greater tendency to express symptoms associated with certain neuropsychiatric diseases.

Some properties of MAO other than activity have been examined in human platelets. Some workers have reported altered substrate affinity and the presence of endogenous inhibitors of MAO in platelets from schizophrenic, as compared to control, individuals (Berretini and Vogel, 1978), but this has not been observed by others (Murphy, 1978). Further, Murphy et al. (1976*a*) found no apparent difference in the thermal stability of platelet MAO in control and schizophrenic populations. Attempts to analyze electrophoretic properties of platelet MAO have been uninformative due to incomplete solubilization of the enzyme (Belmaker, Ebstein, Rimon, Wyatt, and Murphy, 1976). Thus, although studies in platelets have established the inherited basis of human MAO activity and highlighted the potential role of varying levels of activity in human behavior, they have not elucidated the molecular basis for these variations in activity.

Polymorphisms of A Activity in Fibroblasts

Another approach to analysis of the inheritance of MAO in the human population is to study enzyme activity in cultured human skin fibroblasts. These cells offer some advantages over platelets. Fibroblasts possess predominantly type A activity (Roth et al., 1976; Groshong et al., 1977), and since the A and B forms of the enzyme appear to be under different genetic controls (see above), they offer the opportunity to evaluate the inheritance of another type of activity. Since fibroblasts can be studied as actively metabolizing cells under controlled conditions of culture, the inherent cellular regulation of activity can be assessed apart from variations due to the physiologic state of the patient. The primary disadvantages of fibroblasts are that the more intrusive nature of obtaining skin biopsies makes it harder to procure donors, and the time-consuming nature of cell culture limits the number of lines that can be examined. Still, once the lines have been established, substantial amounts of tissue can be grown, and the cells can be stored viably in the frozen state for further study independent of patient contact.

Monoamine oxidase has a wide range of activities in human fibroblasts (see above). We have begun to assess the molecular basis of these variations in activity and their role in certain inherited neurologic diseases. In order to determine whether MAO from fibroblasts with low and high activity differs in structure, we have studied several other properties of the enzyme in a small number of lines (Costa et al., 1979). Monoamine oxidase measurements in homogenates with high and low activity did not differ in affinity for tryptamine, suggesting no alteration in structure at the substrate binding site and no endogenous competitive inhibitors. However, two lines with very low activity (<2 pmol/min/mg protein) were observed to have a more thermostable form of the enzyme as compared to lines with higher activity. This was measured as a twofold increase in the time required to inhibit half the enzyme activity at 56° C. This difference in thermal stability could be due to differences in the amount of the enzyme protein, its structure, or its association with other cellular components, e.g., lipids. Electrophoretic analysis of the enzyme from lines with high and low activity has been limited by the small amount of enzyme present in these cells.

Another explanation for variations in levels of enzyme activity would be differences in the number of active enzyme molecules present per cell. The most direct way to assess this would be to titrate the amount of enzyme by using specific antibodies; however, such antibodies are not available to us. Alternatively, we have attempted to titrate the amount of enzyme using clorgyline, a drug that binds specifically, irreversibly, and stoichiometrically to the A form of the enzyme (Rando, 1974; Egashira, Ekstedt, Kinemuchi, Wiberg, and Oreland, 1976; Fowler and Callingham, 1978). Using a constant amount of homogenate protein per assay, the amount of clorgyline required to completely inhibit enzyme activity was determined by a plot of percent inhibition versus clorgyline concentration (Costa et al., 1979). Five lines varying 50-fold in activity required essentially the same amount of clorgyline (pmol/mg homogenate protein) to completely inhibit activity. The two lines with the lowest activity (<2 pmol/min/mg protein) bound approximately half as much as the other lines, but this was not proportional to their level of enzyme activity. Several explanations could account for the observation that the amount of clorgyline needed to inhibit activity does not appear to reflect the level of enzyme activity. Lines with very low activity may be homozygous for a gene resulting in a structurally altered enzyme molecule in which the reaction velocity proceeds more slowly, so that the same number of active enzyme molecules produces a slower reaction rate. Another possibility would be that a large portion of the clorgyline binds nonspecifically to other cellular components and that this large background reduces our sensitivity in measuring the number of enzyme molecules. Further, if lines with very low activity take a longer time to form an irreversible bond between clorgyline and MAO, this might cause us to overestimate the number of MAO molecules in low lines. We have shown that lines with low activity do not have proportionally more of the B form of the enzyme (Roth et al., 1976).

Together, these studies suggest that polymorphisms for the A form of MAO do exist in the human population. One theory would be that two common alleles at one gene locus determine levels of MAO activity in these cells. Individuals homozygous for the allele conferring low activity would be distinguished by having an activity in the range of 0.5–2 pmol/min/mg protein and a greater thermal stability. Individuals homozygous for the allele conferring high activity would have an activity of 20–40 pmol/min/mg protein and a lower thermal stability. Individuals heterozygous for these alleles would have intermediate levels of activity, 2–15 pmol/min/mg protein; thermal stability characteristics would be similar to the homozygous high population, since activity contributed by the high allele would be over 10 times that contributed by the low allele. It is tempting to speculate that this allele is the structural gene for the A form of MAO, but further work is required to evaluate this possibility.

Studies using monozygotic twins and family pedigrees are under way to establish the inherited basis of levels of MAO activity in fibroblasts. Information is available now from only one small pedigree consisting of paternal grandparents, parents, and two daughters (Figure 6). The pattern of inheritance is consistent with activity being determined by two alleles, with members of this family being either homozygous for the high allele or heterozygous for the high and low alleles. One of the daughters has an activity (3 pmol/min/mg protein) on the borderline between low and intermediate levels. Her activity, however, has a thermal stability typical of high and intermediate types of activity and clearly distinguishable from the low type. It is apparent from the wide range of activities within the intermediate category (3–15 pmol/

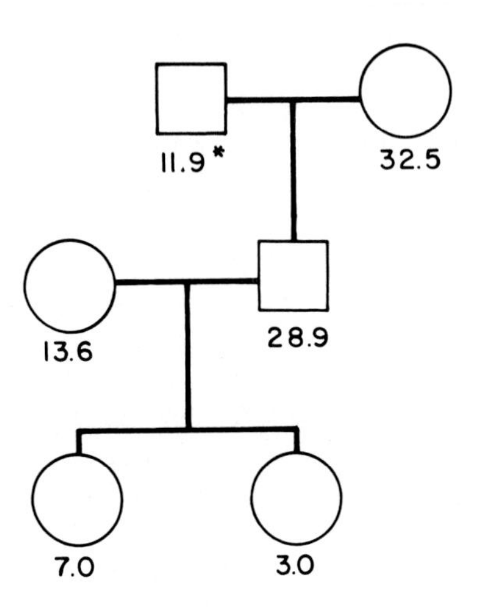

FIGURE 6. MAO activity in fibroblasts from a control pedigree. Cell lines were obtained from the Institute for Medical Research in Camden, New Jersey. Activity was measured in cell homogenates against tryptamine as described in the legend to Figure 5. The value presented for each cell line is the mean of the activities measured in three subcultures of that line. The ages of the family members at the time that skin biopsies were taken were as follows: grandfather, aged 70 years; grandmother, 71; father, 37; mother, 37; daughter at left, 13; daughter at right, 11.

min/mg protein) that even if two alleles primarily control activity, other factors must also modulate it. Since these categories of activity undoubtedly overlap, it will be necessary to monitor other properties of the enzyme, such as thermal stability and electrophoretic mobility.

To date, only a limited number of studies have been carried out to determine the levels of MAO activity in fibroblasts from individuals with neurologic and psychiatric diseases. The largest study has been a comparison of MAO activity in fibroblasts from 20 Lesch-Nyhan patients and 24 age-, sex- and race-matched controls (Figure 5, Costa et al., 1979). The Lesch-Nyhan syndrome is an X-linked inherited disease with loss of hypoxanthine-guanine phosphoribosyltransferase

(HPRT) activity causing an overproduction of uric acid (Lesch and Nyhan, 1964; Seegmiller, Rosenbloom, and Kelley, 1967). In addition to gout and kidney failure (treatable by inhibition of xanthine oxidase), these patients manifest profound neurologic symptoms, including spasticity, choreoathetosis, mental retardation, and self-mutilation (which are not amenable to treatment). Other studies with HPRT-deficient neuroblastoma cells (Breakefield, Castiglione, and Edelstein, 1976) led us to study MAO activity in this disease. Fibroblasts from the typical Lesch-Nyhan patients have MAO activities exclusively in the low and intermediate control range. Expressed as total population means, the Lesch-Nyhan population has about half the activity of the control population ($P < .05$). No biochemical link has been established between loss of HPRT activity and decrease in MAO activity (Breakefield, Edelstein, and Costa, 1979). Low MAO activity in the Lesch-Nyhan population might be explained by random sampling from a small population, such that patients with high MAO activity were missed by chance. However, another finding suggests that MAO activity may play some role in the neurologic manifestations of this disease. A variant patient, with mild symptoms (no mental retardation, choreoathetosis, or self-mutilation) and <3% of normal HPRT activity in blood cells (Manzke, 1976), had very high MAO activity in his fibroblasts (Edelstein, Castiglione, and Breakefield, 1978). If type A MAO activity is also high in the nervous tissue of this individual, it may compensate for some of the metabolic derangements produced by HPRT deficiency. If this theory is true, fibroblasts may be obtained more often from HPRT-deficient individuals with low or intermediate MAO activity because these individuals are more severely affected than those with high MAO activity and thus more frequently identified for study. In order to resolve this question, fibroblasts need to be obtained from a large number of HPRT-deficient individuals with varying severity of symptoms. As proposed by Murphy et al. (1974) to explain low normal platelet MAO in chronic schizophrenics, the physiologic expression or penetrance of certain inherited lesions may be modulated by normal variations in the products of other gene loci.

Levels of MAO activity have been measured in a small number of fibroblast lines from patients with a number of other inherited diseases with neural involvement, including dystonia musculoram deformans (autosomal recessive and dominant forms), familial dysautonomia (autosomal recessive), and Duchenne's muscular dystrophy (X-linked recessive) (Edelstein, Costa, Castiglione, and Breakefield, 1979). Several other disease populations have also been examined in which the pattern of inheritance is not clear-cut and symptoms span both neurologic and psychiatric categories, e.g., the Gilles de la Tourette syndrome, autism (Giller, Breakefield, Young, and Cohen, in preparation), and chronic schizophrenia (Groshong et al., 1978). In all cases, activity fell within the normal range; the small sample sizes preclude the possibility of establishing the significance of minor deviations. It is not known whether a major functional alteration in MAO would be compatible with life and/or produce neurologic dysfunction. It is useful, then, to screen individuals with inherited neurologic diseases for altered MAO. To accurately evaluate the role of variations in MAO activity in any disease, large numbers of affected individuals and their family members must be analyzed.

MONOAMINE OXIDASES AS ISOZYMES

Numerous examples exist of multiple molecular forms of the same enzyme (isozymes) (Markert and Moller, 1959). The genetic basis of these variations in protein structure has been eloquently and extensively reviewed by Harris (1977). To paraphrase his interpretations, such variations can occur by the presence of multiple gene loci coding for similar proteins, multiple allelism at one or more gene loci, and/or posttranslational modification of proteins. In this discussion we will concentrate on the possibility that different gene loci code for the A and B forms of the enzyme.

There are many examples of evolutionarily related gene loci giving rise to isozymic forms of enzymes that differ in their tissue distribution, developmental expression, drug sensitivity, and substrate affinity (Harris, 1977). During the course of evolution, certain gene loci have duplicated, and these duplicated loci have undergone independent mutational events leading to divergence in amino acid sequence. Since the evolutionary history of related genes varies between species, the same ancestral gene may appear later in evolution in varying numbers, lengths, and nucleotide sequences in different species, and even in different individuals within a species, as illustrated by the hemoglobin gene loci (see Harris, 1977). This may account for the fact that some of the classic characteristics of A and B types of activity break down in comparisons between species. Thus, although many species may express two distinct forms of MAO, some may have only one. Further, the properties for each form may vary between species, so that, for example, type A activity might be somewhat different in humans and rodents.

The variation in expression of MAO activity in different tissues at different times in development is again reminiscent of the situation described for other isozymes. For example, lactate dehydrogenase is a tetramer composed of polypeptides coded for by three separate gene loci (see Harris, 1977). These polypeptide subunits have similar molecular weights but differ in amino acid sequence; this in turn affects their interaction with substrates and cofactor and their electrophoretic mobility. Some tissues express almost exclusively one form of lactate dehydrogenase, while others express other forms composed of combinations of the polypeptide subunits. The relative amounts of each type of activity are thought to result from differences in the rate of synthesis and degradation of the subunits within a given cell. Further, different forms of lactate dehydrogenase are expressed in the same tissue at different stages of development. Control of activity by temporal regulation of gene expression is also illustrated by fetal and adult hemoglobins.

The molecular differences between A and B types of activity are not clear. The products of genetically related gene loci frequently retain the same chain length while differing in amino acid sequence. This would be compatible with the similar molecular weights seen for A and B forms of MAO in SDS polyacrylamide gels. No information is available as to possible differences in the charge of these forms under fully solubilizing conditions.

The relative expression of the A and B types of MAO activity within a given tissue could result from differential and temporal gene expression and/or cell-specific differences in synthesis, degradation, or processing. The same and/or different gene loci may be involved in controlling the structure and expression of the A and B types of MAO activity. At present, more information is needed to evaluate the role of genetic elements in determining levels and types of MAO activity.

ACKNOWLEDGMENTS

We thank John Pintar for critical evaluation of this paper, Carmela M. Castiglione for expert assistance and advice, and Lisa Poteet and Regina Gambardella for skilled preparation of the manuscript. This work was supported by United States Public Health Service grant GM20124 and by the National Foundation-March of Dimes. M.R.C.C. was funded through a James H. Brown Fellowship from Yale University and the Dystonia Medical Research Foundation; M.H. was funded by National Science Foundation Faculty Fellowship GZ-3723 and National Institute of General Medical Sciences MARC Fellowship GM05913.

REFERENCES

Belmaker, R. H., R. Ebstein, R. Rimon, R. J. Wyatt, and D. L. Murphy (1976). Electrophoresis of platelet monoamine oxidase in schizophrenia and manic-depressive illness, *Acta Psychiatr. Scand.* **54**:67–72.

Berger, P. A. (1977). Antidepressant medications and the treatment of depression, pp. 174–207 in *Psychopharmacology,* Barchas, J. D., P. A. Berger, R. D. Ciaranello, and G. R. Elliott, eds. Oxford University Press, Oxford.

Berretini, W. H. and W. H. Vogel (1978). Evidence for an endogenous inhibitor of platelet MAO in chronic schizophrenia, *Am. J. Psychol.* **135**:605–607.

Berretini, W. H., W. H. Vogel, and R. Clouse (1977). Platelet monoamine oxidase in chronic schizophrenia, *Am. J. Psychiatry* **134**:805–806.

Bond, P. A. and R. L. Cundall (1977). Properties of monoamine oxidase (MAO) in human blood platelets, plasma, lymphocytes and granulocytes, *Clin. Chim. Acta* **80**:317–326.

Bourgoin, S., F. Artaud, J. Adrieu, F. Hery, J. Glowinski, and M. Hamon (1977). 5-Hydroxytryptamine catabolism in the rat brain during ontogenesis, *J. Neurochem.* **28**:415–422.

Breakefield, X. O., C. M. Castiglione, and S. B. Edelstein (1976). Monoamine oxidase activity decreased in cells lacking hypoxanthine phosphoribosyltransferase activity, *Science* **192**:1018–1020.

Breakefield, X. O., S. B. Edelstein, and M. R. C. Costa (1979). Genetic analysis of neurotransmitter metabolism in cell culture: studies on the Lesch-Nyhan syndrome, pp. 197–234 in *Neurogenetics: Genetic Approaches to the Nervous System*, Breakefield, X. O., ed. Elsevier/North Hollland, New York.

Buchsbaum, M. S., R. D. Coursey, and D. L. Murphy (1976). The biochemical high-risk paradigm: behavioral and familial correlates of low platelet monoamine oxidase, *Science* **194**:339–341.

Buchsbaum, M. S., R. J. Harir, and D. L. Murphy (1977). Suicide attempts, platelet monoamine oxidase and the average evoked response, *Acta Psychiatr. Scand.* **56**:69–79.

Campbell, I. C., D. W. Gallager, D. L. Murphy, and J. F. Tallman (1978). Effects of chronic monoamine oxidase inhibitors on central noradrenergic systems, *Soc. Neurosci. Abstr.*, Vol. 4, p. 269.

Chuang, H. Y. K., D. R. Patek, and L. Hellerman (1974). Mitochondrial monoamine oxidase. Inactivation by pargyline. Adduct formation, *J. Biol. Chem.* **249**:2381–2384.

Collins, G. G. S. and M. B. H. Youdim (1975). The binding of [^{14}C]phenethylhydrazine to rat liver monoamine oxidase, *Biochem. Pharmacol.* **24**:703–706.

Costa, M. R. C. and X. O. Breakefield (1979). Electrophoretic characterization of monoamine oxidase by ^{3}H-pargyline binding in rat hepatoma cells with A and B activity, *Mol. Pharmacol.*, in press.

Costa, M. R. C., S. B. Edelstein, C. M. Castiglione, H. Chao, and X. O. Breakefield (1979). Properties of monoamine oxidase in control and Lesch-Nyhan fibroblast, *Am. J. Hum. Genet.*, submitted.

Dennick, R. G. and R. J. Mayer (1977). Purification and immunochemical characterization of monoamine oxidase from rat and human liver, *Biochem. J.* **161**:167–174.

Diez, J. A. and J. L. Maderdrut (1977). Development of multiple forms of mouse brain monoamine oxidase in vivo and in vitro, *Brain Res.* **128:**187–192.
Donnelly, C. H. and D. L. Murphy (1977). Substrate- and inhibitor-related characteristics of human platelet monoamine oxidase, *Biochem. Pharmacol.* **26:**853–856.
Donnelly, C. H., E. Richelson, and D. L. Murphy (1976). Properties of monoamine oxidase in mouse neuroblastoma N1E-115 cells, *Biochem. Pharmacol.* **25:**1639–1643.
Edelstein, S. B., C. M. Castiglione, and X. O. Breakefield (1978). Monoamine oxidase activity in normal and Lesch-Nyhan fibroblasts, *J. Neurochem.* **31:**1247–1254.
Edelstein, S. B., M. R. C. Costa, C. M. Castiglione, and X. O. Breakefield (1979). Monoamine oxidase activity in fibroblasts from control individuals and patients with inherited neurologic diseases, in *Catecholamines: Basic and Clinical Frontiers,* Usdin, E., ed. Plenum Press, New York, in press.
Edwards, D. J. and S.-S. Chang (1975). Evidence for interacting catalytic sites of human platelet monoamine oxidase, *Biochem. Biophys. Res. Commun.* **65:**1018–1025.
Egashira, T., B. Ekstedt, H. Kinemuchi, A. Wiberg, and L. Oreland (1976). Molecular turnover numbers of different forms of mitochondrial monoamine oxidase in rat, *Med. Biol. (Helsinki)* **54:**272–277.
Ekstedt, B. and L. Oreland (1976). Effect of lipid-depletion on the different forms of monoamine oxidase in rat liver mitochondria, *Biochem. Pharmacol.* **25:**119–124.
Erwin, V. G. and R. A. Deitrich (1971). The labelling in vivo of monoamine oxidase by ^{14}C-pargyline: a tool for studying the synthesis of the enzyme, *Mol. Pharmacol.* **7:**219–228.
Erwin, V. G. and L. Hellerman (1967). Mitochondrial monoamine oxidase. I. Purification and characterization of the bovine kidney enzyme, *J. Biol. Chem.* **242:**4230–4238.
Fowler, C. J. and B. A. Callingham (1978). The inhibition by clorgyline of 5-hydroxytryptamine deamination by the rat liver, *J. Pharm. Pharmacol.* **30:**304–309.
Fowler, C. J., B. A. Callingham, T. J. Mantle, and K. F. Tipton (1977). Monoamine oxidase A and B: a useful concept? *Biochem. Pharmacol.* **27:**97–101.
Groshong, R., R. J. Baldessarini, A. Gibson, J. F. Lipinski, D. Axelrod, and A. Pope (1978). Activities of types A and B MAO and catechol-*O*-methyltransferase in blood cells and skin fibroblasts of normal and chronic schizophrenic subjects, *Arch. Gen. Psychiatry* **35:**1198–1205.
Groshong, R., D. A. Gibson, and R. J. Baldessarini (1977). Monoamine oxidase activity in cultured human skin fibroblasts, *Clin. Chim. Acta* **80:**113–120.
Harris, H. (1977). *The Principles of Human Biochemical Genetics.* North Holland Publishing Company, New York.
Hawkins, M., Jr. and X. O. Breakefield (1978). Monoamine oxidase A and B in cultured cells, *J. Neurochem.* **30:**1391–1397.
Hawkins, M., Jr., M. R. C. Costa, and X. O. Breakefield (1978). Distinct forms of monoamine oxidase expressed in hepatoma and HeLa cells in culture, *Biochem. Pharmacol.* **28:**525–528.

Hellerman, L. and V. G. Erwin (1968). Mitochondrial monoamine oxidase. II. Action of various inhibitors for the bovine kidney enzyme. Catalytic mechanism, *J. Biol. Chem.* **243**:5234–5243.

Hollunger, G. and L. Oreland (1970). Preparation of soluble monoamine oxidase from pig liver mitochondria, *Arch. Biochem. Biophys.* **139**:320–328.

Houslay, M. D. (1977). A model for the selective mode of action of the irreversible monoamine oxidase inhibitors clorgyline and deprenyl, based on studies of their ability to activate Ca^{++}-Mg^{++} ATPase in defined lipid environments, *J. Pharm. Pharmacol.* **29**:664–669.

Houslay, M. D. and K. F. Tipton (1973). The nature of electrophoretically separable forms of rat liver monoamine oxidase, *Biochem. J.* **135**:173–186.

Houslay, M. D., K. F. Tipton, and M. B. H. Youdim (1976). Multiple forms of monoamine oxidase: fact and artifact, *Life Sci.* **19**:467–478.

Johnston, J. P. (1968). Some observations upon a new inhibitor of monoamine oxidase in brain tissue, *Biochem. Pharmacol.* **17**:1285–1297.

Kandaswami, C., J. M. Diaz Borges, and A. D'Iorio (1977). Studies on the fractionation of monoamine oxidase from rat liver mitochondria, *Arch. Biochem. Biophys.* **183**:273–280.

Knoll, J. and K. Magyar (1972). Some puzzling pharmacological effects of monoamine oxidase inhibitor, *Adv. Biochem. Psychopharmacol.* **5**:393–408.

Lesch, M. and W. L. Nyhan (1964). A familial disorder of uric acid metabolism and central nervous system function, *Am. J. Med.* **36**:561–570.

MacPike, A. D. and H. Meier (1976). Genotype dependence of monoamine oxidase in inbred strains of mice, *Experientia* **32**:979–980.

Manzke, H. (1976). Variable Expressivität der Genwirkung beim Lesch-Nyhan-Syndrom, *Dtsch. Med. Wochenschr.* **101**:428–429.

Markert, C. L. and F. Moller (1959). Multiple forms of enzymes: tissue, ontogenetic, and species specific patterns, *Proc. Natl. Acad. Sci. USA* **45**:753–763.

McCauley, R. (1976). 7[^{14}C]Pargyline binding to mitochondrial outer membranes, *Biochem. Pharmacol.* **25**:2214–2216.

McCauley, R. (1978). Properties of monoamine oxidase from rat liver mitochondrial outer membranes, *Arch. Biochem. Biophys.* **189**:8–13.

McCauley, R. and E. Racker (1973). Separation of two monoamine oxidases from bovine brain, *Mol. Cell. Biochem.* **1**:73–81.

Murphy, D. L. (1978). Substrate-selective monoamine oxidases—inhibitor, tissue, species and functional differences, *Biochem. Pharmacol.* **27**:1889–1893.

Murphy, D. L., R. H. Belmaker, M. Buchsbaum, N. F. Martin, R. Ciaranello, and R. J. Wyatt (1977). Biogenic amine-related enzymes and personality variations in normals, *Psychol. Med.* **7**:149–157.

Murphy, D. L., R. Belmaker, and R. Wyatt (1974). Monoamine oxidase in schizophrenia and other behavioral disorders, *J. Psychol. Res.* **11**:221–247.

Murphy, D. L., C. H. Donnelly, L. Miller, and R. J. Wyatt (1976*a*). Platelet monoamine oxidase in chronic schizophrenia. Some enzyme characteristics relevant to reduced activity, *Arch. Gen. Psychiatry* **33**:1377–1381.

Murphy, D. L., C. H. Donnelly, and E. Richelson (1976*b*). Substrate and inhibitor-related characteristics of monoamine oxidase in C6 rat glial cells, *J. Neurochem.* **26:**1231–1235.

Murphy, D. L. and R. Weiss (1972). Reduced monoamine oxidase activity in blood platelets from bipolar depressed patients, *Am. J. Psychiatry* **128:** 1351–1357.

Murphy, D. L., C. Wright, M. Buchsbaum, A. Nichols, J. L. Costa, and R. J. Wyatt (1976*c*). Platelet and plasma amine oxidase activity in 680 normals: sex and age differences and stability over time, *Biochem. Med.* **16:**254–265.

Nara, S., B. Gomes, and K. T. Yasunobu (1966). Amine oxidase. VII. Beef liver mitochondrial monoamine oxidase, a copper-containing protein, *J. Biol. Chem.* **241:**2774–2780.

Neff, N. H. and H.-Y. T. Yang (1974). Another look at the monoamine oxidases and the monoamine oxidase inhibitor drugs, *Life Sci.* **14:**2061–2074.

Nies, A., D. S. Robinson, K. R. Lamborn, and R. P. Lampert (1973). Genetic control of platelet and plasma monoamine oxidase activity, *Arch. Gen. Psychiatry* **28:**834–838.

Powell, J. F. and I. W. Craig (1978). The characterization of monoamine oxidase (E.C. 1.4.3.4) in cultured mammalian cells, *J. Neurochem.* **32:**521–527.

Rando, R. R. (1974). Chemistry and enzymology of K_{cat} inhibitors, *Science* **185:**320–324.

Roth, J. A., X. O. Breakefield, and C. M. Castiglione (1976). Monoamine oxidase and catechol-*O*-methyltransferase activities in cultural human skin fibroblasts, *Life Sci.* **19:**1705–1710.

Seegmiller, J. E., F. M. Rosenbloom, and W. N. Kelley (1967). Enzyme defect associated with a sex-linked human neurological disorder and excessive purine synthesis, *Science* **155:**1682–1684.

Sottocasa, G. L., B. Kuylenstierna, L. Ernster, and A. Bergstrand (1967). An electron transport system associated with the outer membrane of liver mitochondria. A biochemical and morphologic study, *J. Cell Biol.* **32:**415–438.

Sullivan, J. L., C. N. Stanfield, and C. Dackis (1977). Platelet MAO activity in schizophrenia and other psychiatric illnesses, *Am. J. Psychol.* **134:**1098–1103.

Suzuki, O. and K. Yagi (1977). Multiple forms of monoamine oxidase in the human cerebral cortices at different ages, pp. 100–107 in *Maturation of Neurotransmission,* Vernadakis, A. and G. Filogamo, eds. S. Karger, Basel.

Tipton, K. F., M. D. Houslay, and T. J. Mantle (1976). The nature and locations of the multiple forms of monoamine oxidase, pp. 5–31 in *Monoamine Oxidase and Its Inhibition,* Wolstenholme, G. W. E. and J. Knight, eds. Elsevier/North Holland, New York.

Walker, W. H., E. B. Kearney, R. L. Seng, and T. P. Singer (1971). The covalently-bound flavin of hepatic monoamine oxidase. 2. Identification and properties of cysteinyl riboflavin, *Eur. J. Biochem.* **24:**328–331.

Weinshilboum, R. M. (1979). Catecholamine biochemical genetics in human populations, pp. 257–282 in *Neurogenetics: Genetic Approaches to the Nervous System*, Breakefield, X. O., ed. Elsevier/North Holland, New York.

White, W. L. and A. T. Glassman (1977). Multiple binding sites of human brain and liver monoamine oxidase: substrate specificities, selective inhibitors and attempts to separate enzyme forms, *J. Neurochem.* **29**:987–997.

Wyatt, R. J. and D. L. Murphy (1976). Low platelet monoamine oxidase activity and schizophrenia, *Schizophrenia Bull.* **2**:77–89.

Youdim, M. B. H. (1976). Rat liver mitochondrial monoamine oxidase—an iron-requiring flavoprotein, pp. 593–609 in *Flavins and Flavoproteins,* Singer, T. P., ed. Elsevier Scientific Publishing Company, New York.

Youdim, M. B. H. and M. Sandler (1968). The effect of prenylamine on monoamine oxidase, *Biochim. Appl.* **14**:175–184.

GENETIC REGULATION OF CATECHOL-*O*-METHYLTRANSFERASE

Richard M. Weinshilboum

Mayo Medical School, Rochester, Minnesota

INTRODUCTION

The catecholamines are an important class of neurotransmitters. Variation in the function of dopaminergic, noradrenergic, and adrenergic neurons may be involved in the pathogenesis of many neuropsychiatric and cardiovascular diseases and in variations in response to drugs used to treat these diseases. Very little is known of the possible role of inheritance in the regulation of catecholamine neurotransmitter systems in man. One reason for the lack of human catecholamine biochemical genetic information is the difficulty of obtaining neural tissue from a large number of subjects. A strategy that might be used as a first step in biochemical genetic studies of this neurotransmitter system is the measurement of the activities of enzymes involved in catecholamine biosynthesis and metabolism in nonneural tissue. Underlying such experiments would be the assumptions that the enzymes in nonneural tissue are biochemically identical with and are regulated in parallel with those in neural tissue. These assumptions must eventually be tested experimentally. The following review will describe experiments in which this approach has been used to study the biochemical genetics of the catecholamine metabolic enzyme catechol-*O*-methyltransferase.

Catechol-*O*-methyltransferase (COMT, EC 2.1.1.6) is involved in the metabolism of catecholamines and a variety of catechol drugs (Axelrod and Tomchick, 1958). COMT is present in many tissues of man and experimental animals, including the erythrocyte (RBC) (Axelrod and Cohn, 1971). RBC COMT is immunologically and biochemically similar to the enzyme in the brain, heart, and liver (Assicot and Bohuon, 1969; Creveling, Borchardt, and Isersky, 1973; Quiram and Weinshilboum, 1976*a*). The presence of COMT in blood, an easily obtained human tissue, made it possible to begin to study the role of inheritance in the regulation of this enzyme activity in man.

BIOCHEMICAL GENETICS OF COMT IN MAN

COMT catalyzes the *O*-methylation of catechol compounds with *S*-adenosyl-L-methionine as a methyl donor. Magnesium is required for enzyme activity (Axelrod and Tomchick, 1958). The *O*-methylation of norepinephrine by COMT to yield normetanephrine is shown in Figure 1. Calcium has been found to be a potent, reversible, noncompetitive inhibitor of COMT (Quiram and Weinshilboum, 1976*b*; Weinshilboum and Raymond, 1976). This observation made it possible to develop an accurate and reproducible COMT assay in which calcium was removed from tissue homogenates and RBC lysates (Raymond and Weinshilboum, 1975). When this assay was used to measure RBC COMT activity in blood samples from 376 randomly selected subjects aged 16–18 years, there was a significant sibling-sibling correlation of the enzyme activity among the 56 sibling pairs studied ($r = 0.49, P < .001$) (Weinshilboum, Raymond, Elveback, and Weidman, 1974). If a biochemical trait is entirely determined by heredity, it would be expected that the correlation coefficient for siblings and dizygotic twins would be 0.5 and that for monozygotic twins would be 1.0 (Cavalli-Sforza and Bodmer, 1971). An independent evaluation of RBC COMT also demonstrated a significant familial aggregation of the RBC enzyme activity with a sibling-sibling correlation coefficient of 0.53 ($P < .05$) (Gershon and Jonas, 1975), and a study of RBC COMT activity in twins yielded correlation coefficients of 0.37 for dizygotic and 0.90 for mono-

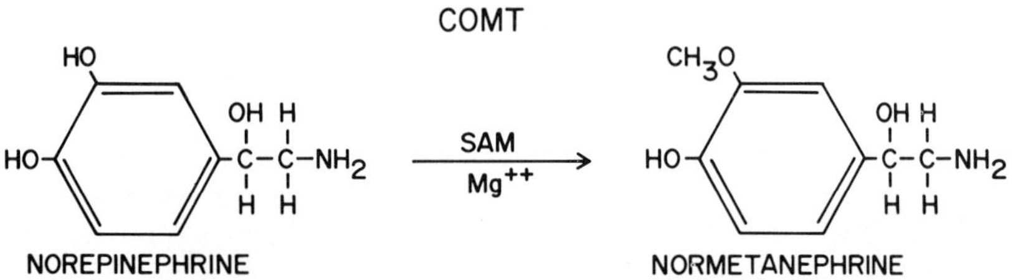

COMT

NOREPINEPHRINE —(SAM, Mg^{++})→ NORMETANEPHRINE

FIGURE 1. COMT reaction. The conversion of norepinephrine to normetanephrine by COMT with *S*-adenosyl-L-methionine (SAM) as a methyl donor is shown.

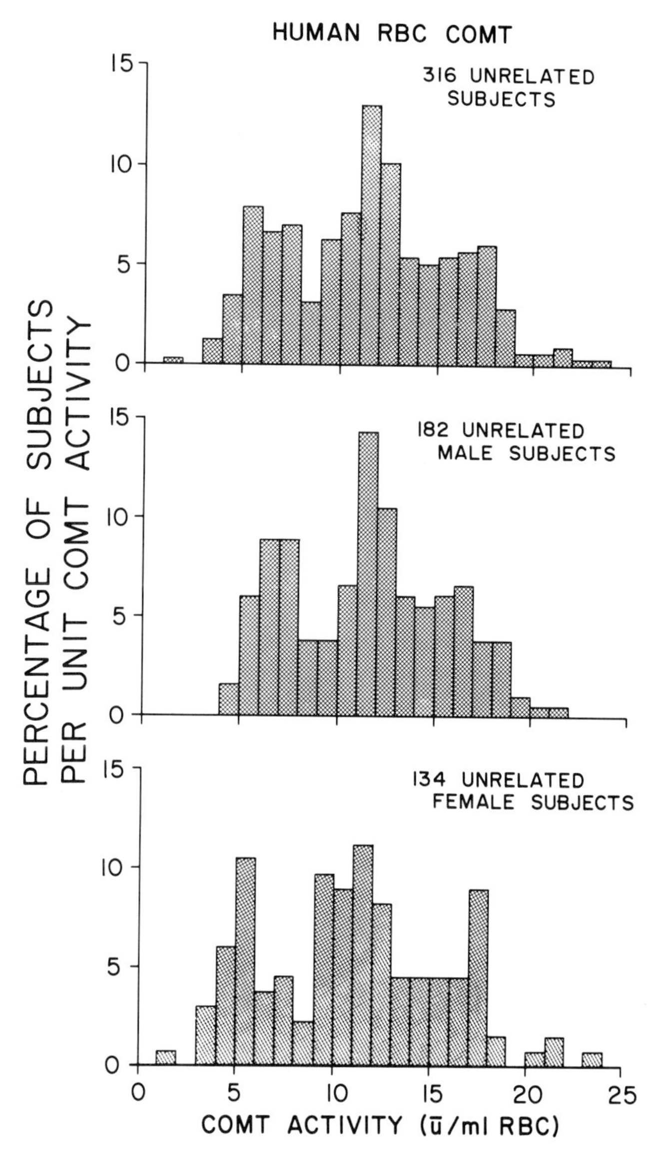

FIGURE 2. Frequency distribution of RBC COMT activity. The frequency distribution of RBC COMT activity in blood samples from 316 unrelated, randomly selected subjects is shown. One unit of enzyme activity (ū) represents the formation of one nmol of 4-methoxy-3-hydroxybenzoic acid from 3,4-dihydroxybenzoic acid per hour of incubation.

zygotic twin pairs (Grunhaus, Ebstein, Belmaker, Sandler, and Jonas, 1976). The data from twins were used to calculate estimates of the heritability of RBC COMT that ranged from 68–100% depending on the method of analysis used (Grunhaus et al., 1976).

Even though sibling and twin studies indicated that human RBC COMT activity was controlled primarily by inheritance, it was not possible to determine from those results whether inheritance was monogenic (mendelian) or polygenic. This was an important question, since the biochemical basis of a monogenically inherited trait might be elucidated more easily than that of a trait controlled by a number of loci. However, the frequency distribution of RBC enzyme activities in the study of the 376 adolescent subjects described above was bimodal, and approximately 25% of the subjects were included in a subgroup with low enzyme activity (<8 units/ml RBC). A bimodal frequency distribution of RBC COMT activity has been found in all studies in which calcium has been removed from erythrocyte lysates (Weinshilboum et al., 1974; Weinshilboum and Raymond, 1977*a*; Scanlon, Raymond, and Weinshilboum, 1979). The distribution of RBC COMT activities from a recent study of 316 randomly selected adult subjects is shown in Figure 2 (Scanlon et al., 1979). About one-quarter of these subjects also fell into a subgroup with low COMT (<8 units/ml). To test the possibility of monogenic inheritance of the trait low RBC COMT, blood samples were obtained from first-degree relatives of subjects with low enzyme activity in 48 families (Weinshilboum and Raymond, 1977*a*). The results of pedigree and segregation analyses of the data from these family studies were all compatible with the autosomal recessive inheritance of the trait of low RBC COMT (Weinshilboum and Raymond, 1977*a*). Subjects with low activity were homozygous for this trait. The gene frequency of the allele for low activity was approximately 0.5, an observation that indicated that about half of the population was heterozygous at the locus *COMT*. It has been proposed that the allele for low human RBC COMT be designated $COMT^L$ and that the alternative allele, the allele for high activity, be designated $COMT^H$ (Scanlon et al., 1979; Weinshilboum, 1979). These designations are in accordance

FIGURE 3. Lineweaver-Burk plots of the effects on COMT activity of dihydroxybenzoic acid (DBA), *S*-adenosyl-L-methionine (SAM), and magnesium. COMT activity was measured in RBC lysates from three subjects with low (○---○) and three subjects with high (●——●) enzyme activity in the presence of varying concentrations of DBA (top), SAM (middle), and magnesium (bottom).

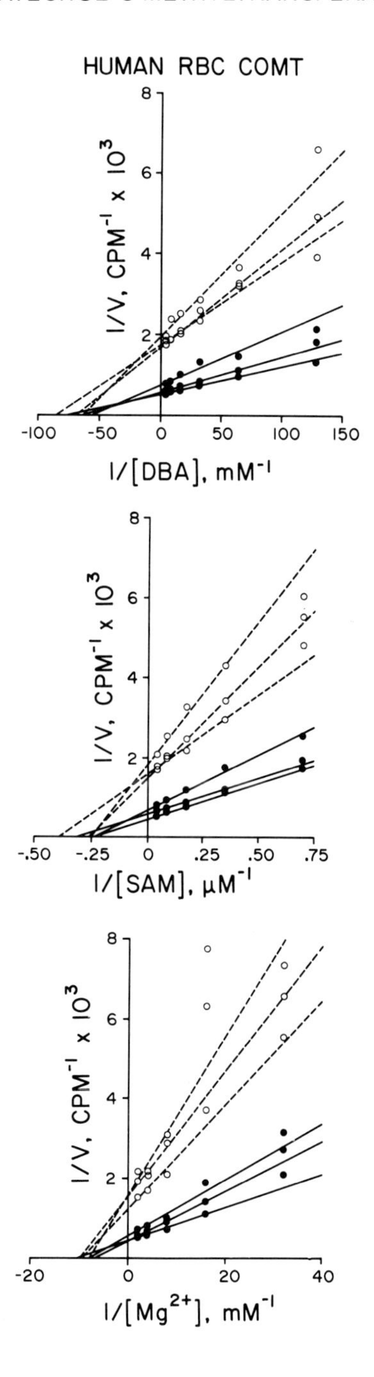
HUMAN RBC COMT
I/V, CPM^{-1} x 10^3
8
6
4
2
-100
-50
0
50
100
150
I/[DBA], mM^{-1}
I/V, CPM^{-1} x 10^3
8
6
4
2
-.50
-.25
0
.25
.50
.75
I/[SAM], µM^{-1}
I/V, CPM^{-1} x 10^3
8
6
4
2
-20
0
20
40
I/[Mg^{2+}], mM^{-1}

with the recommendations of the Committee on Nomenclature of the Third International Workshop on Human Gene Mapping (1976).

One possible explanation for the effect of the allele pair $COMT^L$ and $COMT^H$ is that their presence results in biochemically different forms of the enzyme in RBCs and possibly in other tissues. To test this hypothesis, several biochemical characteristics of COMT in blood from three subjects homozygous for the allele for low enzyme activity (group mean ± SEM = 6.0 ± 0.3) were compared with those of the enzyme in RBCs of three subjects with much higher enzyme activity (group mean = 19.2 ± 2.4 units/ml) (Scanlon et al., 1979). There were no differences between the two groups in apparent Michaelis-Menten (K_m) values for the two cosubstrates of the reaction, dihydroxybenzoic acid and *S*-adenosyl-L-methionine, or for the enzyme activator, magnesium (Figure 3) (Scanlon et al., 1979). Neither were there differences between the groups in the concentrations of three different COMT inhibitors (tropolone, *S*-adenosyl-L-homocysteine, and calcium) at which 50% inhibition of the enzyme occurred (Figure 4) (Scanlon et al., 1979). Relative thermal stability, a sensitive indicator of differences in protein structure (Paigen, 1971), was also tested. There was a significant difference between the thermal stability of COMT in a lysate from a subject with low activity (6.2 units/ml) when compared with a lysate from a subject with high activity (23.4 units/ml). When aliquots of these samples were heated at various temperatures for 15 minutes, the enzyme from the subject with low activity was 50% inactivated at 47.8° C, while that from the subject with high activity was 50% inactivated at 48.9° C (Figure 5) (Scanlon et al., 1979). These results were confirmed in experiments with lysates from the three subjects with low and the three subjects with high activity used for the kinetic studies described above. The half-life of COMT during incubation at 48° C was significantly different in these two groups, 21.2 ± 1.4 minutes (mean ± SEM) in lysates with high activity and 12.5 ± 0.9 minutes in lysates with low activity ($P < .01$) (Scanlon et al., 1979).

These experiments were then expanded to include determinations of the thermal stability of COMT in blood from the 316 randomly se-

---

FIGURE 4. Effects of tropolone, *S*-adenosyl-L-homocysteine (SAH), and calcium on COMT activity. Inhibition by increasing concentrations of tropolone (top), SAH (middle), and calcium (bottom) with RBC lysates from three subjects with low (○ - - - ○) and three subjects with high (● —— ●) enzyme activity are shown. The results are expressed as a percentage of the activity in the absence of the inhibitor.

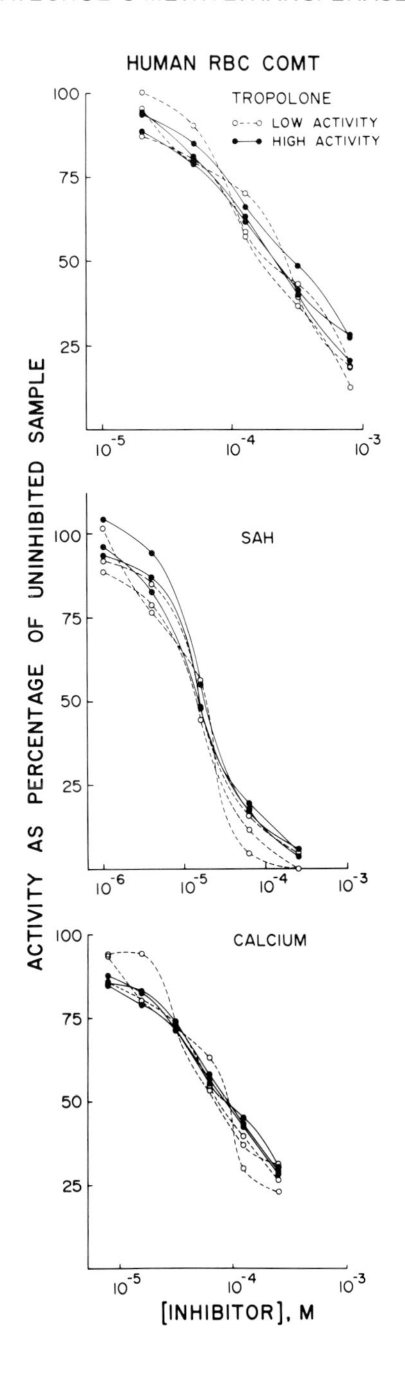
HUMAN RBC COMT
TROPOLONE
LOW ACTIVITY
HIGH ACTIVITY
SAH
CALCIUM
ACTIVITY AS PERCENTAGE OF UNINHIBITED SAMPLE
[INHIBITOR], M
100
75
50
25
10^{-5}
10^{-4}
10^{-3}
10^{-6}

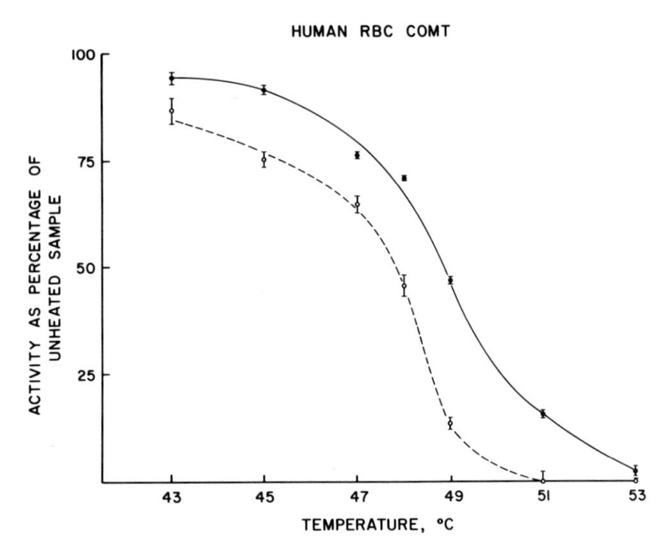

FIGURE 5. Thermal inactivation of COMT. COMT activity was measured after incubation at various temperatures for 15 min in RBC lysates from one subject with low (○ - - - ○) and one with high (● —— ●) enzyme activity. The results are expressed as a percentage of the activity in an unheated sample. Each point represents the mean ± SEM of three determinations.

lected subjects whose basal enzyme activity is shown in Figure 2. COMT activity in these samples was measured after heating the lysate at 48° C for 15 minutes (Figure 6). The ratio of COMT activity in the heated lysate to that in the unheated lysate (heated divided by control or H/C ratio) was used as a measure of thermal stability. The H/C ratios were significantly lower for subjects homozygous for the allele $COMT^L$ (H/C = 0.47 ± 0.01, mean ± SEM, n = 84) than for subjects with high RBC COMT (H/C = 0.65 ± 0.01, n = 232, $P < .001$) (Figure 7) (Scanlon et al., 1979). Experiments in which lysates were mixed and experiments in which partially purified rat liver COMT was added to lysates demonstrated that the differences in thermal stability between subjects with high and subjects with low enzyme activity were not due

---

FIGURE 6. Frequency distribution of RBC COMT activity after thermal inactivation. The frequency distribution of COMT activities in RBC lysates heated at 48° C for 15 min from 316 unrelated, randomly selected subjects is shown. These are the same subjects for whom basal enzyme activities are shown in Figure 2. The meaning of the symbol ū is the same as in Figure 2.

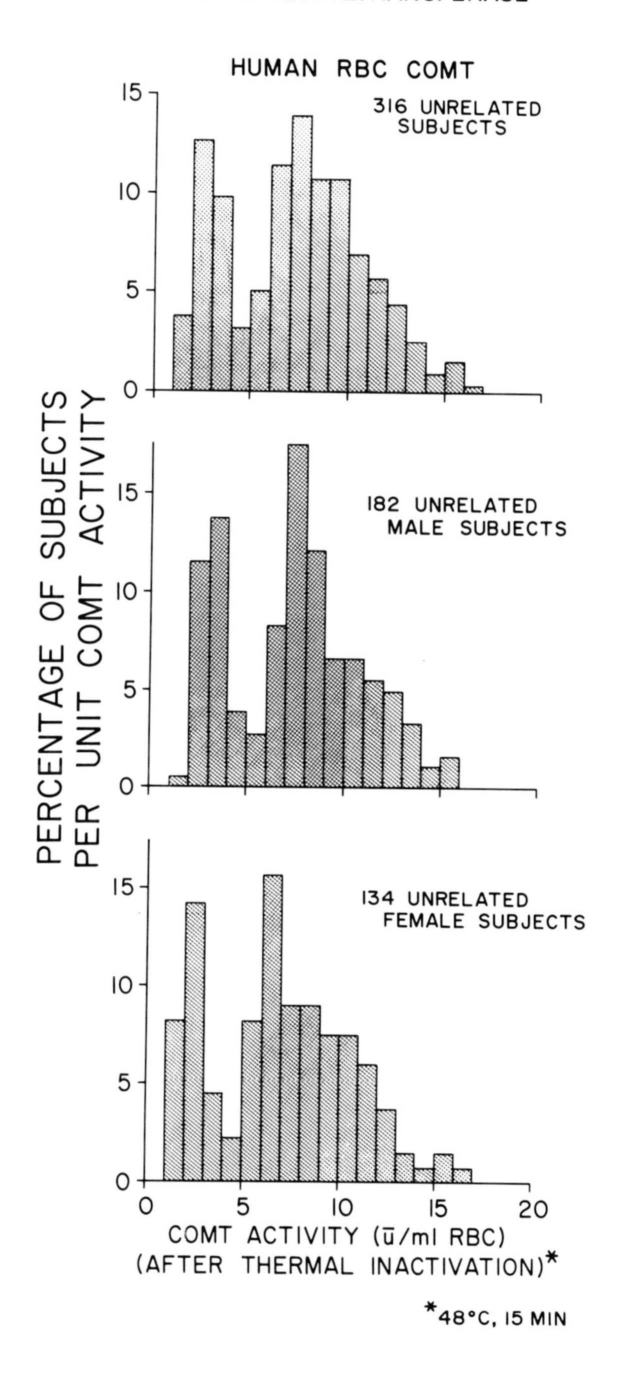
HUMAN RBC COMT
316 UNRELATED SUBJECTS
182 UNRELATED MALE SUBJECTS
134 UNRELATED FEMALE SUBJECTS
PERCENTAGE OF SUBJECTS PER UNIT COMT ACTIVITY
COMT ACTIVITY (ū/ml RBC)
(AFTER THERMAL INACTIVATION)*
*48°C, 15 MIN

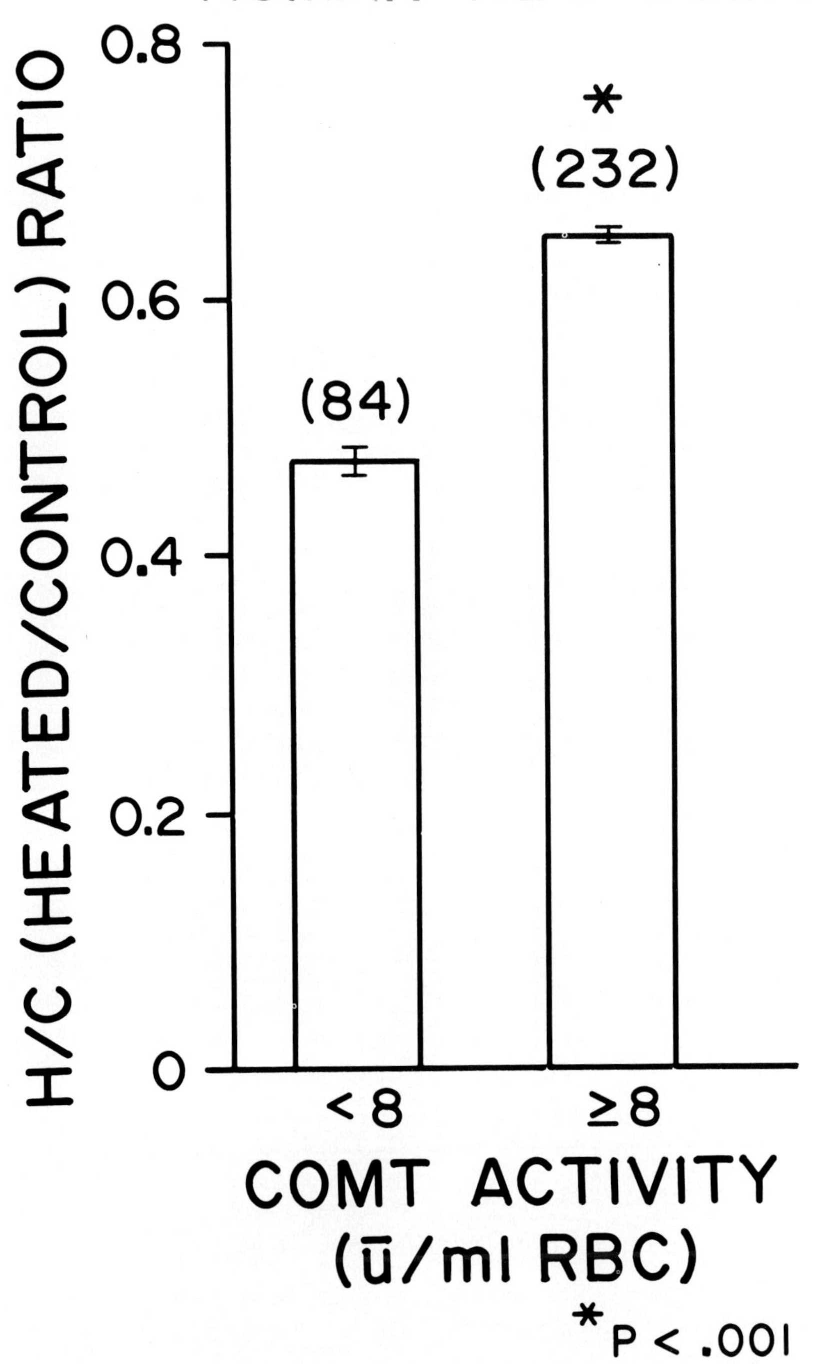

HUMAN RBC COMT
H/C (HEATED/CONTROL) RATIO
0.8
0.6
0.4
0.2
0
*
(232)
(84)
< 8
≥8
COMT ACTIVITY
(ū/ml RBC)
*P < .001

to differences in the thermal stability of endogenous enzyme inhibitors, activators, or COMT degrading enzymes. Therefore, there is a difference in the biochemical properties of COMT in the RBCs of subjects homozygous for the allele $COMT^L$ as compared with RBCs with higher levels of enzyme activity. These results raise the possibility that the locus *COMT* might represent the structural gene for the enzyme in man. However, genetically determined posttranslational modification of the enzyme cannot be ruled out as an explanation for these findings.

To determine whether the relative COMT activity in blood reflects the relative enzyme activity in other tissues, COMT was measured in blood and in lung or kidney tissue from 29 patients who underwent pneumonectomy and 12 patients who underwent nephrectomy for the removal of tumors (Weinshilboum, 1978). The enzyme activity was measured in tissue as far removed from the tumor as possible. It was also important to remove calcium from homogenates of these organs prior to determination of COMT activity. For example, the activity in five lung samples was reduced an average of 50% and the activity in five kidney samples was reduced an average of 20% if calcium was not removed from the tissue homogenate. There was a significant positive correlation of relative RBC COMT activity with the enzyme activity in human lung and kidney from the same subjects. The correlation coefficient for lung and RBC COMT was 0.59 ($n = 29$, $P < .001$) and for kidney and RBC was 0.81 ($n = 12$, $P < .005$) (Weinshilboum, 1978).

BIOCHEMICAL GENETICS OF COMT IN EXPERIMENTAL ANIMALS

Many biochemical genetic and pharmacogenetic experiments cannot be performed in man, and it would be very useful if there were an animal model for genetic differences in COMT activity. Two groups have reported wide variations of COMT activities in different inbred strains of mice (Schlesinger, Harkins, Deckard, and Paden, 1975; Gershon and Jonas, 1976). One study demonstrated that brain COMT is twice as great in C57BL/6J mice as in DBA/2J mice (Schlesinger et al., 1975). The other group of investigators reported that brain COMT is 40% higher in C57BL/6J mice than it is in the C3HF strain (Gershon and

---

FIGURE 7. H/C ratios for RBC COMT activity in lysates from randomly selected subjects with less than and greater than 8 units of enzyme activity per ml RBC (see text for details). The error bars represent SEM; *$P < .001$.

Jonas, 1976). Breeding experiments were performed by both groups, but were only carried through the F_1 (hybrid) generations. In both cases the enzyme activities in the brains of F_1 animals were closer to those in the high COMT parental strain than to those in the low COMT parental strain, an indication of dominance of the high enzyme phenotype. However, since the breeding experiments did not include F_2 generation and backcross animals, it was not possible to determine whether the inheritance of COMT in these mice was monogenic or polygenic.

In a study of COMT in nine strains of rat, it was found that hepatic and renal activities in Sprague-Dawley and Fischer-344 (F-344) rats were only about 50–60% as great as those in the same organs of the other strains tested (Weinshilboum and Raymond, 1977*b*). Wistar strains, including Wistar-Furth (W-F) rats, had much higher levels of liver and kidney activity than did Sprague-Dawley or F-344 rats (Figure 8). There was much less difference among strains in the COMT activities of heart, brain, intestine, lung, and blood. These initial experiments were followed with breeding studies of F-344 and W-F rats. The breeding experiments involved the assay of tissue from a total of 400 parental, F_1, F_2, and backcross rats (Weinshilboum, Raymond, and Frohnauer, 1979). The results of these studies were compatible with the autosomal

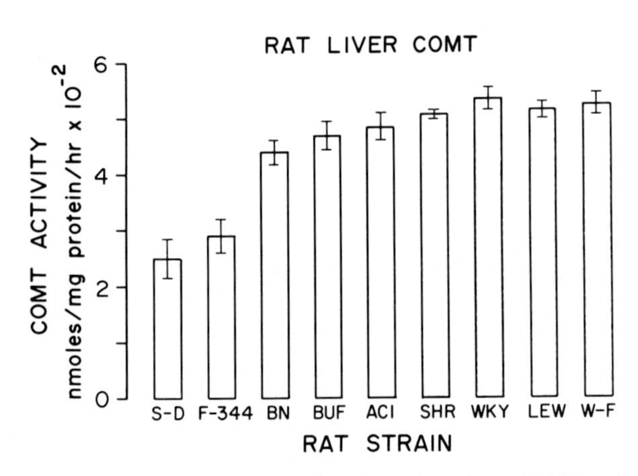

FIGURE 8. Rat liver COMT activity in a variety of strains. COMT activities in the livers of male rats 9–11 weeks old are shown. Abbreviations for strain names include the following: S-D, Sprague-Dawley; F-344, Fischer-344; BUF, Buffalo; SHR, spontaneously hypertensive rats of Okamoto; WKY, Wistar Kyoto; LEW, Lewis Wistar; and W-F, Wistar-Furth. All values are mean ± SEM for determinations in five animals.

TABLE 1. *Estimated gene number for rat COMT activity*

Sex	Organ	Estimated number of genes
Male	Kidney	1.387
	Liver	0.737
Female	Kidney	1.096
	Liver	1.248

Estimates of number of genes responsible for inheritance of COMT activity in kidneys and livers of F-344 and W-F rats. Gene number was estimated by the method of Falconer (1963).

recessive (monogenic) inheritance of the trait low liver and kidney COMT in F-344 and W-F rats. This conclusion was based on the observation that a subgroup of approximately 25% of the F_2 rats had low enzyme activity in these two organs, and that estimates of the number of genes involved in the determination of liver and kidney COMT activities, estimates based on a comparison of the variance of enzyme activities in F_1 and F_2 animals, were close to a gene number of 1.0 (Table 1). It has been proposed that the locus responsible for this genetic difference in rat COMT be referred to as *Co* and that the alleles for low and high activity be designated Co^l and Co^h respectively (Weinshilboum et al., 1979). This nomenclature is analogous to that suggested for biochemical genetic variants in the mouse by the Committee on Standardized Genetic Nomenclature for Mice (1973).

When comparisons were made of the biochemical properties of COMT in liver homogenates of F-344 and W-F rats, no differences were found in the apparent K_m values for 3,4-dihydroxybenzoic acid, *S*-adenosyl-L-methionine, or magnesium. Neither were there differences in the concentrations of three different COMT inhibitors (tropolone, *S*-adenosyl-L-homocysteine, and calcium) that were required to inhibit the enzyme 50%. The R_f values of rat liver COMT after disc gel electrophoresis were identical in the two strains, and there was no difference in thermal stability of the enzyme in liver homogenates from W-F and F-344 rats (Weinshilboum et al., 1979). On the basis of these preliminary data, the biochemical basis of the monogenically inherited difference in liver and kidney COMT between F-344 and W-F rats does not appear to be analogous to the genetic difference in human RBC COMT. However, these animals will still be useful for future biochemical genetic and pharmacogenetic experiments.

CONCLUSIONS

Much has been learned about the biochemical genetics of COMT in man and experimental animals. Polymorphism at a single locus, *COMT*, is responsible for most of the individual variation in enzyme activity in the human erythrocyte. The trait of low RBC COMT activity is inherited as an autosomal recessive trait. The gene frequency of the allele for low activity, $COMT^L$, is about 50%, and approximately one-quarter of the population is homozygous for this allele. The enzyme in RBCs of individuals homozygous for $COMT^L$ differs biochemically from that in RBCs with higher activity. These results are compatible with the conclusion that the locus *COMT* may be the structural gene for the enzyme in man. In addition, the relative enzyme activity in the human RBC reflects the relative activity in at least two other tissues, the lung and kidney. These observations have immediate practical implications with regard to the possibility that individual variations in the metabolism of catechol drugs such as L-dopa and isoproterenol may be predicted by measurement of RBC COMT activity. There have already been preliminary reports of a significant positive correlation between relative RBC COMT activity and the formation of 3-*O*-methyldopa in patients treated with L-dopa (Reilly and Rivera-Calimlim, 1978). Whether individual variations in the metabolism of endogenous catecholamine neurotransmitters result from genetic variations in COMT activity is not known.

There are wide variations in COMT activity among inbred strains of mice and rats. Breeding studies with W-F and F-344 rats have shown that the trait of low liver and kidney enzyme activity in these strains is inherited in an autosomal recessive fashion. However, no difference in the biochemical characteristics of COMT between these strains has been demonstrated, so the biochemical basis of the genetic differences between these animals appears to differ from the situation in man. Further studies of the biochemical basis of this inherited variation in enzyme activity will have to be performed. In addition, these animals may prove useful for pharmacogenetic experiments designed to study the functional significance of differences in COMT activity with respect to drug and neurotransmitter metabolism.

COMT represents one example in which the application of biochemical genetic techniques has increased understanding of individual variation in a neurotransmitter system. Many similar examples will undoubtedly be described in the future. It is anticipated that the application of genetic techniques to the study of other neurotransmitter systems may help make it possible to understand and predict individual variations in the function of these systems.

ACKNOWLEDGMENTS

Supported in part by National Institutes of Health grants 11014 and HL17487. R.M.W. is an Established Investigator of the American Heart Association. Thanks to Luanne Wussow for her assistance with the preparation of this manuscript.

REFERENCES

Assicot, M. and C. Bohuon (1969). Production of antibodies to catechol-*O*-methyltransferase in rat liver, *Biochem. Pharmacol.* **18:**1893–1898.

Axelrod, J. and C. K. Cohn (1971). Methyltransferase enzymes in red blood cells, *J. Pharmacol. Exp. Ther.* **176:**650–654.

Axelrod, J. and R. Tomchick (1958). Enzymatic *O*-methylation of epinephrine and other catechols, *J. Biol. Chem.* **233:**702–705.

Cavalli-Sforza, L. L. and W. F. Bodmer (1971). Pp. 532–537, 573–597 in *The Genetics of Human Populations.* W. H. Freeman and Co., San Francisco.

Committee on Nomenclature of the Third International Workshop on Human Gene Mapping (1976). Report of the committee on nomenclature, *Birth Defects Orig. Artic. Ser.* **12:**65–74.

Committee on Standardized Genetic Nomenclature for Mice (1973). Guidelines for nomenclature of genetically determined biochemical variants on the house mouse, *Mus musculus*, *Biochem. Genet.* **9:**369–374.

Creveling, C. R., R. T. Borchardt, and C. Isersky (1973). Immunological characterization of catechol-*O*-methyltransferase, pp. 117–119 in *Frontiers in Catecholamine Research*, Usdin, E. and S. Snyder, eds. Pergamon Press, New York.

Falconer, D. S. (1963). Quantitative inheritance, pp. 193–216 in *Methodology in Mammalian Genetics*, Burdette, W. J., ed. Holden-Day, Inc., San Francisco.

Gershon, E. S. and W. Z. Jonas (1975). Erythrocyte soluble catechol-*O*-methyltransferase activity in primary affective disorder, *Arch. Gen. Psychiatry* **32:**1351–1356.

Gershon, E. S. and W. Z. Jonas (1976). Inherited differences of brain and erythrocyte soluble catechol-*O*-methyltransferase activity in two mouse strains, *Biol. Psychiatry* **11:**641–645.

Grunhaus, L., R. Ebstein, R. Belmaker, S. G. Sandler, and W. Jonas (1976). A twin study of human red blood cell catechol-*O*-methyltransferase, *Br. J. Psychiatry* **128:**494–498.

Paigen, W. (1971). The genetics of enzyme realization, pp. 1–44 in *Enzyme Synthesis and Degradation in Mammalian Systems,* Rechcigl, M., Jr., ed. University Park Press, Baltimore.

Quiram, D. R. and R. M. Weinshilboum (1976*a*). Catechol-*O*-methyltransferase in rat erythrocyte and three other tissues: comparison of biochemical properties after removal of inhibitory calcium, *J. Neurochem.* **27**:1197–1203.

Quiram, D. R. and R. M. Weinshilboum (1976*b*). Inhibition of rat liver catechol-*O*-methyltransferase by lanthanum, neodymium and europium, *Biochem. Pharmacol.* **25**:1727–1732.

Raymond, F. A. and R. M. Weinshilboum (1975). Microassay of human erythrocyte catechol-*O*-methyltransferase: removal of inhibitory calcium with chelating resin, *Clin. Chim. Acta* **58**:185–194.

Reilly, D. K. and L. Rivera-Calimlim (1978). Red blood cell catechol-*O*-methyltransferase, plasma 3-*O*-methyldopa and dyskineasias, *Pharmacologist* **20**:156.

Scanlon, P. D., F. A. Raymond, and R. M. Weinshilboum (1979). Catechol-*O*-methyltransferase: thermolabile enzyme in erythrocytes of subjects homozygous for the allele for low activity, *Science* **203**:63–65.

Schlesinger, K., J. Harkins, B. S. Deckard, and C. Paden (1975). Catechol-*O*-methyltransferase and monoamine oxidase activities in brains of mice susceptible and resistant to audiogenic seizures, *J. Neurobiol.* **6**:587–596.

Weinshilboum, R. M. (1978). Human erythrocyte catechol-*O*-methyltransferase: correlation with lung and kidney activity, *Life Sci.* **22**:625–630.

Weinshilboum, R. M. (1979). Catecholamine biochemical genetics in human populations, pp. 257–282 in *Neurogenetics: Genetic Approaches to the Nervous System,* Breakefield, X. O., ed. Elsevier/North Holland, New York.

Weinshilboum, R. M. and F. A. Raymond (1976). Calcium inhibition of rat liver catechol-*O*-methyltransferase, *Biochem. Pharmacol.* **25**:573–579.

Weinshilboum, R. M. and F. A. Raymond (1977*a*). Inheritance of low erythrocyte catechol-*O*-methyltransferase activity in man, *Am. J. Hum. Genet.* **29**:125–135.

Weinshilboum, R. M. and F. A. Raymond (1977*b*). Variations in catechol-*O*-methyltransferase activity in inbred strains of rats, *Neuropharmacology* **16**:703–706.

Weinshilboum, R. M., F. A. Raymond, L. R. Elveback, and W. H. Weidman (1974). Correlation of erythrocyte catechol-*O*-methyltransferase activity between siblings, *Nature (London)* **252**:490–491.

Weinshilboum, R. M., F. A. Raymond, and M. Frohnauer (1979). Monogenic inheritance of catechol-*O*-methyltransferase activity in the rat: biochemical and genetic studies, *Biochem. Pharmacol.* **28**:1239–1248.

PLASTICITY OF VISUAL CORTICAL DEVELOPMENT

THE DEVELOPMENT OF OCULAR DOMINANCE COLUMNS IN THE CAT

Simon LeVay and Michael P. Stryker

Harvard Medical School, Boston, Massachusetts

In the striate cortex of cats and monkeys, cells responding preferentially to stimulation of one or the other eye are grouped together in columns that extend vertically through all the cortical layers (Hubel and Wiesel, 1965, 1968). The anatomical basis for the columnar arrangement is a segregation, in layer IV, of the terminals of the geniculocortical afferents serving the two eyes. These afferents arise in different laminae of the lateral geniculate nucleus. The segregation has been demonstrated by a number of techniques (Hubel and Wiesel, 1972; Wiesel, Hubel, and Lam, 1974; LeVay, Hubel, and Wiesel, 1975), of which the most useful has been the autoradiographic method, which depends on the transneuronal transport of tritiated proline injected into one eye (Wiesel et al., 1974). In surface view, the columns have the shape of bands or patches, about 400 microns wide, that alternate with each other to form a rather constant overall pattern in area 17 (LeVay et al., 1975; Shatz, Lindström, and Wiesel, 1977).

Until quite recently, it was thought that the ocular dominance system was already established in newborn animals and closely resembled that of adults (Hubel and Wiesel, 1963; Wiesel and Hubel, 1974; Blakemore, Van Sluyters, and Movshon, 1975). The first indication that this might not be the case came from autoradiographic observations on a monkey that had received an eye injection on the second day of life: instead of clear bands of label in layer IV, separated by unlabeled gaps, the label was distributed continuously on each side of the brain, with slight, wave-like fluctuations in labeling density visible on the ipsilateral side only (Hubel, Wiesel, and LeVay, 1977). Physiologically, too, the columnar pattern in layer IV was somewhat blurred in a week-old monkey, as compared with the very precise segregation found in adults. At the same time, Rakic (1976) injected the eyes of fetal monkeys of various gestational ages and found completely uniform distribution of label in layer IV up until about 3 weeks before birth. These findings suggested that when the left- and right-eye afferents first grow into the cortex, they intermingle freely with each other and only later segregate out into the columnar arrangement.

TIME COURSE OF SEGREGATION

Because of its immaturity at birth, the cat is a more convenient animal in which to study this developmental process. As may be seen from Figure 1, eye injections produced continuous, uniform labeling in layer IV up until 2 weeks of postnatal age. At 3 weeks, undulations in labeling density became apparent; at 39 days these undulations were sharper; and by 3 months the adult columnar pattern had been reached (LeVay, Stryker, and Shatz, 1978).

The autoradiographic picture is contaminated, to some extent, by an artifact that we refer to as "spillover." This is the leakage of radioactivity between laminae of the lateral geniculate nucleus, its uptake by the wrong set of geniculate neurons, and subsequent transport to their terminals in the cortex. We have measured spillover and calculated its effect on the cortical labeling pattern at each age. The results of these calculations indicate that, although the early uniform distribution of label reflects a genuine overlap of left- and right-eye fibers in the cortex, the process of segregation occurs faster than the autoradiographs suggest, being largely complete by about 6 weeks of age. The later increase in clarity of the columns is due almost entirely to a reduction in spillover.

Intermingling of left- and right-eye inputs is reflected in the physiology of the postsynaptic neurons (LeVay et al., 1978). Recordings made during long, tangential penetrations through layer IV in several 2-week-old kittens gave no hint of the regular alternations in eye preference that are so characteristic of this layer in adult cats (Shatz and Stryker, 1978). Instead, most cells were about equally influenced by the two eyes (Figure 2). This observation, besides supporting the anatomical evidence for an intermingling of the afferents, suggests that they form functional connections with cortical neurons in their early, overlapping pattern. This in turn suggests that, during the process of segregation, functional connections from the left eye are broken in the forming right-eye columns, and vice versa. Physiologically, the progression from

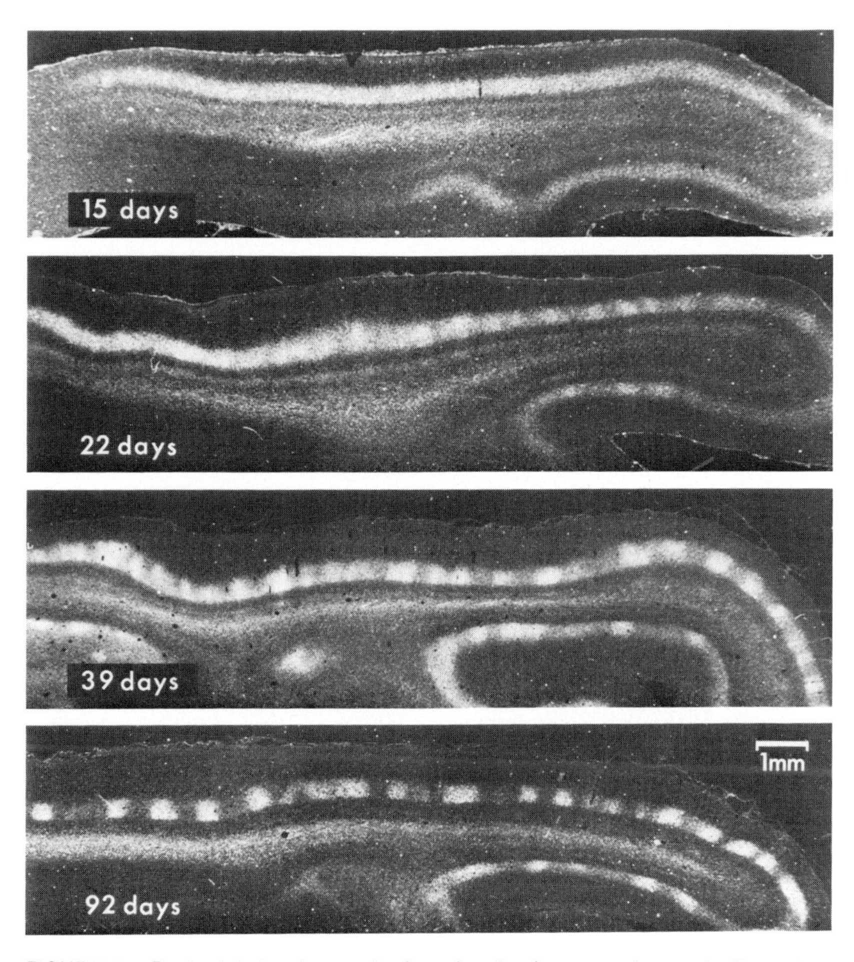

FIGURE 1. Postnatal development of ocular dominance columns in the cat as shown by transneuronal transport of [^{3}H]proline injected into one eye. These are darkfield autoradiographs of the visual cortex at four different ages, ipsilateral to the injected eye. Horizontal sections, midline at the top of each figure, anterior to the left. At 15 days of age the afferents are spread uniformly along layer IV, completely intermingled with the (unlabeled) afferents serving the contralateral eye. At later ages the afferents progressively aggregate into clumps—the anatomical basis for the physiologically defined ocular dominance columns. The gaps are occupied by unlabeled afferents serving the other eye. (From LeVay, S., M. P. Stryker, and C. J. Shatz [1978]. Ocular dominance columns and their development in layer IV of the cat's visual cortex: a quantitative study, *J. Comp. Neurol.* **179:**223–244. With permission of The Wistar Press.)

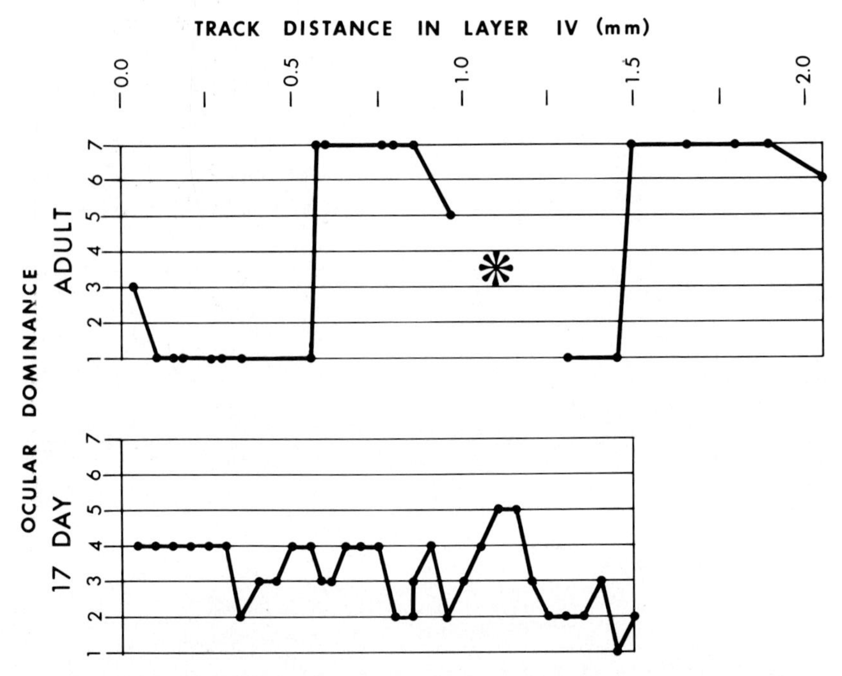

FIGURE 2. Ocular dominance distribution of cells encountered during the passage of an electrode tangentially through the fourth layer in two cats: an adult (above) and a 17-day-old kitten (below). Each point represents an isolated single unit whose ocular dominance, on the 7-point scale of Hubel and Wiesel (1962), is plotted against its position along the electrode track. In the kitten most cells could be driven from either eye, with on the whole a slight preference for the contralateral eye. In the adult there is a periodic variation from exclusively contralateral (Group 1) to exclusively ipsilateral (Group 7) eye dominance. The asterisk marks the position of a lesion made for the purpose of the histological reconstruction of the track; cells could not be recorded for a short distance after the lesion. The transitions in ocular dominance are more abrupt in this penetration than is commonly observed: more often, one or several binocular neurons are encountered at column borders. (From LeVay, S., M. P. Stryker, and C. J. Shatz [1978]. Ocular dominance columns and their development in layer IV of the cat's visual cortex: a quantitative study, *J. Comp. Neurol.* **179:**223–244. With permission of The Wistar Press.)

early binocularity to more monocular responses was seen not only in layer IV but also in the other cortical layers, although in these layers the degree of ocular segregation in normal adult cats is always less than that found in layer IV. Columnar development in these layers probably depends both on the columnar segregation of the direct geniculate inputs to these layers (LeVay and Gilbert, 1976) and on the increasing monocularity of the input relayed through layer IV.

STRUCTURE OF AXONS BEFORE SEGREGATION

The changing autoradiographic picture seen in layer IV between 2 weeks and maturity seems likely to result from roughly synchronous changes in the arborizations of thousands of overlapping geniculocortical axons. An idea of the cytological processes involved can be obtained by reconstructing the entire arborizations of single afferent axons, before and after columnar segregation. In young kittens, when the afferents are not yet myelinated, it is possible to impregnate them in their entirety with the Golgi method. Figure 3 is an example of one such arborization in camera lucida reconstruction. It is one of the larger type of geniculocortical afferents, which terminates in the upper half of layer IV with some extension into the bottom of layer III. Most impressive about this arborization is its size: it extends for over 2 millimeters in both the dorsoventral and the anteroposterior directions. When one considers that the kitten's cortex is about 30% smaller, in linear dimensions, than the adult cat's, and that ocular dominance columns are not more than 0.5 mm wide in adults, it is clear that this arborization extends over a territory that is destined to become a number of columns. Within this area, the axon branches profusely, without any obvious local clumping.

Looked at more closely (Figure 4), the individual branches of the axonal tree had an irregular outline, marked by countless bumps and bulges, but with relatively few obvious boutons. It was hard to guess, from the Golgi picture, where the synapses might be located. In order to answer this question, we have injected the lateral geniculate nucleus with tritiated proline and studied the cortex with electron microscopic (EM) autoradiography. As in the Golgi picture, afferents identified in this way had a notably irregular outline as they coursed through the neuropil (Figure 5). It could now be seen that synapses were formed with neighboring dendritic elements along the course of the axon. These synapses were, for the most part, quite elementary in construction, consisting of apposed membrane thickenings and a very small number of synaptic vesicles. The glial wrapping, which is prominent in the adult geniculocortical synapse, was either absent or poorly developed. Although the EM autoradiographic method does not permit us to recon-

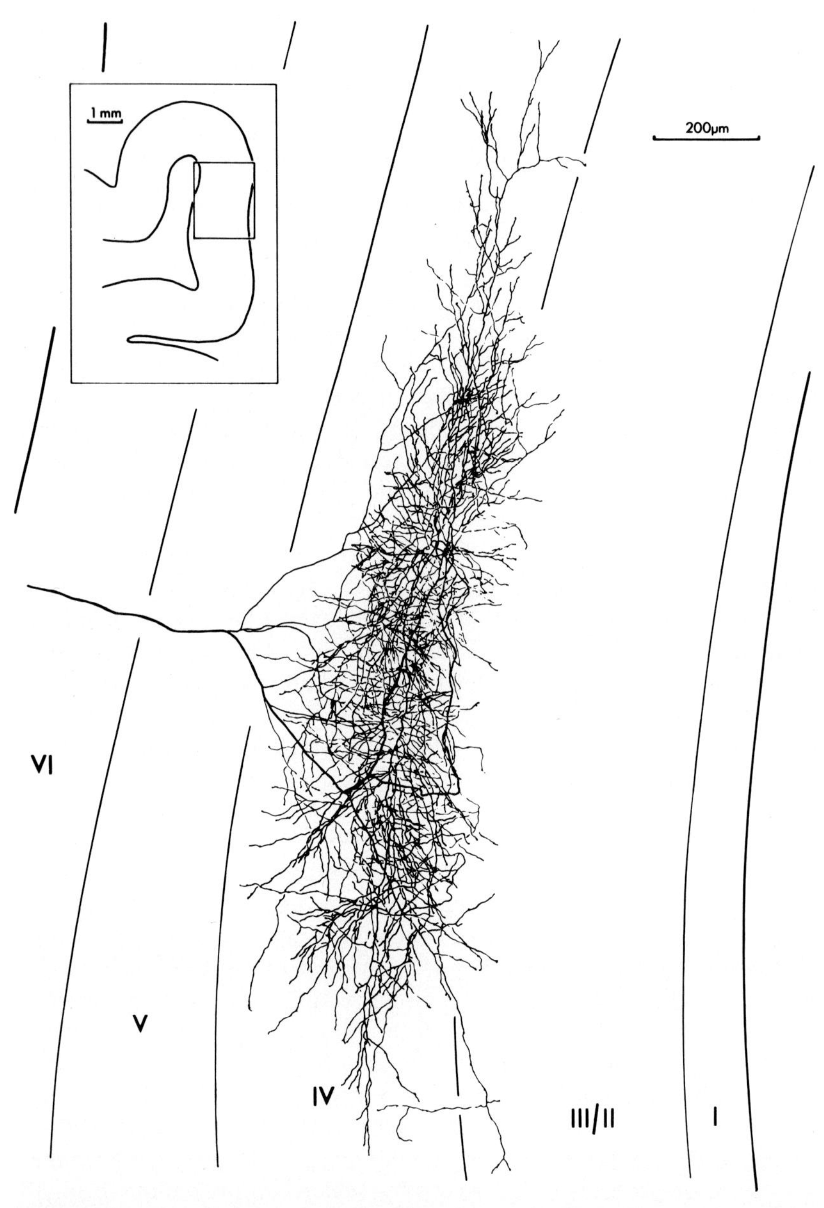

FIGURE 3. Arborization of a single geniculocortical afferent in the visual cortex of a 17-day-old kitten, i.e., just prior to the beginning of columnar segregation. This is a camera lucida reconstruction made from 25 successive coronal sections, each 100 μm thick, of an axon impregnated with the rapid Golgi method and em-

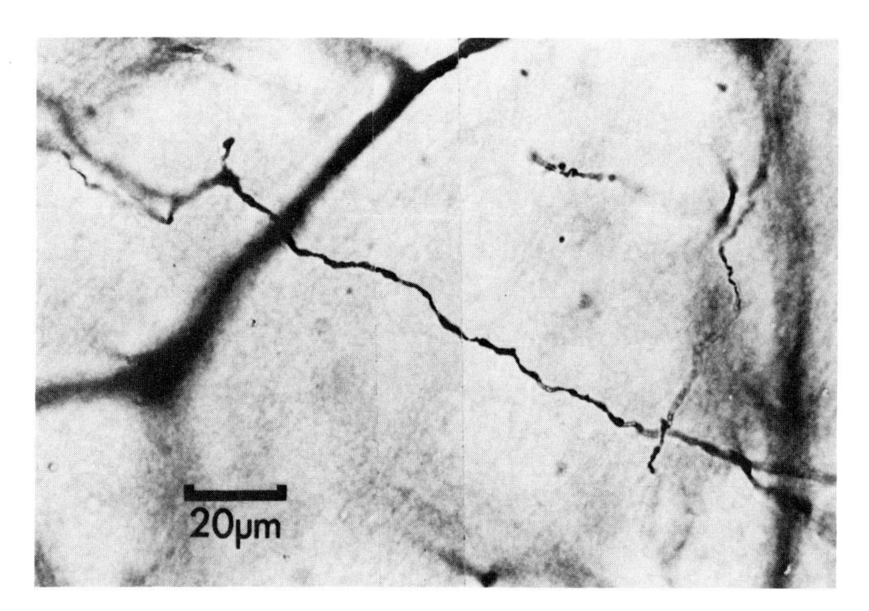

FIGURE 4. Detail from a Golgi-impregnated arborization in a 17-day-old kitten, similar to the one reconstructed in Figure 3. The immaturity of the arborization is revealed by a lack of clear differentiation into boutons (synaptic region) and connecting segments (compare with the adult pattern illustrated in the inset to Figure 6). A single bouton on a side twig is seen at left.

struct the synapses over a whole axonal tree, it seems likely, taking the EM and Golgi pictures together, that these rather simple *en passage* synapses are found over the entire arborization.

STRUCTURE OF AXONS IN ADULT

In the adult, the Golgi method was no longer successful in revealing entire arborizations. Instead, Ferster and LeVay (1978) were able to fill afferents with horseradish peroxidase (HRP) from an extracellular injection in the optic radiation. Figure 6 is an example of one of the larger afferents, which arborizes in the upper half of layer IV and the bottom of layer III, in an adult cat. As in the young kitten, the extent

---

bedded in Epon according to the method of Nevin, Tanaka, and Cruce (1978). The axon arborizes profusely and uniformly over a disc-shaped area that is more than 2 mm in diameter. The entire arborization is unmyelinated; the myelin sheath of the afferent trunk probably leaves off in layer VI at the point where the impregnation begins.

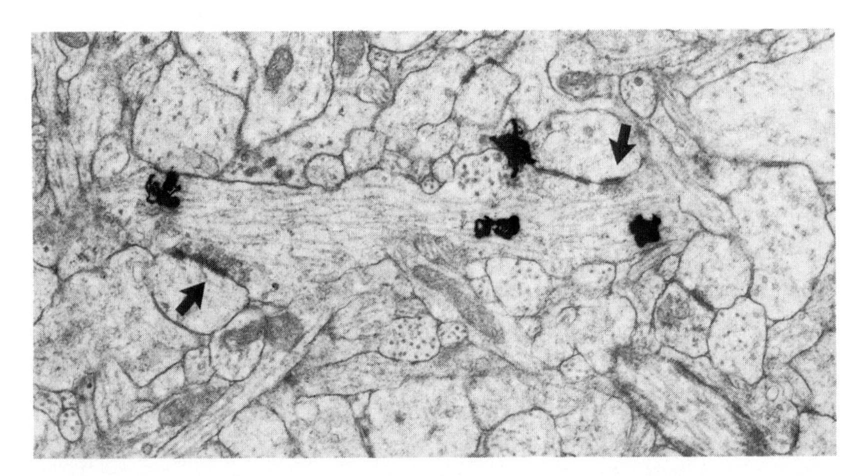

FIGURE 5. Part of a geniculocortical arborization in cortical layer IV from a 17-day-old kitten, identified by EM autoradiography after injection of [^{3}H]proline into the A laminae of the lateral geniculate nucleus. The axon forms two *en passage* synapses with processes that are probably dendritic spines (arrows). Note the small number of synaptic vesicles and the lack of mitochondria in the axon, which is mainly filled with microtubules. ×30,000.

of the arborization in layer IV is quite large—a little under 2 mm in both directions. In the adult, however, the terminal arborization is broken up into a number of clumps, connected by thicker axonal branches that run tangentially within layer IV. These clumps and gaps, though somewhat variable in size, have the appropriate dimensions to correspond to individual ocular dominance columns. Although this conclusion has not been verified by recording or by combining the HRP fillings with transneuronal autoradiography, it is hard to doubt that the clumps of dense arbor do correspond to columns serving the same eye.

Besides this clumping of their terminals, the adult arborizations differ from those seen in young kittens in being myelinated. Myelin ensheaths all but the finest, terminal-bearing segments of the arborizations. The axonal branch-points occur, of course, at nodes of Ranvier. A notable difference between the axonal branching patterns in adult cats and in kittens is that whereas in the kitten the branching is mostly dichotomous, in the adult many branch-points, particularly the higher-order ones, involve a splitting of the parent axon into a spray of three to six daughters. A curious feature of the daughter branches is that they often hug the outside of the myelin sheath of the parent trunk for a variable distance before turning off sharply in a new direction (see inset to Figure 6). It seems likely that this configuration, as well as the multiple branching, arises from a tendency of the myelinating oligodendrocytes to sweep the randomly located branchpoints to common sites—the definitive nodes—as sketched in Figure 7.

The boutons on the mature arborizations are very distinct bulbous structures that occur either as varicosities or as side twigs on the smooth, very fine preterminal axons (see inset to Figure 6). It is a simple if tedious task to count them: one axon of the same type as illustrated in Figure 6 possessed 1,543 boutons in layers IV and III (Ferster and LeVay, 1978). The boutons are, of course, the sites of synaptic specialization. This was shown by autoradiography (Figure 8). The swellings were packed with synaptic vesicles and mitochondria, formed synapses with one or several postsynaptic elements, and were enveloped in an astrocytic lamella. No synapses were found on any other parts of the axon tree (LeVay and Gilbert, 1976).

MYELINATION

The intermediate stages of columnar development have not yet been studied in detail. One aspect of cortical maturation that has been followed, however, is the process of myelination (Figure 9). At 22 days, when the segregation of the afferents is getting under way, there was no myelin in the cortex, except for a very occasional trunk in layer VI. At 49 days, when segregation is essentially complete, there were lightly myelinated axons running vertically and obliquely through layers VI and V. These were probably the parent trunks of the geniculocortical arborizations. There was still virtually no myelin in layer IV. By 3 months, there was a lightly myelinated tangential plexus in layer IV (which probably included the arborizations of the geniculate afferents) and in layer I. The density of myelinated fibers was still far lighter than in the adult, however, particularly in the upper cortical layers. It is clear then that the myelination of the afferent arborizations occurs *after* their segregation into ocular dominance columns is complete.

DISCUSSION

It has been suggested previously that, in forming ocular dominance columns, the geniculocortical afferents express a compromise between two opposing instructions—one for both sets of afferents to occupy the entire visual cortex according to a single retinotopic map, and the

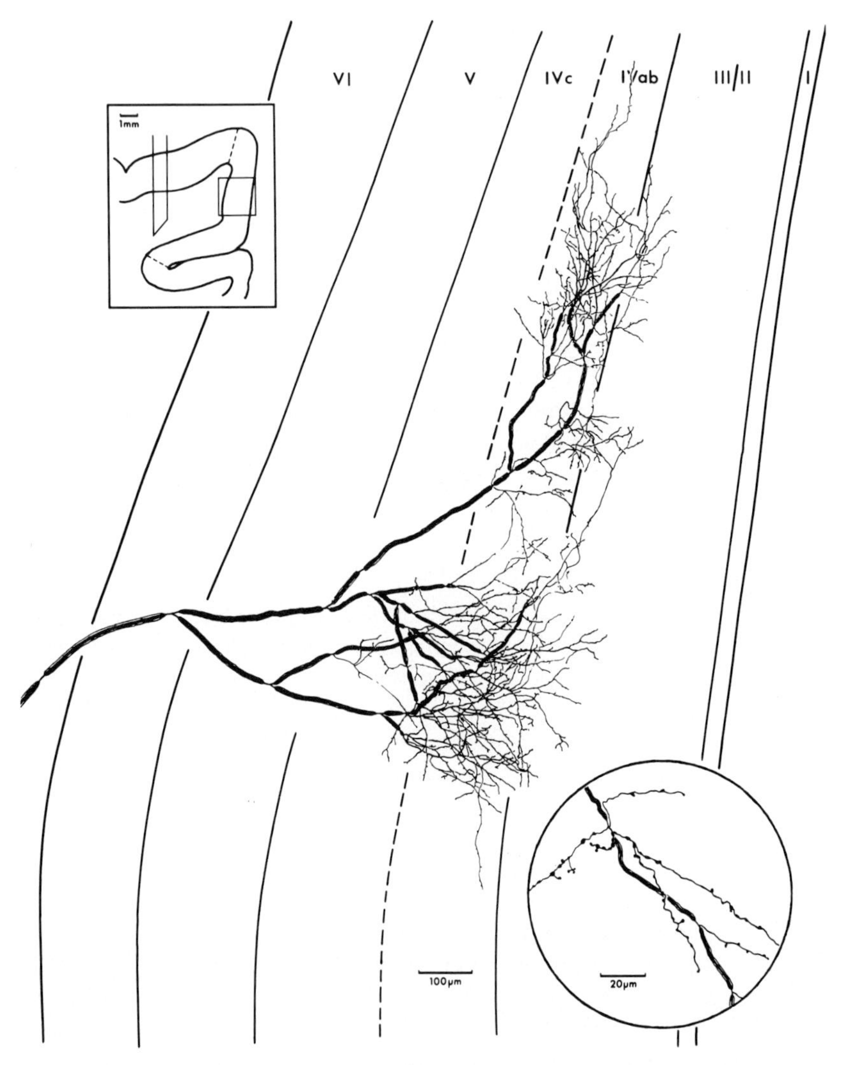

FIGURE 6. Arborization of a large geniculocortical afferent in an adult cat. This axon was filled with horseradish peroxidase from an extracellular injection into the optic radiation (see inset, top left). Camera lucida reconstruction from 16 successive coronal sections, each 100 μm thick. In its overall form and laminar distribution, this axon resembles the one reconstructed from a young kitten (Figure 3). It differs from it in four respects: (i) the arborization is not uniform but is divided into four clumps (two are superimposed in the lower part of the reconstruction), (ii) the arborization is myelinated, (iii) the mode of branching is no longer mainly dichoto-

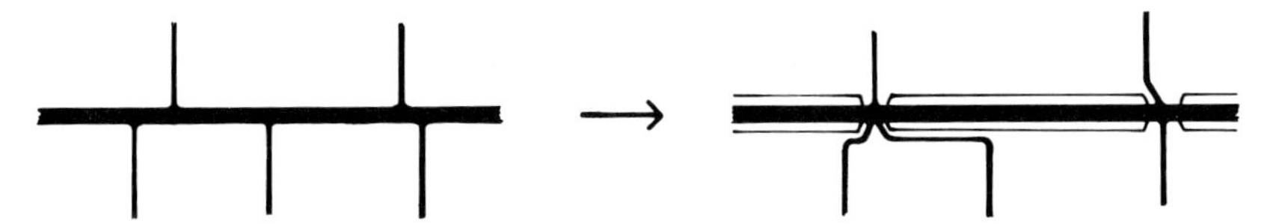

FIGURE 7. Typical axonal branching patterns in young kittens (left) and adult cats (right), arranged to illustrate the hypothesis that axonal branch-points are translocated to nodes of Ranvier during the process of myelination (see text).

other for fibers serving one eye to remain apposed to each other, rather than intermingling with those for the other eye (LeVay et al., 1975). These two instructions will generate a pattern of alternating bands, if the numbers of fibers serving the two eyes are about equal, because such an arrangement minimizes the length of interface between the two sets of fibers. The finding that the two sets of afferents are at first intermingled suggests that these two instructions are expressed at different stages of development. The first, retinotopic innervation of the cortex, is expressed, one may guess, during the initial ingrowth of the geniculocortical afferents. The second, sorting according to eye preference, is not expressed until 2–3 weeks postnatally in the cat. What the signal for this sorting process may be remains to be determined. In the monkey, at least, it does not seem to be dependent on visually evoked activity in the retina, since the process of segregation begins prenatally (Rakic, 1976) and goes to completion postnatally, even in conditions of dark-rearing or binocular lid-suture (Wiesel and Hubel, 1974; Hubel, Wiesel, and LeVay, unpublished).

The reconstruction of complete terminal arbors of geniculocortical afferents, before and after columnar development, has strengthened the autoradiographic evidence for a sorting process. More importantly, it has given us a clearer idea of the anatomical changes that may be involved (Figure 10). For the larger axons, at least, columnar development does not seem to require any major movement of entire arbors. Since, before segregation begins, afferent arbors span a cortical terri-

---

mous: as seen in the inset, lower right, several daughter branches can arise from each node of Ranvier, and (iv) the boutons—either of the terminal or *en passage* variety—are easily recognizable specializations on the terminal axon branches. (From Ferster, D. and S. LeVay [1978]. The axonal arborizations of lateral geniculate neurons in the striate cortex of the cat, *J. Comp. Neurol.* **182**:923–944. With permission of The Wistar Press.)

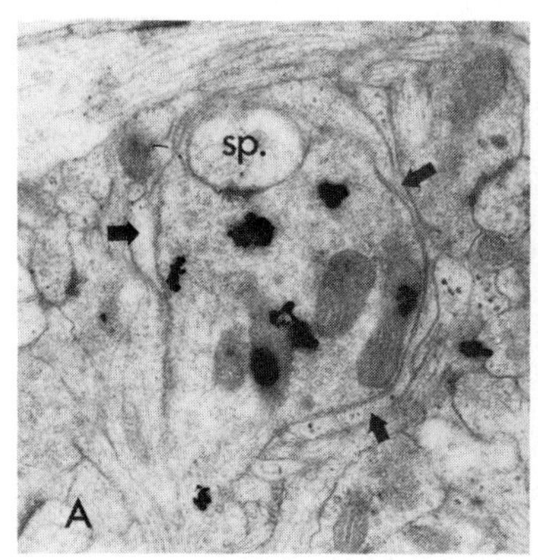

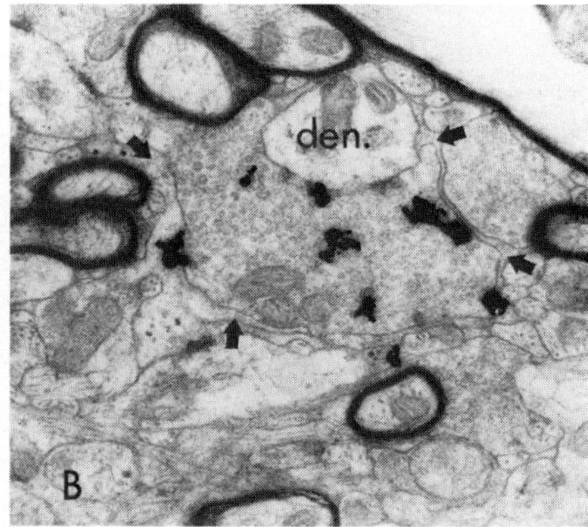

FIGURE 8. Geniculocortical terminals in an adult cat, identified by EM autoradiography. In contrast to the early synapses (Figure 5), those in the adult are situated at specialized swellings (boutons) that are ensheathed in astrocytic lamellae (arrows). The boutons contain large numbers of synaptic vesicles and mitochondria, but few or no microtubules. The synapses are made onto spines (sp.) or small dendritic shafts (den.). ×30,000. (From LeVay, S. and C. D. Gilbert [1976]. Laminar patterns of geniculocortical projection in the cat, *Brain Res.* **113**:1–19. With permission of Elsevier/North Holland Biomedical Press.)

tory that is large enough to house a number of neighboring columns, the sorting process may involve nothing more than local changes in density of the terminal branches on a static arbor.

The physiological observations suggest that the left- and right-eye afferents establish functional connections with cortical neurons in their early, overlapping pattern, and hence that synapses in the appropriate columns are likely to be broken during the process of segregation. It should not be thought, however, that these temporary connections have the elaborate structure of mature geniculocortical synapses. On the contrary, the EM autoradiographic observations show that most of these connections are very simple *en passage* contacts, with only small numbers of synaptic vesicles, just as has been described for early synapses in other parts of the developing nervous system. We have not yet applied the EM autoradiographic method to a study of the geniculocortical synapses during the process of segregation. Given the elementary structure of the contacts and the obvious limitations of electron microscopy in the third and fourth dimensions, it will not be easy to catch any synapses unmistakably in the act of dissolution, even though it is probably a commonplace event.

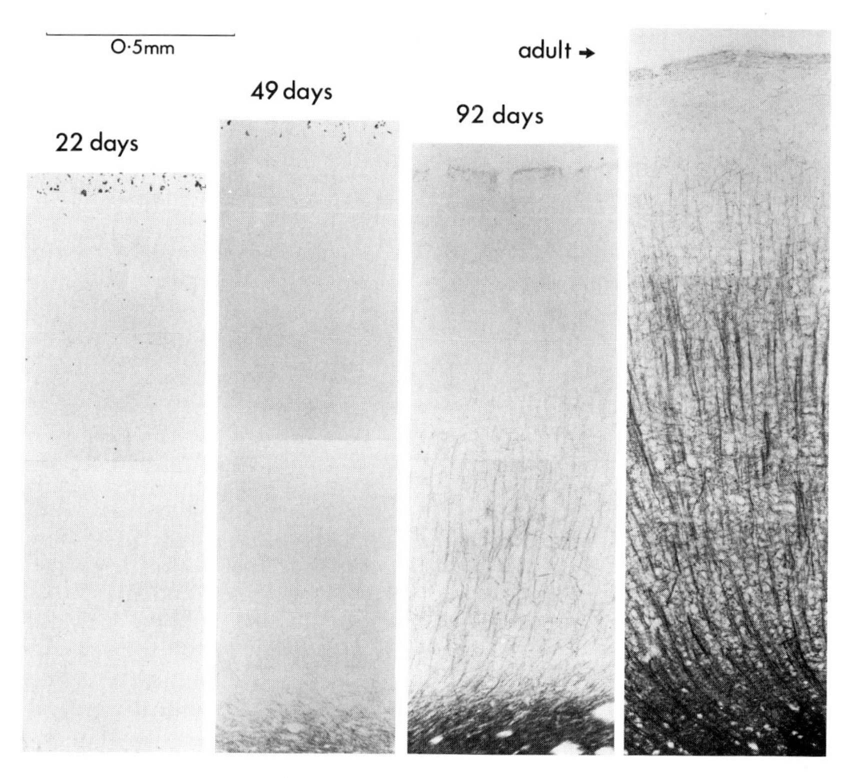

FIGURE 9. Time course of myelination in cat visual cortex. These are coronal sections stained by the method of Jebb and Woolsey (1977). The series illustrates that most of the myelin—except for that ensheathing large axonal trunks in the lower layers —is laid down after completion of columnar development (compare with Figure 1).

The rearrangement of axonal arbors in normal development envisioned here takes place during a "critical period" when the visual cortex and its afferents are susceptible to the effects of monocular deprivation (Wiesel and Hubel, 1963). During the fourth week of life in the cat (Hubel and Wiesel, 1970), or perhaps slightly earlier (Van Sluyters and Freeman, 1977; Van Sluyters, personal communication), a few days of monocular suture lead to a reduced influence of the deprived eye on cortical cells. Similar deprivation at younger ages does not reduce the influence of the deprived eye. Thus, in the cat, the beginning of the critical period roughly coincides with the onset in normal development of columnar segregation. Columnar segregation has begun by

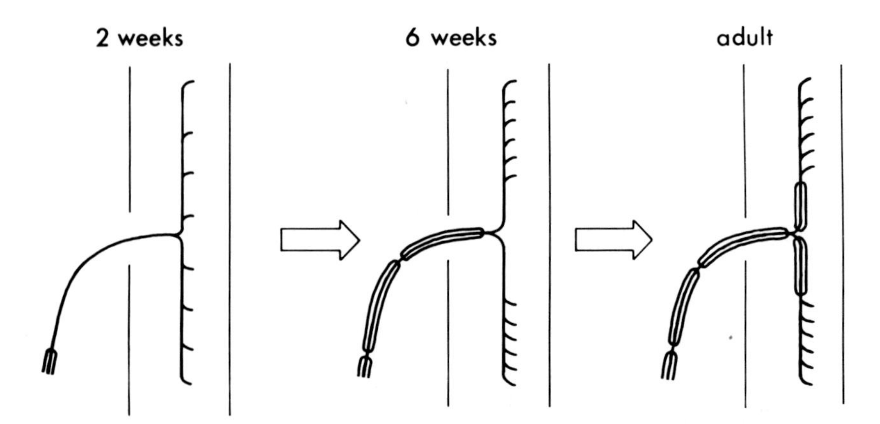

FIGURE 10. Probable sequence of events in the maturation of a large geniculocortical afferent. Each part of the drawing represents an axonal trunk entering the cortex and arborizing in layer IV. At 2 weeks the terminal branches are distributed uniformly. By 6 weeks branches in the inappropriate columns have degenerated or have been reabsorbed, and branches in the appropriate columns strengthened, but the overall structure and position of the axonal tree is unchanged. After completion of columnar development the arborization is myelinated (see text).

birth in the monkey (Rakic, 1976; Hubel et al., 1977), and the critical period is advanced accordingly (Hubel, Wiesel, and LeVay, unpublished). The end of the critical period for the cortex as a whole occurs well after the end of columnar segregation in layer IV in both the monkey (Hubel, Wiesel, and LeVay, unpublished) and in the cat (Hubel and Wiesel, 1970; LeVay et al., 1978). In the monkey, however, the critical period for layer IV ends earlier than for the other layers, although its exact timing with respect to the end of the columnar segregation is not yet clear, and similar data are as yet lacking for the cat. If plasticity does persist in the afferent axons for a period after the end of columnar segregation, it will be tempting to blame the end of this plastic period on some later-occurring process such as myelination.

For the maturation of the afferents is by no means complete with the completion of columnar segregation. A number of events follow, of which the myelination of the arborizations is the most dramatic. This is not merely the addition of myelin to an axon. As argued above, myelination seems likely to involve some local restructuring of the axonal branching pattern. More significantly, there is likely to be a further loss of synapses associated with myelination. Since the early synapses appear to be situated throughout the arborization of an axon, and since, in the adult, all but the finest terminal branches are myelinated, and hence nonsynaptic, it seems probable that the contacts made by the more proximal branches on the connecting trunks are either removed prior to myelination or are actually stripped off by the myelinating oligodendrocytes. Unlike the situation during columnar segregation, where synapses lost by one axon are likely to be reformed by another, breakage of connections during myelination would lead to a net reduction in synaptic number. Cragg (1972) has indeed reported an apparent drop in total synaptic number in the cat's visual cortex between 6 weeks and maturity, the period during which most of the myelination occurs.

ACKNOWLEDGMENTS

We thank E. Coughlin for technical assistance, M. Peloquin for photography, and S. Wilson for typing the manuscript. This study was supported by National Institutes of Health grant R01 EY01960, and by a grant from the Milton Fund.

REFERENCES

Blakemore, C., R. C. Van Sluyters, and J. A. Movshon (1975). Synaptic competition in the kitten's visual cortex, *Cold Spring Harbor Symp. Quant. Biol.* **40:**601–609.

Cragg, B. G. (1972). The development of synapses in cat visual cortex, *Invest. Ophthalmol.* **11:**377–384.

Ferster, D. and S. LeVay (1978). The axonal arborizations of lateral geniculate neurons in the striate cortex of the cat, *J. Comp. Neurol.* **182:**923–944.

Hubel, D. H. and T. N. Wiesel (1962). Receptive fields, binocular interaction and functional architecture in the cat's visual cortex, *J. Physiol. (London)* **160:**106–154.

Hubel, D. H. and T. N. Wiesel (1963). Receptive fields of cells in striate cortex of very young, visually inexperienced kittens, *J. Neurophysiol.* **26:**994–1002.

Hubel, D. H. and T. N. Wiesel (1965). Binocular interaction in striate cortex of kittens reared with artificial squint, *J. Neurophysiol.* **28:**1041–1059.

Hubel, D. H. and T. N. Wiesel (1968). Receptive fields and functional architecture of monkey striate cortex, *J. Physiol. (London)* **195:**215–243.

Hubel, D. H. and T. N. Wiesel (1970). The period of susceptibility to the effects of unilateral lid closure in kittens, *J. Physiol. (London)* **206:**419–436.

Hubel, D. H. and T. N. Wiesel (1972). Laminar and columnar distribution of geniculo-cortical fibers in the macaque monkey, *J. Comp. Neurol.* **146:** 421–450.

Hubel, D. H., T. N. Wiesel, and S. LeVay (1977). Plasticity of ocular dominance columns in monkey striate cortex, *Philos. Trans. R. Soc. London B Biol. Sci.* **278**:377–409.

Jebb, A. H. and T. A. Woolsey (1977). A simple stain for myelin in frozen sections: a modification of Mahon's method, *Stain Technol.* **52**:315–318.

LeVay, S. and C. D. Gilbert (1976). Laminar patterns of geniculocortical projection in the cat, *Brain Res.* **113**:1–19.

LeVay, S., D. H. Hubel, and T. N. Wiesel (1975). The pattern of ocular dominance columns in the monkey striate cortex revealed by a reduced-silver stain, *J. Comp. Neurol.* **159**:559–576.

LeVay, S., M. P. Stryker, and C. J. Shatz (1978). Ocular dominance columns and their development in layer IV of the cat's visual cortex: a quantitative study, *J. Comp. Neurol.* **179**:223–244.

Nevin, T. T., D. Tanaka, and W. L. R. Cruce (1978). A modified Epon embedding and counterstaining procedure for light microscopic examination of Golgi impregnated neurons in vertebrates, *Neurosci. Lett.* **7**:307–312.

Rakic, P. (1976). Prenatal genesis of connections subserving ocular dominance in the rhesus monkey, *Nature (London)* **261**:467–471.

Shatz, C. J., S. H. Lindström, and T. N. Wiesel (1977). The distribution of afferents representing the right and left eyes in the cat's visual cortex, *Brain Res.* **131**:103–116.

Shatz, C. J. and M. P. Stryker (1978). Ocular dominance in layer IV of the cat's visual cortex and the effects of monocular deprivation, *J. Physiol. (London)* **281**:267–283.

Van Sluyters, R. C. and R. D. Freeman (1977). The physiological effects of monocular deprivation in very young kittens, *Soc. Neurosci. Abstr.*, Vol. 3, p. 433.

Wiesel, T. N. and D. H. Hubel (1963). Single-cell responses in striate cortex of kittens deprived of vision in one eye, *J. Neurophysiol.* **26**:1003–1017.

Wiesel, T. N. and D. H. Hubel (1974). Ordered arrangement of orientation columns in monkeys lacking visual experience, *J. Comp. Neurol.* **158**:307–318.

Wiesel, T. N., D. H. Hubel, and D. M. K. Lam (1974). Autoradiographic demonstration of ocular-dominance columns in the monkey striate cortex by means of transneuronal transport, *Brain Res.* **79**:273–279.

MODIFICATION OF CORTICAL BINOCULAR CONNECTIVITY

Max S. Cynader, Michael J. Mustari, and Jill C. Gardner

Dalhousie University, Halifax, Nova Scotia, Canada

A number of animals, including men, monkeys, and cats, have two frontally located eyes. Although such species lose the advantage of panoramic vision, they gain extra capacities for locating objects in three-dimensional space. The differences in the retinal images of the two eyes provide the basis for stereoscopic depth perception, a mechanism by which frontal-eyed animals can determine not only the location of an object in depth, but also its trajectory. Stereoscopic capacities are dependent on visual experience (Duke-Elder and Wybar, 1973), and accumulating evidence indicates that atypical patterns of cortical binocular connectivity induced by abnormal visual exposure underlie anomalies of binocular vision (Wiesel and Hubel, 1963, 1965; Hubel and Wiesel, 1965; Packwood and Gordon, 1975). In this paper we will discuss the modifications of cortical binocular connectivity that occur following strabismus and monocular deprivation.

In strabismus, the visual axes of the two eyes are misaligned, and so the opportunity to correlate images in the two eyes is lost. When Hubel and Wiesel (1965) studied the effects of strabismus, surgically induced in young kittens, on the development of the striate cortex, they found that this noncorrelated input from the two eyes resulted in profound alterations in cortical binocular connectivity. Their results, shown in Figure 1, illustrate the relative effectiveness of inputs from the two eyes on a population of cells studied in normal (left-hand side) and strabismic (right-hand side) cat striate cortex (area 17). The ordinates of the ocular dominance distributions of Figure 1 represent the number of cells encountered; the abscissae represent a 7-point scale of ocular dominance. Units in Group 1 receive excitation exclusively through the eye contralateral to the hemisphere under study, and units in higher ocular dominance groups receive successively more

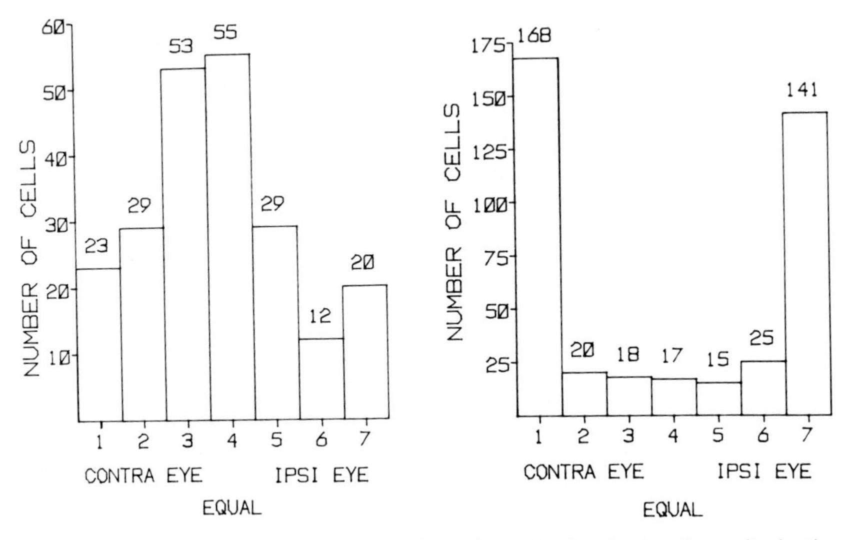

FIGURE 1. On the left is the distribution of ocular dominance for units in the striate cortex (area 17) of normal cats (Hubel and Wiesel, 1962); the right-hand side illustrates the comparable distribution for strabismic cats (Hubel and Wiesel, 1965). The ordinate represents the number of cells in each ocular dominance group, and the abscissa is the 7-point scale of Hubel and Wiesel (1962). On this scale, the numbers from 1 to 7 represent a trend from complete excitatory dominance by the contralateral eye (Group 1) to complete dominance by the ipsilateral eye (Group 7). Units in Group 4 are driven equally by visual stimuli through either eye. It is evident that the number of binocularly driven cells (Groups 2–6) is much reduced in strabismic kittens.

input from the ipsilateral eye. Units in Group 4 are driven equally through the two eyes, and units in Group 7 exclusively through the ipsilateral eye. In normal cats, many units receive strong inputs from both eyes (Groups 3–5), with relatively few monocular cells (Groups 1 and 7). In the striate cortex of strabismic kittens, however, monocularly driven cells form the vast majority of the units encountered. The virtual absence of binocularly driven units in ocular dominance Groups 3–5 indicates that the reduced opportunity for simultaneous binocular vision during development has resulted in a loss of excitatory convergence from both eyes onto single cortical cells.

We have studied the effects of early strabismus induced by surgery on the responses of units in cat parastriate cortex (area 18). We studied area 18 rather than area 17 because of recent evidence indicating that area 18, in both cat and monkey (Hubel and Wiesel, 1970*a*; Tusa,

Palmer, and Rosenquist, 1975; Cynader and Regan, 1978; Cynader, Gardner, and Douglas, 1978), may play a special role in stereoscopic function. Divergent strabismus was induced in normal, 10–14-day-old kittens by severing the medial rectus muscle. The kittens were then allowed visual exposure in a normally lit colony environment until they were 4–6 months old. The ocular dominance distribution for units in area 18 of these strabismic kittens is compared with that of normal cats in Figure 2. It is evident that the incidence of units that receive equal excitatory input from the two eyes is markedly reduced in the strabismic animals. As in area 17, most cortical units can be driven through either one eye or the other, but not both. In normal cats, about 60% of the units encountered receive strong inputs from both eyes (ocular dominance Groups 3–5), while only 12% of the units in the strabismic animals fall into this category. In both the normal and strabismic cats, the influence of the contralateral eye is greater than that of the ipsilateral eye. If one subdivides the ocular dominance

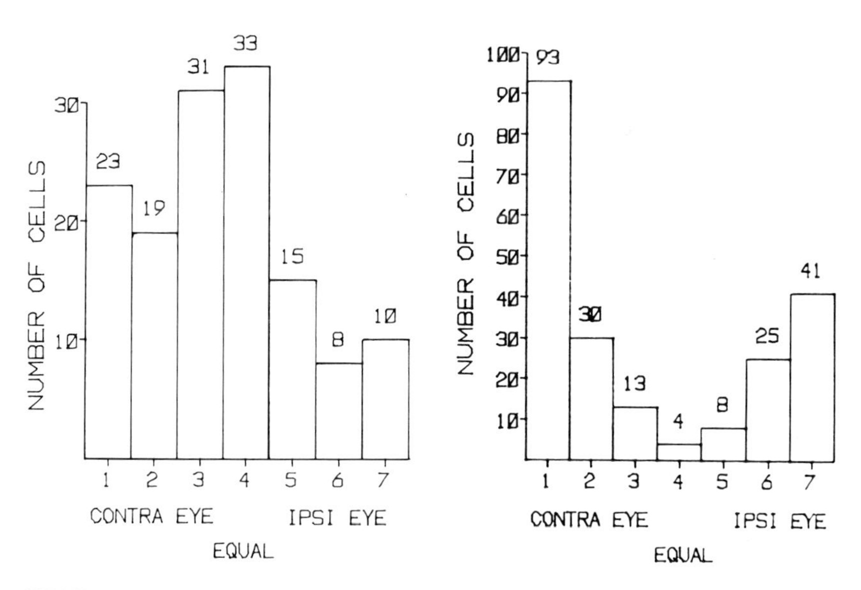

FIGURE 2. On the left is the distribution of ocular dominance for units in area 18 of normally reared cats; the right-hand side illustrates the comparable distribution for area 18 of strabismic cats. The influence of the contralateral eye is stronger than that of the ipsilateral eye in both normal and strabismic cats. In addition, the incidence of cells with binocular influence is much reduced in the strabismic cats. Conventions as in Figure 1.

distribution for the strabismic cats in Figure 2 by cortical hemisphere, one observes that, in addition to the loss of binocularly driven cells, fewer cells are driven through the deviating eye than through the normal eye. Figure 3 shows this trend is very marked in the hemisphere ipsilateral to the deviating eye. Here the trend toward the contralateral eye, observed in both normal and strabismic cats, *sums with* the trend toward the unoperated eye. In the hemisphere contralateral to the deviating eye, the contralateral bias *opposes* the trend toward the normal eye, resulting in approximately equal numbers of cells driven through each eye.

The rare binocularly driven cells observed in these cats tended to occur near the borders between ocular dominance columns (Hubel and Wiesel, 1962). In general, the receptive fields of these binocularly driven units were located on or near corresponding *retinal* points in the two eyes. That is, if the cat displayed a divergent strabismus of 15° by area centralis plots, the two receptive fields of binocular cells were separated by approximately 15°. On a few occasions, however, we observed that the retinal locations of the receptive fields in the

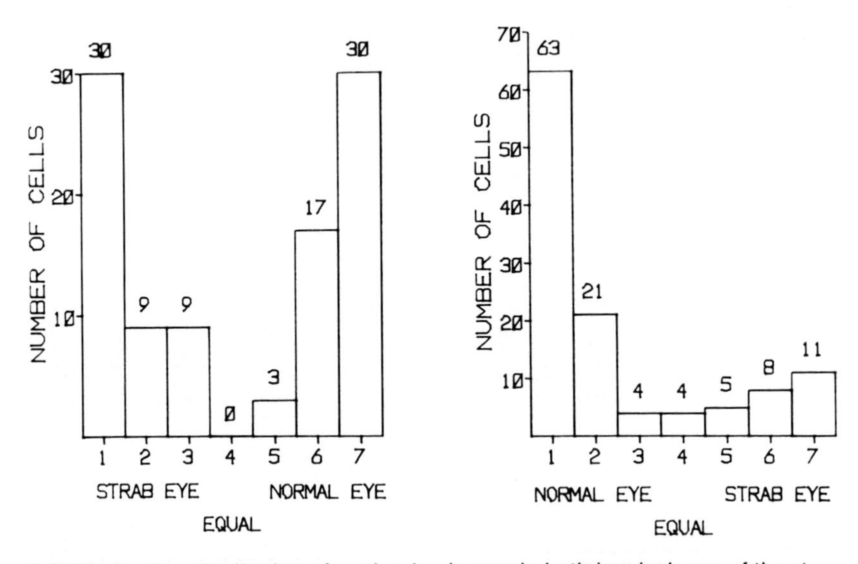

FIGURE 3. The distribution of ocular dominance in both hemispheres of the strabismic cats illustrated in Figure 2. In the hemisphere contralateral to the strabismic eye (left-hand side), the number of cells influenced through either eye alone is equal. In the other hemisphere, the ipsilateral (strabismic) eye influences far fewer units than the normal eye. Conventions as in Figure 1.

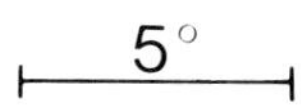

FIGURE 4. The relative positions of the receptive fields from the two eyes in a unit located in the superficial layers of the left area 18 in a strabismic cat. The locations of the areae centrales for the two eyes, as plotted ophthalmoscopically, are indicated by the points labeled A.C. The location and preferred orientation of the receptive fields of the binocularly driven unit are indicated by the two overlapping rectangles. The expected location of the receptive field of the ipsilateral eye if it were in *retinal correspondence* with that of the contralateral eye is indicated by shading in the figure. The receptive fields from the two eyes were nearly superimposed on the tangent screen despite the large divergent strabismus the cat displayed.

two eyes differed. The receptive fields of one of these exceptional units are illustrated in Figure 4. The label indicates the locations of the areae centrales in the two eyes as plotted on the tangent screen. The locations of the receptive fields, as plotted through each eye separately, are indicated by the overlapping rectangles. Since, in this animal, the two eyes were diverged by 22° (by area centralis plots), one would expect that the receptive fields in the two eyes would also be separated by this amount on the tangent screen. Instead, the receptive field plotted through the ipsilateral eye was found 15° ipsilateral to the area centralis, over 20° from its expected location. The shaded rectangle indicates the expected location of the receptive field of the ipsilateral eye. It should be noted that the effect of this anomalous correspondence between the receptive fields in the two eyes is to permit the cat to correlate the images of a single external object in the two eyes *despite* the large divergent strabismus. The mechanisms underlying the anomalous location of the receptive field in the ipsilateral eye remain unclear, but they may be related to the abnormally expanded projection of the corpus callosum in strabismic kittens (Lund, Mitchell, and Henry, 1978; Lund and Mitchell, this volume).

BINOCULAR INTERACTIONS IN NORMAL CATS

The ocular dominance distributions of Figures 1, 2, and 3 indicate the relative effectiveness of excitatory inputs from the two eyes on individual cortical cells as measured with monocular stimulation. However, *interactions* between inputs from the two eyes cannot be observed unless the two eyes are stimulated together. Several earlier workers have shown that, under appropriate circumstances, inputs from the two eyes may result in cortical responses that are greater than the sum of separately measured monocular responses (facilitation). In other circumstances, binocular stimulation may elicit weaker responses than stimulation of one eye alone (inhibition) (Barlow, Blakemore, and Pettigrew, 1967; Pettigrew, Nikara, and Bishop, 1968; Hubel and Wiesel, 1970*a*). Over the last few years, we have developed several new ways to examine the nature of binocular interactions in the visual cortex. In particular, we have been concerned with the effects of binocular stimulation of the two eyes at different spatial and temporal separations.

An example of the response of a unit that shows strong binocular facilitation is illustrated in Figure 5. To derive the three-dimensional plot illustrated in the figure, we presented flashed stimuli to the two eyes with different spatial and temporal separations in the two eyes. The response of the cell, represented on the ordinate, depends critically on the location of binocular stimuli in the two eyes. However, responses at the best location occur only if the *timing* of inputs from the two eyes is appropriate. If stimuli in two eyes are offset in time by more than 20 milliseconds, the response drops back to baseline levels. In the case illustrated, the binocular response is more than 50 times as large as that obtained when each eye is stimulated alone. The nature of the binocular interactions in a given unit cannot be predicted from stimulation of either eye alone. Only when both eyes are stimulated together can one observe binocular interactions of the sort described above. In fact, units that fall into ocular dominance Groups 1 or 7, and that appear to receive monocular input only, may show marked binocular interactions. Figure 6 illustrates the response of an apparently monocular unit (ocular dominance Group 1) to stimulation of the two eyes with different spatial and temporal disparities. To obtain the response plot illustrated in Figure 6, we flashed an optimally oriented bar at one location in the receptive field of the dominant eye and varied the location and timing of stimulation in the apparently silent eye. Stimulation through the silent eye resulted in strong inhibition of the

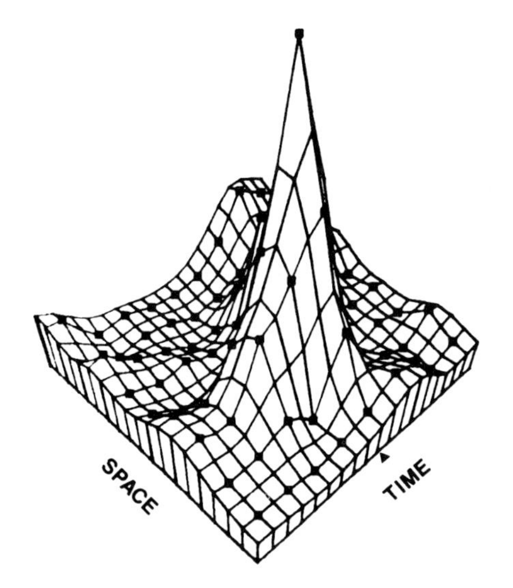

FIGURE 5. The responses of a unit in area 18 that responded poorly to monocularly presented stimuli through either eye, but showed strong facilitation with appropriate binocular stimulation. One horizontal axis of the three-dimensional plot is the spatial separation of stimuli in the two eyes (±6°). The other axis is the delay between stimulation of the two eyes (±40 msec). The ordinate represents the total spikes evoked from the cell under each condition. The baseline surrounding the central peak represents the linear sum of the responses to stimulation of each eye alone. With binocular stimulation of an appropriate spatial and temporal nature, a large facilitation is observed. Stimulus presentation and data collection were performed under control of a PDP 11/34 computer, which interleaved presentation of the 64 stimulus conditions used to generate this plot.

response to the dominant eye, but only when the spatial and temporal characteristics of binocular stimulation were appropriate. The detailed relationship between the characteristics of these binocular interactions and the mechanisms underlying stereoscopic vision are discussed elsewhere (Cynader et al., 1978).

It is also possible to observe binocular interactions in cortical units of normal cats when *moving* stimuli are presented to the two eyes. In this case, one loses information about the spatial and temporal characteristics of the binocular interactions, since different spatial separations of moving stimuli result inevitably in temporal separations as well. However, other characteristics of the interactions between the two eyes can best be observed with moving stimuli (Cynader and Regan, 1978). Figure 7 illustrates the responses of another apparently monocular

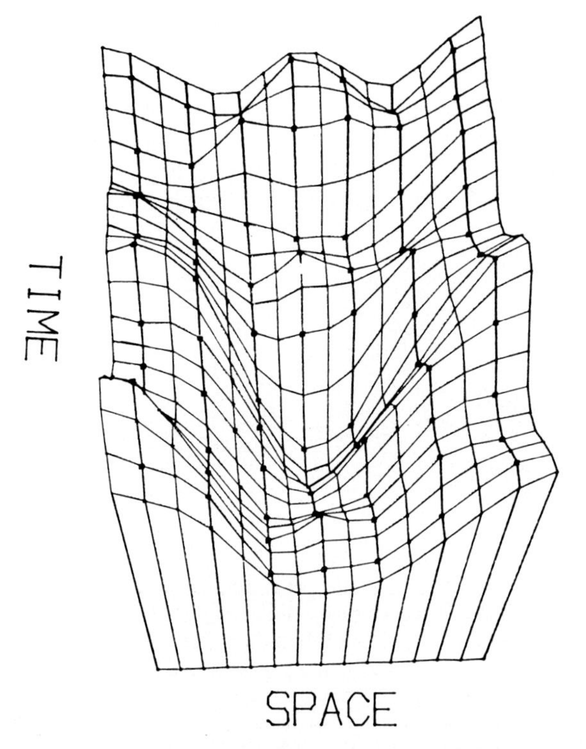

FIGURE 6. A space-time plot of a unit that appeared monocularly driven with conventional testing. For this unit, 7 spatial disparities (±6°) and 7 temporal disparities (±100 msec) were presented in a randomized interleaved fashion, and 49 histograms were collected. As in Figure 5, the ordinate represents the response of the unit, and the abscissae represent spatial and temporal disparities.

When stimuli are presented to the apparently silent eye with appropriate timing and at particular locations, the response to stimulation of the dominant eye (represented on the fringe of the plot) is markedly suppressed.

cell in area 18 of a normal cat to binocularly presented moving stimuli with different disparities. The dotted line represents the response obtained when the dominant eye was stimulated alone. With binocular stimulation, a profound inhibition was observed at a particular disparity. In this case, stimuli moved at the same speed and in the same direction in the two eyes. However, when the nondominant eye was stimulated with a bar that moved in the *opposite* direction at the same speed, binocular inhibition was replaced by binocular facilitation. The lower graph in Figure 8 is the same as that of Figure 7, but the curve marked

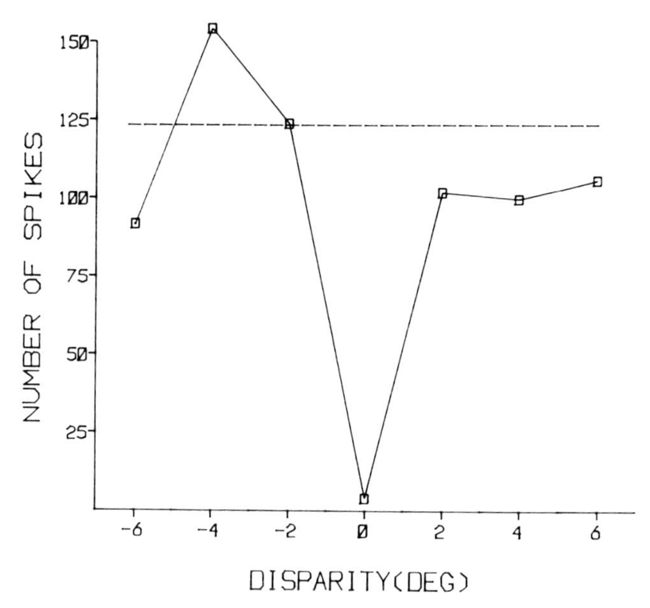

FIGURE 7. The response of an apparently monocular cell in area 18 of a normal cat to moving stimuli at various disparities. The dotted line represents the response to stimulation of the dominant eye alone. Binocular stimulation at a particular disparity (labeled 0) evokes a strong inhibitory influence from the "silent" eye, which abolishes the response to stimulation of the normal eye.

The ordinate represents the number of spikes evoked per 16 stimulus presentations. Stimulus velocity: 20°/sec.

"antiphase" represents the response of this cell to stimuli that moved in opposite directions in the two eyes. These data show that the inhibition from the silent eye illustrated in Figure 7 is a *direction selective* inhibition. With opposite-directed motion in the two eyes, the inhibition is itself suppressed and the binocular response is much enhanced.

Opposite-directed movement in the two eyes has a simple geometric correlate in the movement of visual stimuli in three-dimensional space. A stimulus that is moving in depth directly toward the observer's nose will move at equal and opposite speeds in the two eyes. A stimulus that is moving sideways, rather than toward or away from the observer, will move in the same direction at the same speed in the two eyes. These geometrical considerations show that the behavior of the unit of Figures 7 and 8 may be understood in terms of an inhibition of response to binocular stimuli moving sideways and an enhanced response to

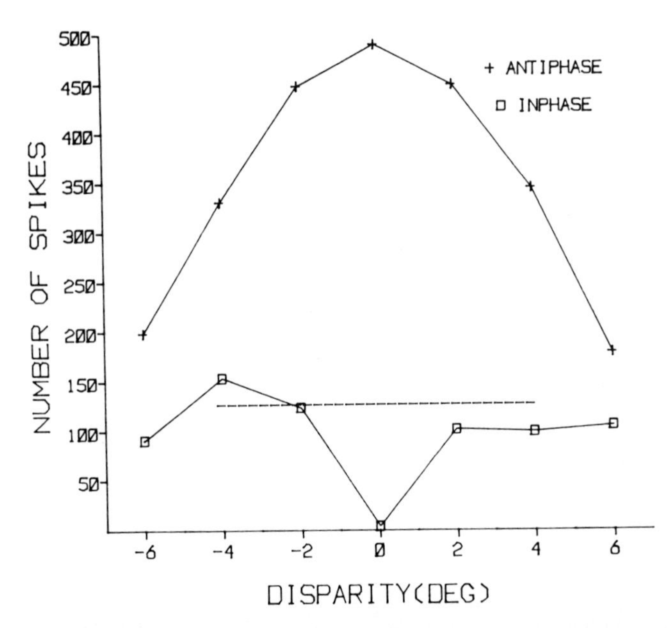

FIGURE 8. Response of the same unit as illustrated in Figure 7 to binocular stimuli moving in the same (□) or opposite (+) directions in the two eyes. As in Figure 7, the dotted line represents the response to stimulation of the dominant eye alone; the lower graph (□) shows that this response is inhibited at a particular disparity when the other eye is stimulated with the same direction of movement. When the apparently silent eye is presented with the opposite direction of stimulus movement (+), the inhibition disappears, and instead the unit gives an enhanced response.

stimuli moving toward (or in other cases, away from) the cat. It is possible to define the directional preferences of a cell for different trajectories of three-dimensional movement by measuring the *velocity* characteristics of the binocular interactions (Cynader and Regan, 1978). It can be shown geometrically that different velocity ratios in the two eyes correspond to different directions of stimulus movement, as viewed from above. By varying both the relative speed and direction of binocular stimulation, we can obtain a profile of the cell's response to stimuli moving in different directions in three-dimensional space. Figure 9 illustrates the response of another apparently monocular unit in area 18 of a normal cat to stimuli moving in different directions in depth (as viewed from above). The two eyes at the bottom of the figure represent the cat's eyes and are included to highlight the distinction be-

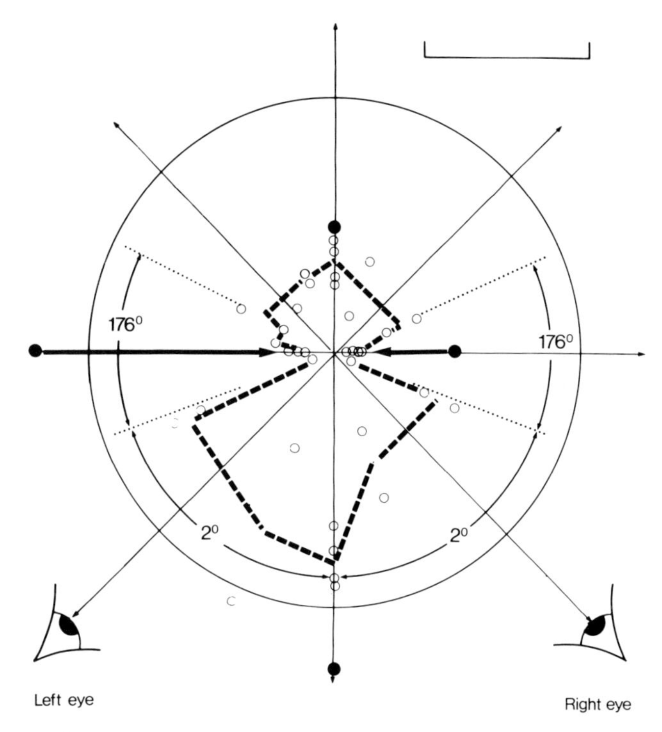

FIGURE 9. The response of an apparently monocular unit to stimuli moving with different velocities in the two eyes. The stimulus to the dominant eye was kept constant (22°/sec movement), while the speed and direction of stimuli presented to the apparently silent eye were varied. The filled circles represent the monocular response and the open circles represent the responses obtained with binocular stimulation. The large inhibition for sideways movement (indicated by arrows) causes strong responses to be restricted to a narrow range of direction (4°) for which the moving target would hit or just miss the animal's body. Scale bar represents five spikes/presentation. For further details, see Cynader and Regan (1978).

tween trajectories that miss the head and trajectories that pass between the animal's eyes. The radial distance of any point from the center of the plot represents the number of spikes elicited per stimulus presentation. The sole visual cue to the direction of motion in depth is the relationship between the velocities of the left and right visual images. The plot of Figure 9 shares the different velocity ratios (from -1.0 to $+1.0$) equally aross the polar plot. However, the relationship between the different directions of stimulus movement in depth and the velocity ratios of the two eyes is markedly nonlinear (Beverley and Regan, 1975). For the viewing distance that we used (145 cm), the four central octants each represent only 0.6° of visual angle. The range of directions *in space* over which this unit responded vigorously, as noted in the figure, comprised only 4° of visual angle.

BINOCULAR INTERACTIONS IN STRABISMIC CATS

The results of the previous section showed that marked facilitatory or inhibitory binocular interactions can be observed when both eyes of normal cats are stimulated together. Moreover, many apparently monocular units may display strong binocular interactions if the characteristics of binocular stimuli are appropriate. Since the major effect of strabismus, assessed qualitatively as in Figures 1–3, is to increase the incidence of units with monocular excitatory drive, it may be that these apparently monocular units show *binocular* interactions just as do many apparently monocular units of normal cats. To examine this question, we undertook a quantitative study of binocular interaction in area 18 of strabismic cats using the same methods as have been described for normal cats.

Figure 10 illustrates the responses of a unit that showed among the strongest binocular interactions we have observed in area 18 of strabismic cats. In this case, the interaction was facilitatory in nature. Appropriate binocular stimulation produced three times as many spikes as did monocular stimulation of the dominant eye alone. Comparison of this figure with Figure 5 shows, however, that the degree of binocular facilitation is much less in this unit of the strabismic cat than that occurring in strongly facilitatory units encountered in normal cats. As in the normal cat, binocular facilitation for the unit in Figure 10 depends on both the spatial and temporal characteristics of binocular stimuli. The spatial extent over which facilitation occurs is comparable in Figures 5 and 10, but in the temporal domain, selectivity appears much broader in the unit of the strabismic cat. This marked reduction in the maximum amplitude of binocular facilitation and the increased duration over which binocular interactions occur distinguish cortical units in the strabismic cats from those of normal cats.

Figure 11 illustrates the effects of moving binocular stimuli on the responses of four apparently monocular units in the cortex of strabismic cats. The unit on the upper right of Figure 11 is representative of over 70% of the monocularly driven units encountered in area 18 of

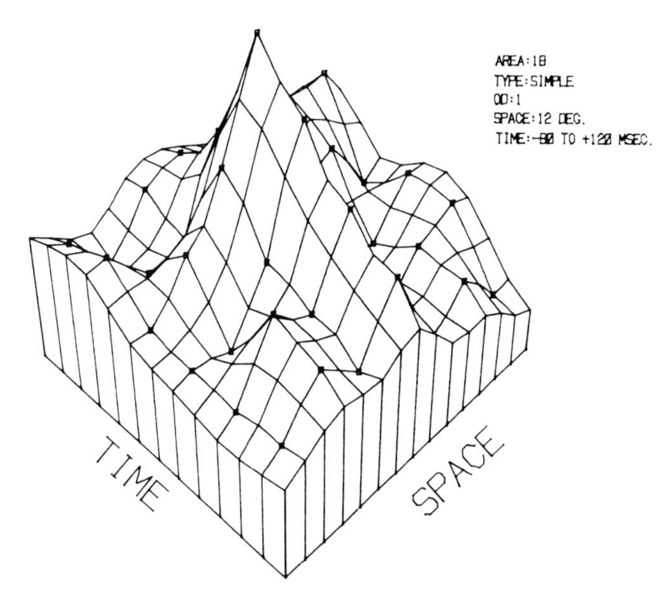

FIGURE 10. A space-time plot showing the degree of binocular interaction in a unit located in area 18 of a strabismic cat. As in Figure 5, the border around the central peak represents the baseline response (in this case, response to the dominant eye alone). The horizontal axis represents space (±6° around 0 disparity) and time (−80 to +120 msec away from simultaneous stimulation of the two eyes). The ordinate represents the number of spikes evoked from the cell under each of the 49 conditions.

The response of the cell is somewhat enhanced, relative to the monocular response, over a fairly narrow range of spatial disparities and over a broad range of interocular delays.

strabismic cats. In these cells, stimulation of the nondominant eye had little or no effect on the responses elicited through the dominant eye. These results show that binocular *interaction* as well binocular excitatory convergence is markedly reduced in strabismic cats. Among the relatively few units encountered in which binocular interactions occurred, several different varieties of response could be observed. These are depicted in the remaining parts of Figure 11. Both inhibitory and excitatory interactions were seen. The largest interactions thus far observed occurred with opposite-directed stimuli in the two eyes (moving toward or away from the body), as illustrated by the unit on the lower left of the figure. Other cells, represented by the unit on the lower right, showed inhibition with either the same or opposite-directed stim-

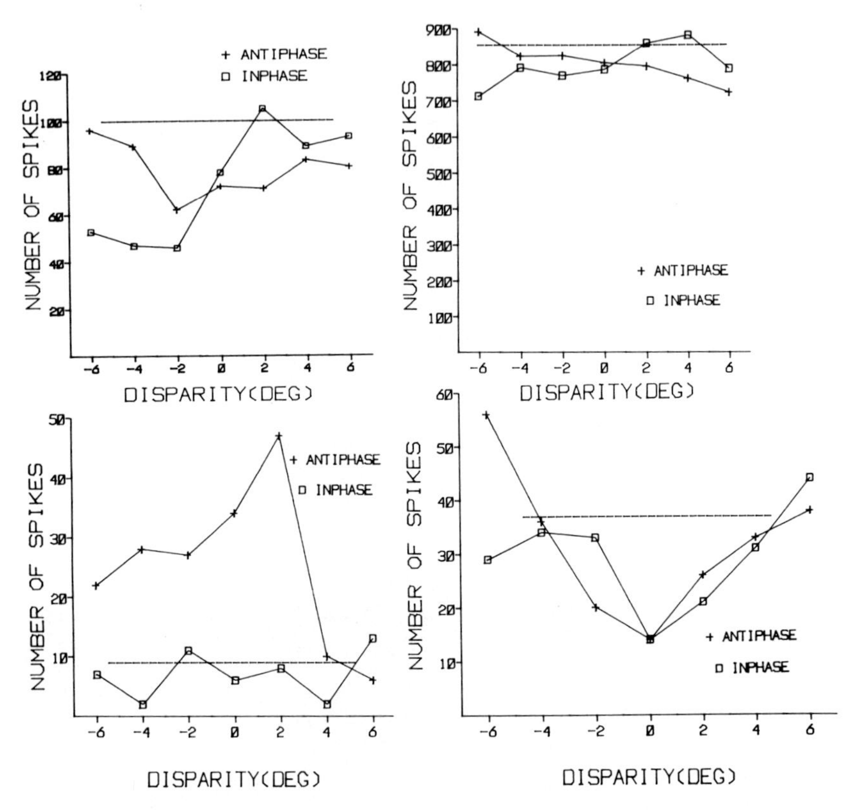

FIGURE 11. The responses of four apparently monocular cells to binocularly presented moving stimuli at different disparities. In all cases, the dotted horizontal line represents the response to stimulation of the dominant eye alone. The graphs labeled with squares represent responses to stimuli moving in the same direction and at the same speed in the two eyes. The graphs labeled (+) represent the responses to stimuli moving in opposite directions at the same speed.

The unit on the top right showed almost no change in its responses with binocular stimulation. Apparently, this cell received little or no input from the "silent" eye. The other three cells in the figure all showed some degree of binocular interaction. The unit on the top left displayed a clear inhibition for moving stimuli with crossed disparities (in front of the fixation plane: $-6°$ to 0), but the monocular responses were relatively unaffected by binocular stimulation at uncrossed disparities (beyond the fixation plane). With opposed motion in the two eyes, a broad but relatively shallow inhibitory trough was observed.

The unit on the lower left displayed little or no change in its firing rate with binocular stimuli moving in the same direction. However, responses were increased by a factor of more than five when stimuli moved in opposite directions in the two eyes. The behavior of this unit is similar to that of the cell illustrated in Figure 8.

ulus movement in the two eyes. In general, these units display the same sorts of binocular interactions as we and others have found in normal cats, but with considerably reduced amplitude.

CONSEQUENCES OF REDUCED BINOCULAR CONNECTIVITY FOR CORTICAL PLASTICITY

The notion of the critical period has become ubiquitous in developmental neurobiology. In the study of postnatal development of the visual system, the term underscores numerous observations indicating that older animals are less vulnerable to environmental alteration of cortical connectivity than younger animals. The mechanisms that permit binocular (and other) cortical connections to be modified by experience in young animals, but not in older animals, remain unknown.

In the previous paper (LeVay and Stryker, this volume), the extent of overlap between lateral geniculate inputs from the two eyes in the striate cortex was shown to diminish markedly as the animal aged. Since the timing of the retraction of cortical afferents representing the two eyes roughly mirrors the decline in sensitivity to monocular deprivation that occurs as the animal ages, it has been suggested that as terminals from the two eyes retract, the probability of binocular competition between inputs from the two eyes is reduced. According to this view, competition is facilitated when fibers representing the two eyes are in close proximity, and is less likely when they are separated.

The effect of strabismus is to *accentuate* the separation between inputs from the two eyes, resulting in reduced binocular input to cortical neurons. Hence, it is possible that strabismic animals might be less vulnerable to subsequent modification of cortical ocular dominance by monocular deprivation than would normal animals of a comparable age. To examine this possibility, divergent strabismus was induced in kittens when they were 10–14 days old (Mustari and Cynader, 1977). They were then reared in a normally lit environment until the age of 45 days, at which time one eyelid was sutured for an additional 1-week

---

The unit illustrated on the lower right showed a clear inhibition of the monocular response when binocular stimuli were presented at a particular disparity. The inhibitory interaction was similar for like-directed and opposite-directed motion in the two eyes. This cell appeared to receive an inhibitory influence from the "silent" eye, but the inhibition was much weaker than that illustrated for the normal cell in Figures 7 and 8, and was not altered by changing the direction of stimulus movement.

period. The effects of 1 week of monocular deprivation in these previously strabismic kittens are compared with those of control normally reared kittens in Figure 12. In the control kittens, 1 week of monocular deprivation causes a marked shift in the ocular dominance distribution as compared with that of normal kittens (Figure 1). The number of binocularly driven units is markedly reduced, and the large majority of the units encountered are driven exclusively through the nondeprived eye. In strabismic kittens that have undergone 1 week of monocular deprivation, binocular units are also rare, but the effect of monocular deprivation is *reduced* relative to that of normal cats. In the normal cats, only 3% of the units encountered receive strong input from the deprived eye (Groups 1 and 2), while 20% of the units encountered in the strabismic kittens fall into this category. This protective effect of the prior strabismus is much stronger in the hemisphere contralateral to the deviating eye than it is in the other hemisphere.

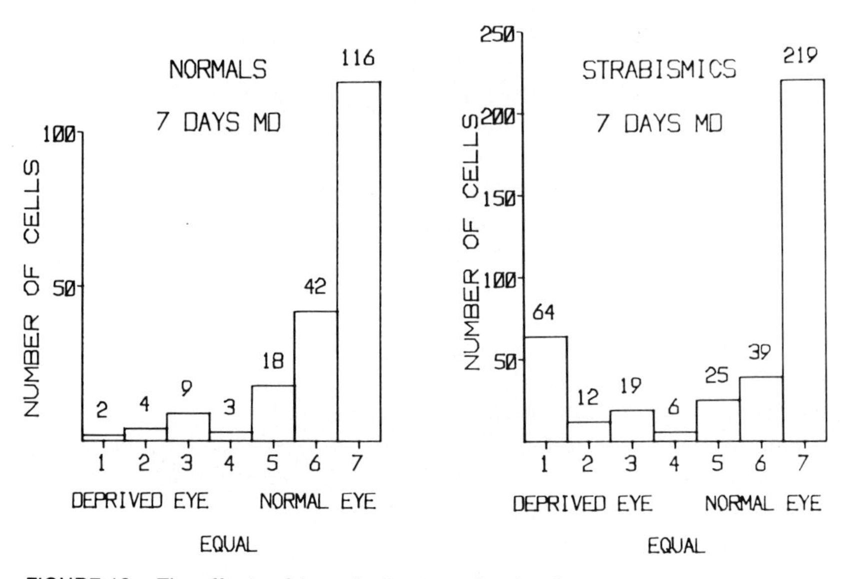

FIGURE 12. The effects of 1 week of monocular deprivation starting at 45 days of age are compared for normal (left-hand side) kittens and kittens in which strabismus was induced at 10–14 days of age (right-hand side). Recording was carried out in both hemispheres, but the data are pooled and displayed as if all units were recorded from the hemisphere contralateral to the deprived eye. The data indicate that monocular deprivation causes most cells in the normal group (left-hand side) to lose excitatory connections from the deprived eye. These effects are reduced in strabismic kittens relative to the normally reared kittens. Conventions as in Figure 1.

Since the effect of neonatally induced strabismus is to accentuate the maturational process by which ocular dominance bands segregate, and since strabismic kittens seem less vulnerable to the effects of monocular eyelid suture, they may, in this sense, be "older" than their normally reared counterparts.

CORTICAL PLASTICITY IN DARK-REARED CATS

Several studies have been made of cortical responses in kittens kept in darkness for prolonged periods of time starting at birth (Pettigrew, 1974; Imbert and Buisseret, 1975; Cynader, Berman, and Hein, 1976). In general, the results of these studies indicate that cortical unit responses in visually deprived kittens become less vigorous and that the sensitivity for stimulus orientation shown by units of normal cats is markedly degraded. The sluggish responses and decreased selectivity of units in dark-reared cats are reminiscent of the weak and relatively unselective responses observed in cortical units of young kittens (Hubel and Wiesel, 1963). Since many aspects of cortical development, including binocular connectivity, may remain in an immature state following visual deprivation, it is possible that visually deprived animals might appear "younger" than their normally reared counterparts with regard to their sensitivity to monocular deprivation, just as strabismic kittens appear "older."

To examine this question, we raised kittens in the dark until they were 4 months of age, beyond the duration of the previously defined critical period (Hubel and Wiesel, 1970*b*; Blakemore and Van Sluyters, 1974), and then brought them into the light and sutured one eyelid shut. The animals were then allowed monocular visual exposure for an additional 1 or 2 months. When we recorded from striate cortex cells of these kittens, we found evidence for marked changes in cortical binocular connectivity. As shown in Figure 13 (left-hand side), the large majority of cortical cells encountered responded to stimuli presented through the nonsutured eye, and only a few cells could be influenced through the deprived eye. These data appeared to provide clear evidence for a prolongation of the critical period in dark-reared cats. Our control experiments were, however, somewhat disappointing. Figure 13 (right-hand side) shows that effects of monocular deprivation can still be observed even in 4-month-old normally reared cats. These experiments indicate that the critical period in normal cats is more prolonged than had previously been thought (Hubel and Wiesel, 1970*b*; Blakemore and Van Sluyters, 1974). While the effects of monocular

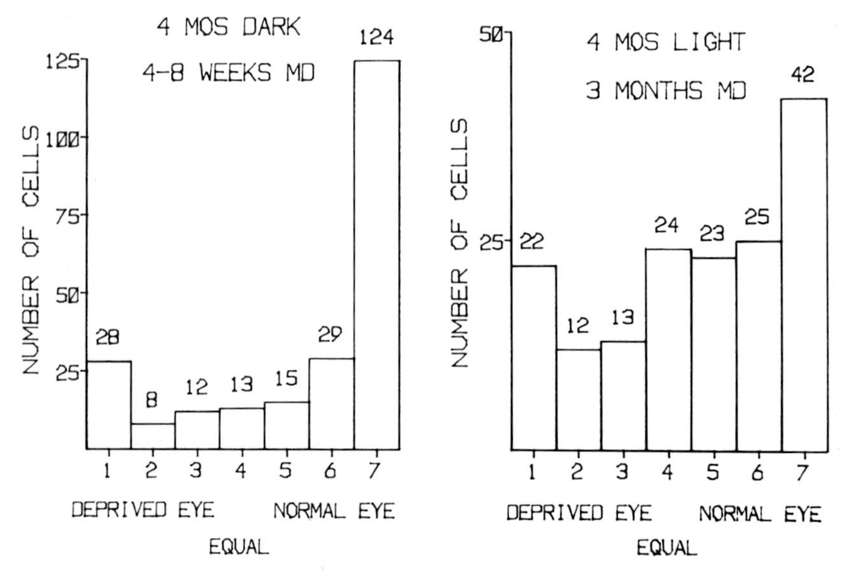

FIGURE 13. On the left is the distribution of ocular dominance for units in the visual cortex of cats reared in the dark for 4 months and then allowed 4–8 weeks of monocular visual exposure. On the right is the comparable distribution for units of cats reared in the light for 4 months and then monocularly sutured for an additional 3-month period. All recordings in these cats were made in the hemisphere contralateral to the sutured eye, so that units dominated by that eye are found in ocular dominance Groups 1–3. In both normal and dark-reared subjects, the deprived eye influences fewer cells than does the nonsutured eye. The effects are, however, much stronger in the dark-reared cats. Conventions as in Figure 1.

deprivation commencing at 4 months of age are much more marked in dark-reared cats than they are in normal cats of the same age, indicating that sensitivity to monocular deprivation has indeed increased, one cannot conclude unequivocally from these data that we have succeeded in extending the period of susceptibility for competitive binocular interactions.

To obtain definitive evidence on this point, we reared kittens in the dark for still longer periods of time and then brought them into the light and sutured one eyelid. The kittens were then allowed monocular visual exposure for an additional 3-month period. The results for these longer deprivations and the associated control data are shown in Figure 14. The data of Figure 14 show that strong monocular deprivation effects can be obtained following dark-rearing of 6, 8, or 10 months.

In all cases, most cortical cells can be driven only through the nonsutured eye. The control data in the lower part of Figure 14 show that some residual cortical plasticity remains even in normally reared 5-month-old kittens, but that no increase in response to the nonsutured eye is observed in the cats allowed 8 or 10 months of normal vision before the monocular deprivation was instituted. The similarities among

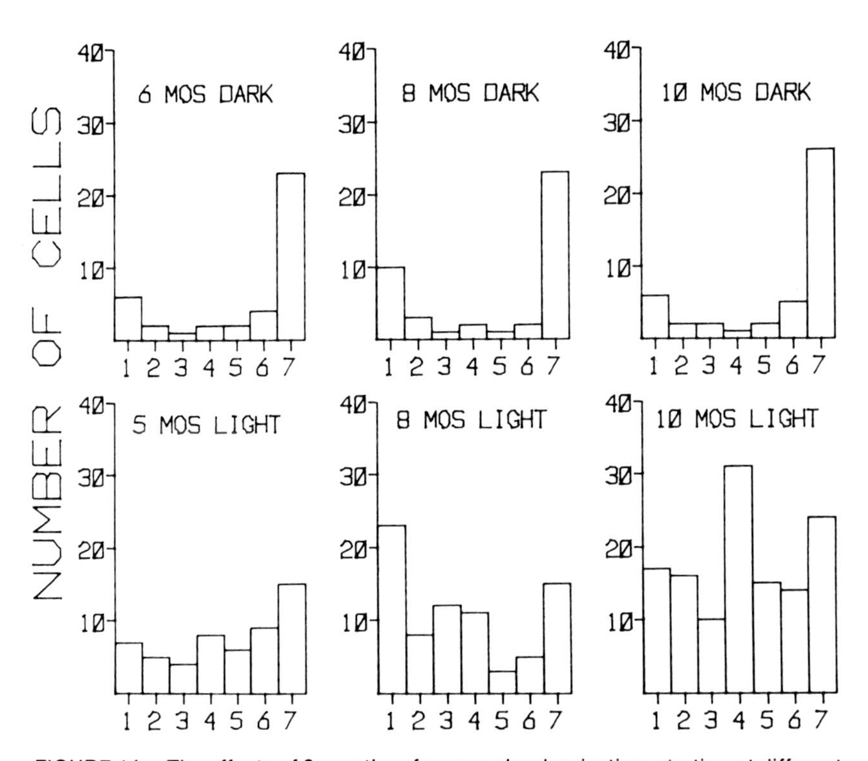

FIGURE 14. The effects of 3 months of monocular deprivation, starting at different ages, on the distribution of ocular dominance in dark-reared and light-reared animals. In all cases, recordings were made in the hemisphere contralateral to the sutured eye, so that the cells in Group 1 are those driven through the previously sutured eye, and those in Group 7 are driven through the normally viewing eye.

In the dark-reared cats, monocular deprivation instituted at any age causes a marked decrease in the relative effectiveness of stimuli presented through the deprived eye. In the normal cats, the deprived eye is somewhat less effective than the nonsutured eye if deprivation is instituted at 5 months, but not at 8–10 months.

the effects of monocular deprivation performed after different periods of dark-rearing, and the difference from the normal cats of comparable ages, indicate that the animal's age is not the sole determinant of its susceptibility to monocular deprivation. Rather, it appears that the type of experience the animal has had and the associated state of cortical maturity determine the susceptibility to later monocular deprivation.

The results of these experiments show that the critical period for modification of binocular connectivity can be extended by prior visual deprivation. The mechanism underlying this enhanced modifiability is not yet known, but a delay in the separation of geniculate terminals representing the two eyes in dark-reared cats must be regarded as a reasonable candidate. Other possibilities certainly exist (e.g., Kasamatsu and Pettigrew, 1976), and it is possible that several underlying processes may interact to control the plastic capabilities of cortical neurons.

CONCLUSIONS

The experiments reported here show that binocular convergence onto single cortical cells is markedly reduced in cats reared with surgically induced strabismus. In normal cats, responses of single cells depend critically on the spatial and temporal characteristics of binocular stimuli, and also on the relative velocities of the images in the two eyes. Many apparently monocular cells in normal cats may show strong binocular interactions with appropriate stimuli. Some binocular interactions, of the same type noted in normal cats, are preserved in the cortex of strabismic cats, but the strength of these interactions is much reduced, and in many cells little or no binocular connectivity can be demonstrated.

This reduced binocular connectivity may play a role in the diminished sensitivity to monocular deprivation that characterizes cats with neonatally induced strabismus. The relative insensitivity of these cats is reminiscent of the decline in sensitivity to monocular deprivation that occurs as kittens grow older. By contrast, dark-reared cats that display markedly enhanced sensitivity to monocular deprivation resemble younger kittens in this regard. Since it appears to be possible to prolong the critical period for the effect of monocular deprivation, apparently indefinitely, by rearing kittens in the dark, or to reduce sensitivity to monocular deprivation in young kittens by making them strabismic early in life, the results show that sensitivity to monocular deprivation depends on the state of the cortex, rather than simply on the age of the animal.

ACKNOWLEDGMENTS

This research was supported by U.S. National Institutes of Health grant EY02248, and by grants from the Medical Research Council (MT5201) and National Research Council (A9939) of Canada.

REFERENCES

Barlow, H. B., C. Blakemore, and J. D. Pettigrew (1967). The neural mechanism of binocular depth discrimination, *J. Physiol. (London)* **193:**327–342.

Beverley, K. I. and D. M. Regan (1975). The relationship between discrimination and sensitivity in the perception of motion in depth, *J. Physiol. (London)* **249:**387–398.

Blakemore, C. and R. C. Van Sluyters (1974). Reversal of the physiological effects of monocular deprivation in kittens: further evidence for a sensitive period, *J. Physiol. (London)* **237:**195–216.

Cynader, M., N. Berman, and A. Hein (1976). Recovery of function in cat visual cortex following prolonged deprivation, *Exp. Brain Res.* **25:**139–156.

Cynader, M., J. Gardner, and R. M. Douglas (1978). Neural mechanisms of stereoscopic depth perception in the cat visual cortex, pp. 373–390 in *Frontiers of Visual Science*, Cool, S. and E. L. Smith III, eds. Springer-Verlag, New York.

Cynader, M. and D. M. Regan (1978). Neurones in cat parastriate cortex sensitive to the direction of stimuli in three-dimensional space, *J. Physiol. (London)* **274:**549–569.

Duke-Elder, S. and K. Wybar, eds. (1973). *System of Ophthalmology*, Vol. 6: *Ocular Motility and Strabismus*. C. V. Mosby Co., St. Louis.

Hubel, D. H. and T. N. Wiesel (1962). Receptive fields, binocular interaction and functional architecture in the cat's visual cortex, *J. Physiol. (London)* **160:**106–154.

Hubel, D. H. and T. N. Wiesel (1963). Receptive fields of cells in striate cortex of very young, visually inexperienced kittens, *J. Neurophysiol.* **26:**994–1002.

Hubel, D. H. and T. N. Wiesel (1965). Binocular interaction in striate cortex of kittens reared with artificial squint, *J. Neurophysiol.* **28:** 1041–1059.

Hubel, D. H. and T. N. Wiesel (1970*a*). Cells sensitive to binocular depth in area 18 of the macaque monkey cortex, *Nature (London)* **225:**41–42.

Hubel, D. H. and T. N. Wiesel (1970*b*). The period of susceptibility to the physiological effects of unilateral eye closure in kittens, *J. Physiol. (London)* **206:**419–436.

Imbert, M. and P. Buisseret (1975). Receptive field characteristics and plastic properties of visual cortical cells in kittens reared with or without visual experience, *Exp. Brain Res.* **22:**25–36.

Kasamatsu, T. and J. D. Pettigrew (1976). Catecholaminergic control of neural plasticity in kitten visual cortex during the critical period, *Soc. Neurosci. Abstr.*, Vol. 2 (2), p. 1120.

LeVay, S. and M. P. Stryker (1979). The development of ocular dominance columns in the cat, *Soc. Neurosci. Symp.* **4:**83–98.

Lund, R. D. and D. E. Mitchell (1979). Plasticity of visual callosal projections, *Soc. Neurosci. Symp.* **4:**142–152.

Lund, R., D. E. Mitchell, and G. H. Henry (1978). Squint induced modification of callosal connection in cats, *Brain Res.* **144:**169–172.
Mustari, M. and M. Cynader (1977). Early strabismus protects neurons in kitten area 17 from the effects of monocular deprivation, *Soc. Neurosci. Abstr.*, Vol. 3, p. 570.
Packwood, J. and B. Gordon (1975). Stereopsis in normal domestic cat, Siamese cat and cats raised with alternating monocular occlusion, *J. Neurophysiol.* **38:**1485–1499.
Pettigrew, J. D. (1974). The effect of visual experience on the development of stimulus specificity by kitten cortical neurones, *J. Physiol. (London)* **237:**49–74.
Pettigrew, J. D., T. Nikara, and P. O. Bishop (1968). Responses to moving slits by single units in cat striate cortex, *Exp. Brain Res.* **6:**373–390.
Tusa, R. J., L. A. Palmer, and A. C. Rosenquist (1975). The retinotopic organization of the visual cortex in the cat, *Soc. Neurosci. Abstr.*, Vol. 1, p. 52.
Wiesel, T. N. and D. H. Hubel (1963). Single cell responses in the striate cortex of kittens deprived of vision in one eye, *J. Neurophysiol.* **26:**1003–1017.
Wiesel, T. N. and D. H. Hubel (1965). Comparison of the effects of unilateral and bilateral eye closure on cortical unit responses in kittens, *J. Neurophysiol.* **28:**1029–1040.

ABNORMAL CONNECTIONS IN THE VISUAL SYSTEM OF SIAMESE CATS

Carla J. Shatz

Stanford University School of Medicine, Stanford, California

Connections in the adult visual system are exquisitely precise. For example, the topographic map of external visual space contained within the primary visual cortex arises from the orderly set of connections that exist between retina, lateral geniculate nucleus (LGN), and cortex. What rules govern the formation of these connections during development? They might be rigidly predetermined—programmed to form in an unalterable pattern. Or, connections might depend to some extent on the identity of input from the periphery—in this case, on information that reaches the brain from the retina. The visual system of the Siamese cat provides an unusual opportunity to examine these alternatives. In these cats, the brain is presented with abnormal input from the retina (Guillery, 1969) as a consequence of a genetic mutation. The question to be considered here is how the rest of the visual system deals with this abnormal input.

NORMAL RETINO-GENICULO-CORTICAL ORGANIZATION

The primary visual cortex, area 17, in each hemisphere contains an orderly binocular map of the contralateral half of the visual field. This arises due to the partial decussation of ganglion cell axons from each retina in the optic chiasm: axons from the nasal half of each retina cross in the chiasm, while those from the temporal half remain uncrossed (Stone, 1966). As shown in the diagram of Figure 1, the right LGN receives input from ganglion cells that, by virtue of their position in the right half of each retina, represent points in the left half (contralateral half) of the visual field.

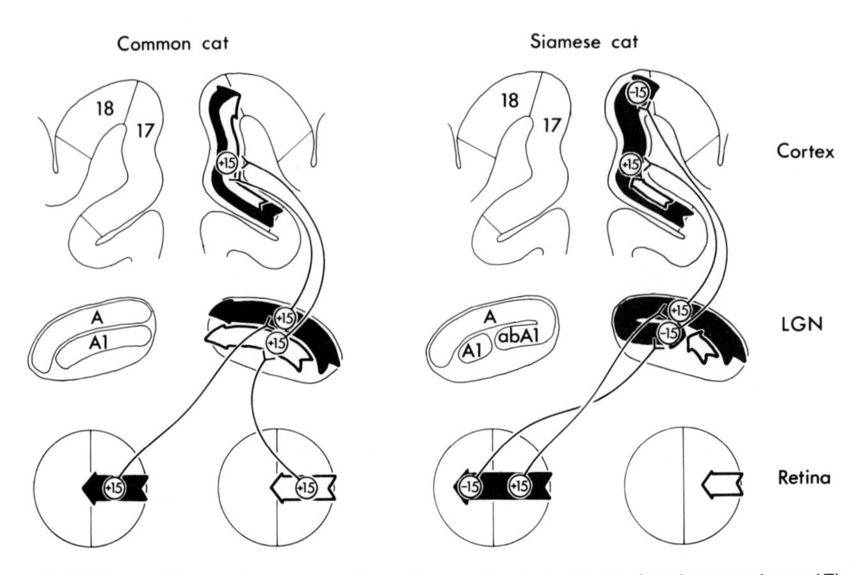

FIGURE 1. Diagram to show pathway from retina to primary visual cortex (area 17) in the common cat and in Siamese cats with the Boston pattern of geniculocortical projection. Only the retinal projections to the right hemisphere are shown, with the contribution from the left retina indicated by a solid arrow and that from the right retina as an open arrow. The line dividing each retina is the representation of the vertical midline of the visual field. The figures +15 and −15 indicate the position of neurons whose receptive fields, with reference to the right side of the brain, are situated 15° contralateral (+15) or 15° ipsilateral (−15) to the vertical midline. Note that in the Siamese cat the LGN lamina A1 is broken into two regions: a medial region (abA1) receiving the abnormal representation of the ipsilateral field, and a lateral region (A1) receiving what remains of the normal contralateral field representation from the ipsilateral eye (open arrow). Other, subsidiary retinal projections to the LGN (to the C laminae and medial interlaminar nucleus) have been omitted for simplicity. (Reprinted with permission from Shatz, C. J. and S. LeVay [1979]. Siamese cat: altered connections of visual cortex, *Science*, in press. Copyright 1979 by the American Association for the Advancement of Science.)

The LGN consists of several cell laminae interleaved by a fiber plexus (Figure 2*C*), and the axons of retinal ganglion cells from each eye terminate separately in alternate laminae (Hayhow, 1958; Laties and Sprague, 1962; Stone and Hansen, 1966; Garey and Powell, 1968; Guillery, 1970). This separation can be demonstrated in the cat's LGN by injecting one eye with tritiated amino acids and allowing sufficient time for axoplasmic transport to convey radioactively labeled protein to the terminals of ganglion cells in the appropriate LGN laminae

(Hickey and Guillery, 1974; Shatz, 1977*a*). The results of such an experiment are shown in the autoradiographs of Figure 2. (In these and all darkfield photographs to follow, radioactively labeled terminals appear white.) For example, the dorsalmost LGN lamina, called lamina A, receives fibers from the contralateral eye; fibers from the ipsilateral eye terminate in lamina A1.

Each LGN lamina contains an orderly map of the contralateral half of the visual field. The maps are in register so that neurons lying opposite one another in adjacent laminae represent the same visual field

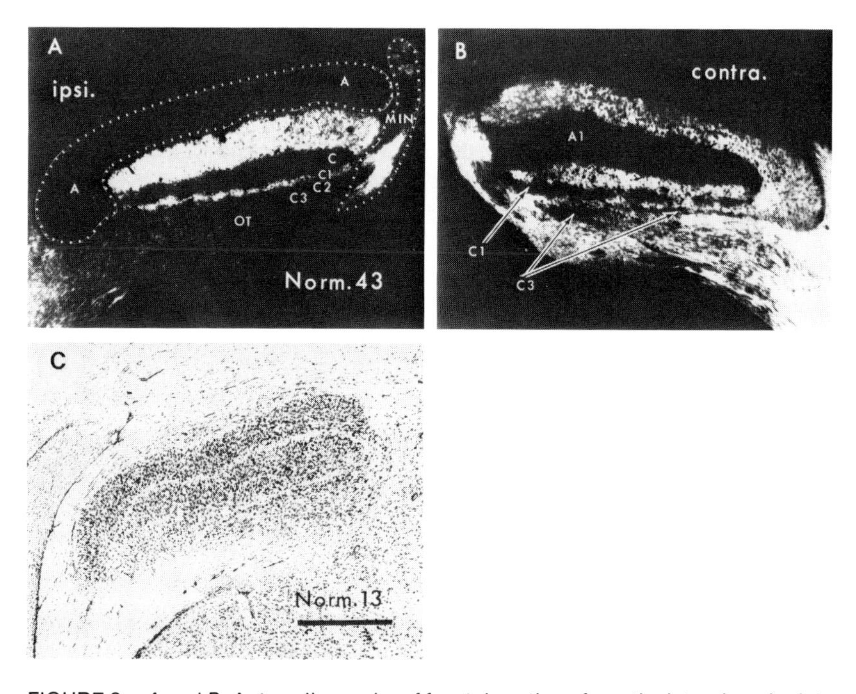

FIGURE 2. *A* and *B*: Autoradiographs of frontal sections from the lateral geniculate nucleus ipsilateral (*A*) and contralateral (*B*) to an injection of tritiated proline into the left eye of a common cat. Darkfield optics. In this and all subsequent darkfield photos, regions with radioactively labeled terminals appear white. Dotted white lines delineate the structure of the unlabeled laminae that receive afferents from the uninjected left eye. Lines were drawn from the brightfield, Nissl-stained view of the same section. OT, Optic tract. *C*: Brightfield, Nissl-stained frontal section through the LGN at a level similar to that of *A* and *B* to show directly the pattern of lamination. Calibration bar for *A*–*C* = 1 mm. (Reprinted with permission from Shatz, C. J. [1977*a*]. A comparison of visual pathways in Boston and Midwestern Siamese cats, *J. Comp. Neurol.* **171**:205–228.)

position in each eye (Hubel and Wiesel, 1961; Sanderson, 1971). As shown in Figure 1, neurons at corresponding positions in each lamina project in turn to the same region of visual cortex (Garey and Powell, 1967; Gilbert and Kelly, 1975). The most medial cells in the LGN represent the vertical midline of the visual field: these project to the border between area 17 and area 18 (the secondary visual area). Cells in the more lateral parts of the LGN, representing more peripheral parts of the visual field, project to regions further within area 17. Such an arrangement ensures that visual field coordinates in the two eyes are precisely aligned and topographically ordered at the cortical level.

ABNORMAL ORGANIZATION IN SIAMESE CATS

In Siamese cats, a genetic mutation at the albino locus, known for producing the characteristic Siamese coat color, also somehow causes an excessive number of ganglion cell axons from each retina to cross in the optic chiasm (Guillery, 1969). Each LGN consequently receives an abnormally large contribution from the contralateral retina. This is evident from Figure 3*B*, a darkfield autoradiograph of the LGN contralateral to an intraocular injection of radioactive amino acids in a Siamese cat (compare with Figure 2*B*). With the exception of two (dark) regions, the LGN is almost entirely occupied by afferents from the contralateral, injected retina.

Microelectrode recordings from the LGN of Siamese cats have indicated that the abnormal retinal projection is supplied largely by ganglion cells located within a region stretching from the vertical midline to 14–20° into the temporal retina (Guillery and Kaas, 1971). This is illustrated in the diagram of Figure 1 (right side). Two recent anatomical studies (Cooper and Pettigrew, 1977; Stone, Campion, and Leicester, 1978) have shown that actually ganglion cells throughout the temporal retina are misrouted, but the majority of those affected are located within the first 20°. The misrouted axons terminate in the medial portion of lamina A1, called the abnormal segment of A1 (Guillery and Kaas, 1971), on the wrong side of the brain (see Figure 1). This gives the LGN a representation of part of the ipsilateral visual field in addition to its normal contralateral field representation in lamina A. The misrouting is orderly: corresponding points in laminae A and A1 now represent mirror-symmetric positions in the contralateral visual field, but in the contralateral eye only.

How is this abnormal representation of the ipsilateral visual field handled by the visual cortex? First suppose that in Siamese cats, the

LGN projects to the cortex in the same pattern as that found in common cats (Figure 1, left). This would have the bizarre effect of conferring onto each cortical location the representation of two mirror-symmetric visual field positions. Such a pattern is indeed found in some Siamese cats, and has been called the "Midwestern" projection pattern because it was first observed by Kaas and Guillery (1973) in Wisconsin.

Physiological mapping experiments suggest that an entirely different pattern of projection, the "Boston" pattern, is present in other Siamese cats (Hubel and Wiesel, 1971; Shatz, 1977*a*). An example of such an

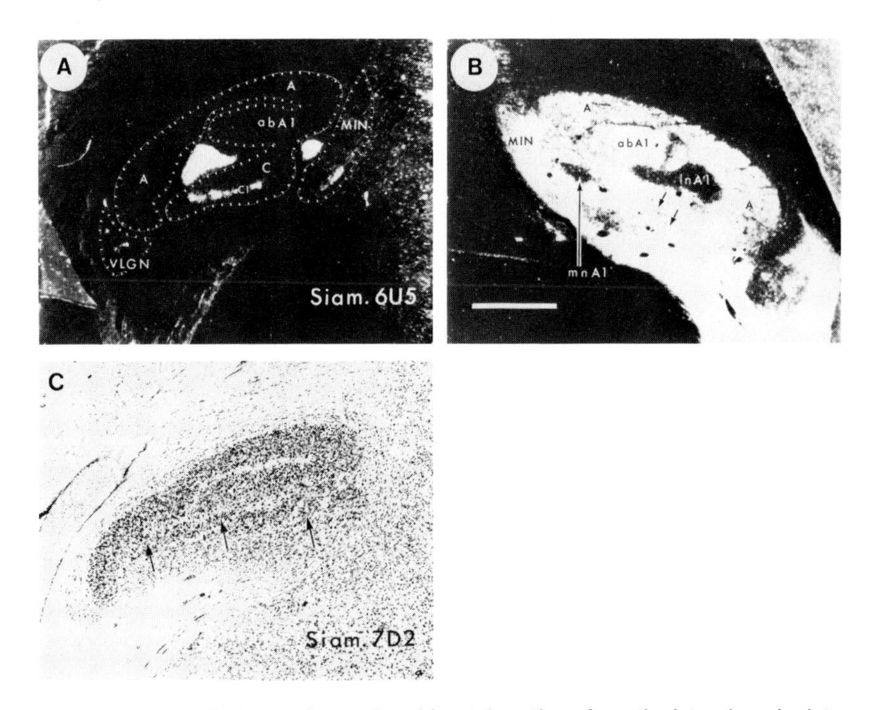

FIGURE 3. *A* and *B*: Autoradiographs of frontal sections from the lateral geniculate nucleus ipsilateral (*A*) and contralateral (*B*) to an eye injection in a Boston Siamese cat. Conventions as in Figure 2. abA1, Abnormal segment of lamina A1; mnA1, medial normal segment of A1; lnA1, lateral normal segment of A1. Calibration 1 mm. *C*: Brightfield, Nissl-stained photograph of a frontal section through the LGN of another Boston Siamese cat to show the abnormal pattern of lamination present in these animals. Note particularly the clear fusion between the medial portion of lamina A and the abnormal segment of lamina A1. This figure should be compared with Figure 2. (Reprinted with permission from Shatz, C. J. [1977*a*]. A comparison of visual pathways in Boston and Midwestern Siamese cats, *J. Comp. Neurol.* **171**:205–228.)

experiment is shown in Figure 4. The microelectrode first entered the cortex in area 18, and next traversed the border between areas 17 and 18, where receptive fields were found abnormally positioned 20° into the ipsilateral half of the visual field (units 6–12). (In common cats, receptive fields would be found at the visual field vertical midline.) As the electrode was further advanced into area 17, receptive fields moved gradually towards, and then crossed, the vertical midline (0°) (units 15–18) and finally moved out into the normal, contralateral half of the visual field (units 19–27). To account for this preservation of continuity in the visual field map at the cortical level, Hubel and Wiesel (1971) proposed that the geniculocortical projection undergoes a remarkable rearrangement. As shown in Figure 1 (right side), the abnormal geniculate laminae representing the ipsilateral visual field now project to their own strip of cortex, which is separate but adjacent to that receiving normal input from geniculate laminae representing the contralateral visual field.

This reorganization has been verified anatomically by means of the retrograde axonal transport of horseradish peroxidase (HRP) injected into the visual cortex through a recording micropipette (Shatz and LeVay, 1979). When injections were made at the 17/18 border (where receptive fields were located 10–20° into the ipsilateral visual field), labeled neurons were found in the portion of LGN lamina A1 that

FIGURE 4. First penetration in the right hemisphere of a Siamese cat, 7D2, exemplifying the "Boston" pattern. Far right of the figure shows the histological reconstruction of the microelectrode track in frontal section. Arrow indicates the anatomical 17/18 border. Position of the units encountered during the penetration is designated by the numbered horizontal bars to the right of the track. L1 and L2 refer to two electrolytic lesions made during the penetration. In diagrams to the left, the position of each receptive field is shown relative to the area centralis, which is indicated for each unit by a horizontal crossbar intersecting the long vertical line. Ipsilateral visual field is to the right, and contralateral to the left, of each line. When possible, orientation preference for each receptive field is indicated by two short lines. As the electrode approached the anatomical 17/18 border, the receptive fields of units moved out 20° into the ipsilateral visual field. When the electrode was advanced further down the medial bank of the lateral gyrus, receptive fields moved uniformly back towards the vertical midline and out into the contralateral visual field. All units were driven exclusively by the contralateral eye, with the exception of units 24, 25, and 27, which preferred the contralateral eye but could be weakly influenced by the ipsilateral eye. A section from the LGN of cat 7D2 is shown in Figure 3*C*. (Reprinted with permission from Shatz, C. J. [1977*a*]. A comparison of visual pathways in Boston and Midwestern Siamese cats, *J. Comp. Neurol.* **171**:205–228.)

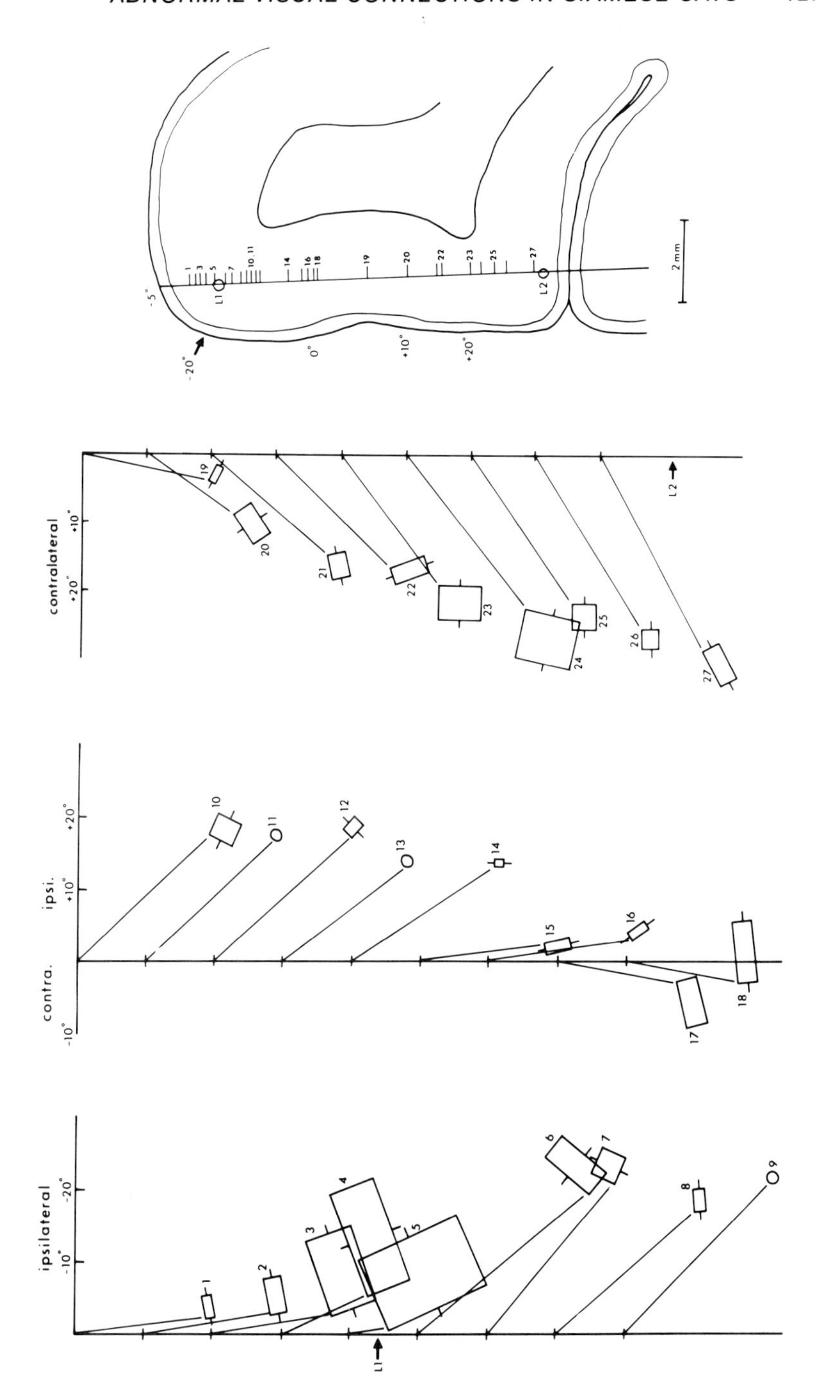
2 mm
L1
L2
contralateral
ipsi.
contra.
ipsilateral

receives the abnormal representation of the ipsilateral visual field (Figure 1: abA1). No labeled cells were found in lamina A. On the other hand, injections of area 17 at positions 12–20° into the contralateral visual field labeled neurons in LGN lamina A, but not those in the abnormal part of A1. This pattern contrasts with that seen in the common cat, in which an HRP injection anywhere in the visual cortex labels cells at corresponding points in both laminae A and A1. However, the finding confirms the suggestion, made from the physiology, that the "Boston" pattern of organization involves an extensive and orderly rearrangement of the geniculocortical projection.

HIGHER ORDER CONNECTIONS IN BOSTON SIAMESE CATS

As a consequence of the geniculocortical rearrangement in Boston Siamese cats, the 17/18 border carries a visual field representation that is different from normal (compare Figure 1, right and left sides). The question arises whether the further connections made by cells in this region are also altered or whether they retain the pattern seen in the common cat.

One set of connections are those of the corpus callosum. In the common cat, physiological (Berlucchi, Gazzaniga, and Rizzolatti, 1967; Hubel and Wiesel, 1967) and anatomical (Hubel and Wiesel, 1965; Wilson, 1968; Fisken, Garey, and Powell, 1975; Shatz, 1977*c*) evidence indicates that these connections are restricted to cortical regions in the two hemispheres that are concerned almost exclusively with the representation of the visual field vertical midline, such as the 17/18 border. This restriction is reflected in the heavy representation of the vertical midline by fibers in the corpus callosum. For example, Figure 5 shows that the receptive fields of callosal fibers recorded in two common cats are located primarily within regions of the visual field close to the vertical midline.

In Boston Siamese cats, the receptive field distribution of callosal fibers is much more widespread, as shown in Figure 6. Nevertheless, a good many receptive fields are located near the vertical midline. This point is clearly illustrated in the histograms of Figure 7. The histograms were constructed by measuring the distance of each callosal receptive field's center from the vertical midline, and combining the results from the Boston Siamese (left) and common (right) cats. Thus the corpus callosum of Boston cats carries a substantial representation of the visual field vertical midline—a remarkable finding, since it implies that the anatomical pattern of callosal connections in these animals is altered.

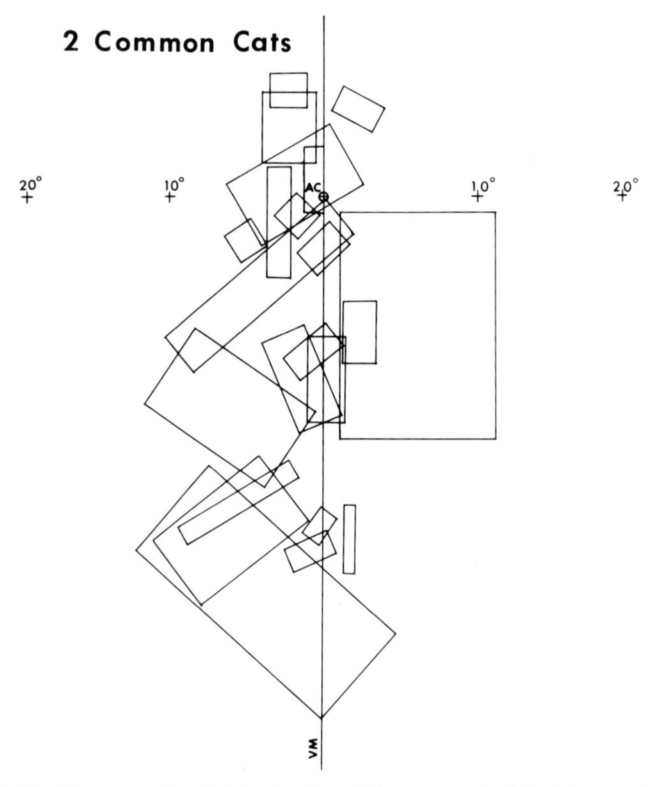

FIGURE 5. The receptive fields of callosal fibers recorded from two common cats. AC, area centralis; VM, vertical meridian. Numbers opposite the small crosses along the horizontal meridian indicate distance from the vertical midline. Thus, most receptive fields here were in the inferior visual field and were located within 10° of the midline. (Reprinted with permission from Shatz, C. J. [1977*b*]. Abnormal interhemispheric connections in the visual system of Boston Siamese cats: a physiological study, *J. Comp. Neurol.* **171**:229–246.)

If the pattern were the same as that seen in common cats, in which only the 17/18 borders in each hemisphere supply fibers, most callosal receptive fields should have been found 15–20° away from the vertical midline, thereby reflecting the altered visual field representation at the border. This possibility is shown in the diagram of Figure 8*A*. The actual distribution seen in Figures 6 and 7*A* suggests that in Siamese cats the 17/18 border makes a definite but small contribution to the corpus callosum. The substantial contribution is supplied by a set of connections—not present in common cats—that originates from sites well within area 17.

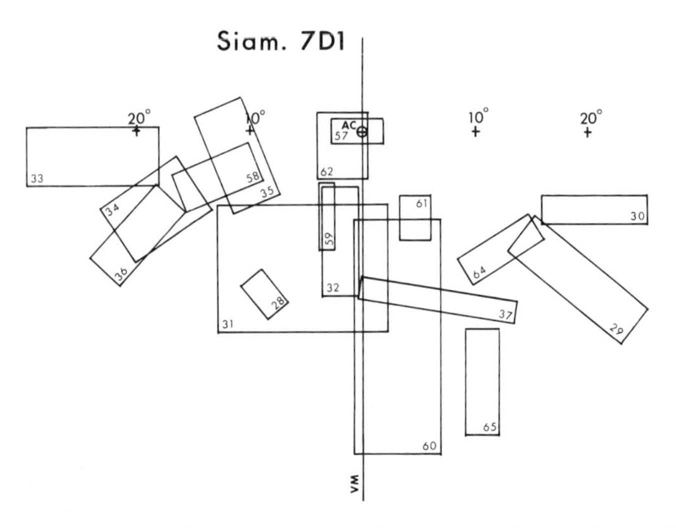

FIGURE 6. Receptive fields recorded from the corpus callosum of a Boston Siamese cat (7D1). Same conventions as in Figure 5. Note the abnormally large spread in the receptive field distribution as compared with that of the common cats of Figure 5. (Reprinted with permission from Shatz, C. J. [1977*b*]. Abnormal interhemispheric connections in the visual system of Boston Siamese cats: a physiological study, *J. Comp. Neurol.* **171**:229–246.)

These suggestions were verified anatomically by using the retrograde transport of HRP injected into one hemisphere to identify neurons in the opposite hemisphere supplying callosal connections to the injection site (Shatz, 1977*c*). In common cats, just as expected, an HRP injection at the 17/18 border labels neurons at the opposite 17/18 border. The results of such an experiment are shown in Figure 9*A*.

One can next ask whether in Boston Siamese cats, as in common cats, the 17/18 border in one hemisphere receives callosal terminals from neurons symmetrically situated at the opposite 17/18 border. If so, then callosal connections would link cortical regions serving two mirror-symmetric visual field locations 20° to either side of the vertical midline. This possibility is shown in the diagram of Figure 8*A*. On the other hand, callosal connections that make sense functionally would be those linking the 17/18 border in one hemisphere with the cortical region in the other serving the same visual field location. As shown in Figure 8*B*, the appropriate region would be one located well within area 17—a region that in common cats is known not to participate in callosal connections.

With these alternatives in mind, HRP was injected at the 17/18 border in several Siamese cats identified as "Boston" by finding receptive fields at the injection site located 15–20° into the ipsilateral hemifield. As shown in Figure 9*B*, the distribution of peroxidase-labeled neurons in the opposite hemisphere differs distinctly from that of the common cat of Figure 9*A*. The majority of labeled neurons were found in regions of areas 17 (arrows) and 18 far from the border, confirming the suggestion (Figure 8*B*) that the border in one hemisphere receives connections from the cortical region in the other representing the identical visual field position.

What about the set of callosal connections supplied by the 17/18 border? These were examined by injecting tritiated proline at the border and using autoradiographic methods to reveal the distribution of radioactively labeled terminals in the opposite hemisphere (Shatz, 1977*c*). In common cats, an injection at one border labels terminals in a restricted zone at the opposite border, as shown in the autoradiograph of Figure 10*A*, thereby confirming previous degeneration studies.

Figure 10*B* shows the results of a similar injection in a Boston Siamese cat. In area 17, the region of heaviest labeling (left arrowhead) was

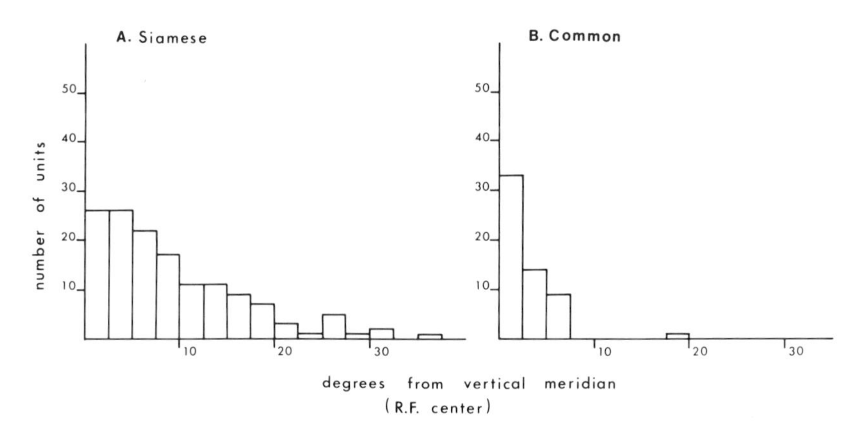

FIGURE 7. Histograms to show the overall distribution of callosal receptive fields combined from 11 Boston Siamese cats (*A*) and from 5 common cats (*B*). To construct the histograms, the distance of each receptive field's center from the vertical midline was measured. Thus in Boston Siamese cats the distribution of callosal receptive fields is more widespread than that in common cats. (Reprinted with permission from Shatz, C. J. [1977*b*]. Abnormal interhemispheric connections in the visual system of Boston Siamese cats: a physiological study, *J. Comp. Neurol.* **171**:229–246.)

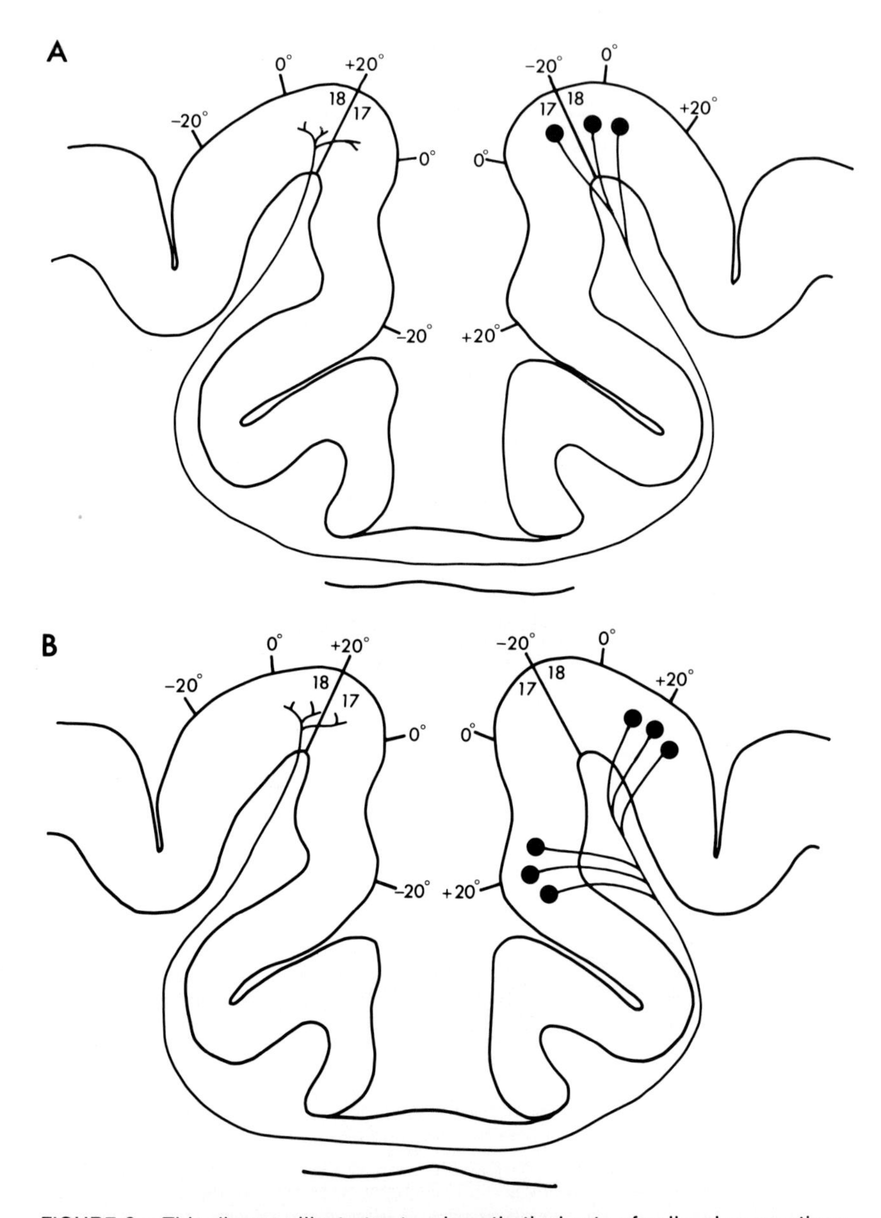

FIGURE 8. This diagram illustrates two hypothetical sets of callosal connections that could exist between the anatomical 17/18 borders in each hemisphere of Boston Siamese cats. One possibility, shown in *A*, is that the two borders are connected as in common cats. (Although, of course, in common cats, the result would be to

far away from the 17/18 border (arrow); a region of labeling was also seen well within area 18 (right arrowhead). These regions are known to carry roughly the same visual field coordinates as those at the injection site (refer to Figure 8*B*), indicating that callosal connections in Boston cats are supplied to visuotopically appropriate targets in the opposite hemisphere.

Taken together, the results of the retrograde and anterograde transport experiments allow a set of callosal connections to be elaborated. In the common cat, the results simply indicate that the 17/18 borders in the two hemispheres are reciprocally interconnected; the rest of area 17 does not participate in callosal connections. In the Boston Siamese cats, unusually large parts of area 17 in the two hemispheres are reciprocally interconnected in such a way as to link cortical regions representing similar visual field coordinates. (The vertical meridian representations within area 17 of each hemisphere are also interconnected [Shatz, 1977*c*].) To accomplish this, the anatomical pattern of callosal connections is profoundly altered in these animals.

The ability, in Boston cats, of neurons to seek out appropriate counterparts elsewhere in the brain that carry similar visuotopic information is evidently not restricted to those of the corpus callosum. Another example is the recurrent projection from area 17 to the LGN. When [^{3}H]proline is injected at the 17/18 border, labeled terminals are seen only within the abnormal portion of lamina A1 (see Figure 1), but not in lamina A, indicating that the cortical and geniculate representations of the ipsilateral visual field are appropriately interconnected (Shatz and LeVay, 1979). In common cats, a similar injection labels terminals in both LGN laminae.

In view of these rearrangements in the corticogeniculate and callosal connections, it was surprising to find, from [^{3}H]proline injections, that the 17/18 border also projects to zones deep within areas 17 and 18 on the same side of the brain (Shatz and LeVay, 1979). An example

---

link regions serving the vertical midline representation.) This pattern, however, would be meaningless functionally, since two locations in the visual field mirror-symmetrically displaced from each other by 20° to either side of the vertical midline would be interconnected. The pattern of connections shown in *B*, although anatomically different from that of common cats, would effectively connect two cortical sites representing identical visual field coordinates (+20°). (Reprinted with permission from Shatz, C. J. [1977*c*]. Anatomy of interhemispheric connections in the visual system of Boston Siamese and ordinary cats, *J. Comp. Neurol.* **173**:497–518.)

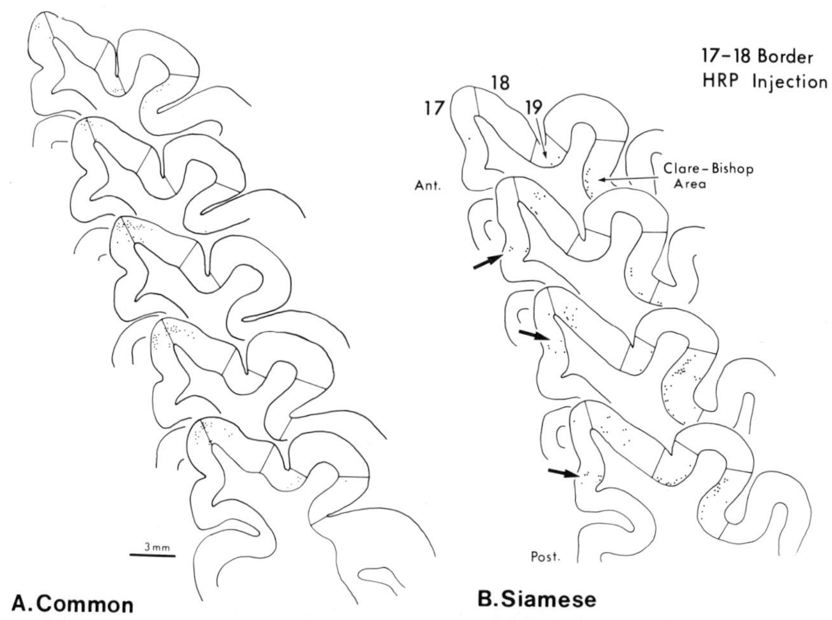

FIGURE 9. Reconstructions following an injection of HRP at the anatomical 17/18 border to compare the distribution of peroxidase-labeled neurons in the opposite hemisphere of Boston Siamese and common cats. Each dot represents a peroxidase-filled cell observed using darkfield optics to improve visibility. Frontal sections were outlined, and cytoarchitectonic borders drawn, from the brightfield thionin-stained view of the same sections. Separation between successive frontal sections is about 160 μm.

A: Peroxidase-labeled callosal neurons in a common cat. Note the accumulation of labeled cells in areas 17 and 18 immediately adjacent to the 17/18 border, and the lateral border of area 19. These positions correspond to the vertical midline representation of the visual field in each cortical area. Receptive fields recorded at the injection site in the opposite hemisphere were also located at the vertical meridian, about 7° below the horizontal meridian.

B: Peroxidase-labeled callosal neurons in a Boston Siamese cat. Note the conspicuous absence of labeled cells at the 17/18 border and their unusual presence well down the medial bank of the lateral gyrus in area 17 (arrows) and also well within area 18. Receptive fields recorded at the injection site were located in the ipsilateral visual field 15° from the vertical midline, and 10° inferior to the horizontal midline. The labeled regions in areas 17 and 18 of the opposite hemisphere represent approximately similar visual field coordinates, suggesting that the pattern of callosal connections in Boston Siamese cats conforms to that postulated in the diagram of Figure 8*B* and therefore differs from that of common cats.

(Reprinted with permission from Shatz, C. J. [1977*c*]. Anatomy of interhemispheric connections in the visual system of Boston Siamese and ordinary cats, *J. Comp. Neurol.* **173:**497–518.)

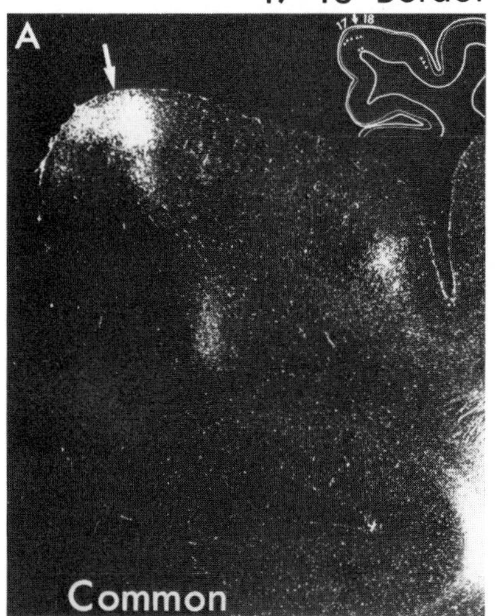

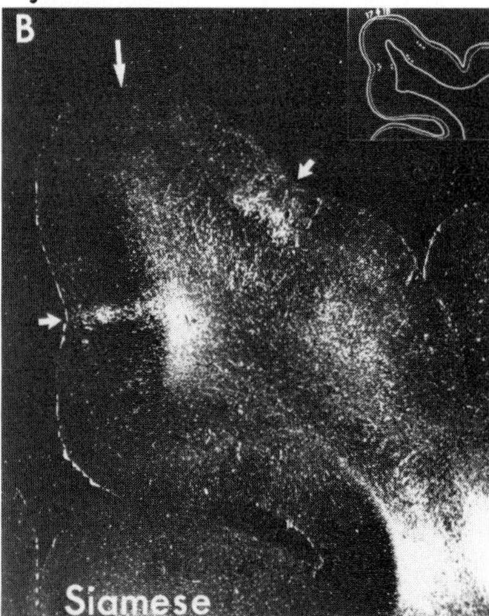

FIGURE 10. Autoradiographs to show the topographic distribution of radioactively labeled terminals in representative frontal sections following an injection of tritiated proline in the opposite hemisphere in common (*A*) and Boston Siamese (*B*) cats. The use of darkfield optics results in the bright appearance of regions containing labeled terminals. Large arrow in each figure indicates the position of the 17/18 border as determined from the brightfield thionin-stained view of the same section. Smaller arrowheads in *B* designate the approximate area receiving the heaviest projection. Inset is an outline of the same region shown in the darkfield photograph. Areas containing labeled terminals, as determined from higher power darkfield optics, are delineated by the triangles in the inset.

A: To show the maximum spread of labeled terminals into area 17 observed in a common cat after an injection at the 17/18 border in the opposite hemisphere. Receptive fields recorded at the injection site were at the vertical midline about 6° inferior to the horizontal midline.

B: Distribution of labeled terminals in a Boston Siamese cat after an injection at the anatomical 17/18 border in the opposite hemisphere. The cortical region receiving the heaviest projection (arrowhead) does not coincide with the 17/18 border, but is located instead well within area 17 in a region that, in Boston cats, represents visual field coordinates similar to those of the injection site. Receptive fields recorded at the injection site were about 11° into the ipsilateral visual field, 9° inferior to the horizontal midline.

(Reprinted with permission from Shatz, C. J. [1977*c*]. Anatomy of interhemispheric connections in the visual system of Boston Siamese and ordinary cats, *J. Comp. Neurol.* **173**:497–518.)

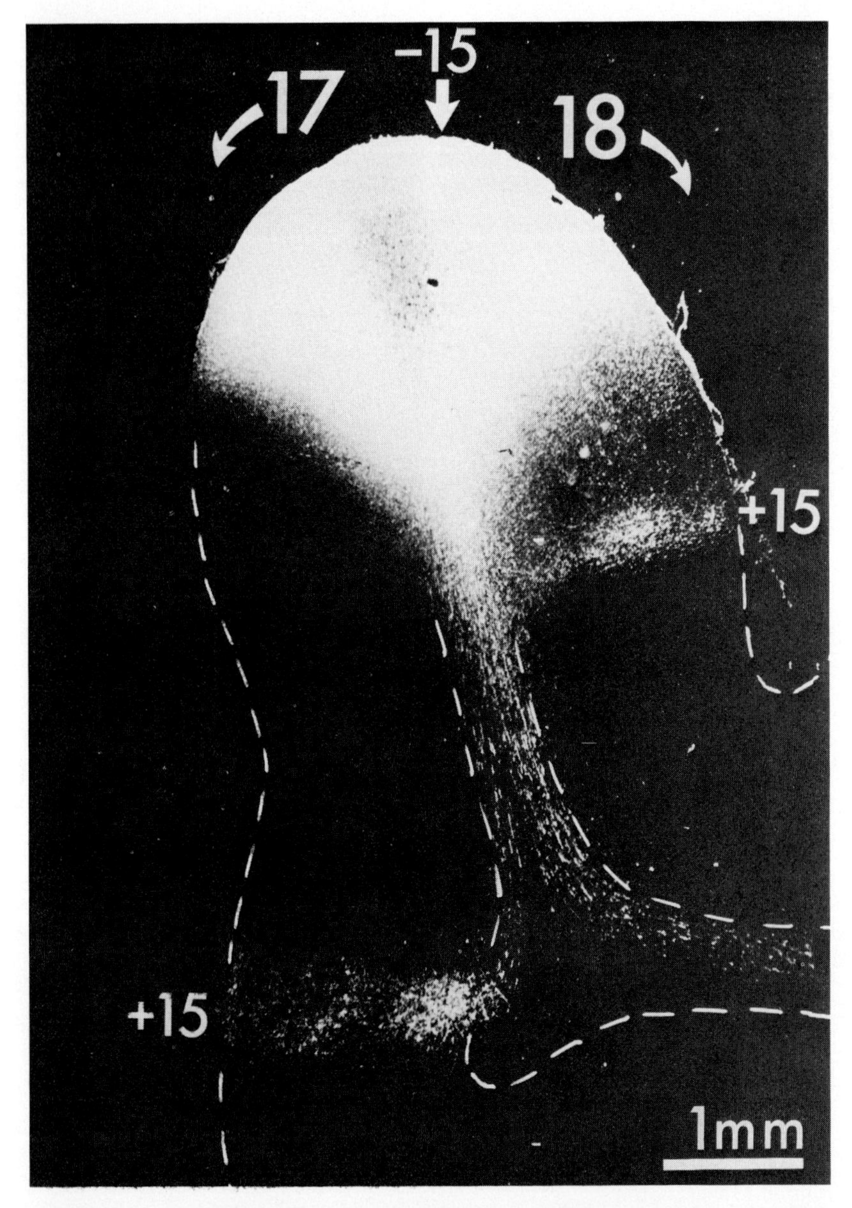

FIGURE 11. Autoradiograph of visual cortex of a Siamese cat that received an injection of [^{3}H]proline at the 17/18 border. In this darkfield micrograph, regions containing silver grains appear white, except for the center of the injection site, which is so heavily labeled as to nullify the darkfield effect. The position of the

is shown in Figure 11. From mapping studies (Hubel and Wiesel, 1971), it is known that the projection zones shown in Figure 11 (+15) represent a region of the contralateral visual field roughly 15° from the vertical meridian. In forming these corticocortical projections, then, there is evidently some confusion about visual field coordinates, since cortical regions representing mirror-symmetric positions are interconnected.

DISCUSSION

In Siamese cats the gene has replaced the scalpel, conveniently operating on the chiasm to reroute optic nerve fibers and thereby confront the brain with abnormal input from the periphery. The cascade-like rewiring of geniculocortical, callosal, corticocortical, and corticogeniculate connections seen in the visual system of Boston Siamese cats indicates that peripheral input, in this case from the retina, must indeed play a role in determining central connections during development.

Just how abnormal retinal input might lead to a rewiring of visual connections is, of course, unknown. But one can construct a scheme based on the assumption that the same rules governing the formation of connections under normal circumstances are also in operation in Siamese cats. Such an assumption requires that the primary effect of the albino gene be confined to the retinal ganglion cells. All subsequent rearrangements of the visual pathways are seen simply as the consequence of normal rules acting on altered inputs. The good correlation found between abnormal optic nerve decussation, on the one hand, and genes —not necessarily at the albino locus—that cause reduced ocular pigmentation (Sanderson, Guillery, and Shackelford, 1974; LaVail, Nixon, and Sidman, 1978; Guillery, Oberdorfer, and Murphy, 1979) makes this assumption a reasonable one.

With this assumption in mind, then, it can be imagined that the retinal afferents themselves convey some sort of positional information

---

17/18 border (vertical arrow) was determined histologically in an adjacent section stained with cresyl violet. A projection site within area 17 is visible (bottom left). From physiological mapping studies, this site is known to represent a visual field position approximately 15° into the contralateral field (+15), i.e., mirror-symmetrical to the field position represented at the injection site (−15). Area 18 carries a second, more compressed representation of the visual field, and an anomalous projection is evident there, too (middle right, +15). Medial is to the left, dorsal is up. (Reprinted with permission from Shatz, C. J. and S. LeVay [1979]. Siamese cat: altered connections of visual cortex, *Science*, in press. Copyright 1979 by the American Association for the Advancement of Science.)

to their postsynaptic targets in the LGN—information that ultimately relates to external visual space, since the visual field is mapped onto each retina. This positional information in turn might be used to determine the pattern of the geniculocortical projection. For example, geniculate neurons receiving input from adjacent retinal positions might project to adjacent cortical positions. In Siamese cats this would give rise to the Boston pattern, since the medial parts of lamina A and the abnormal segment of lamina A1 receive input from adjacent regions of the retina. (For further discussion, see Hubel and Wiesel, 1971; Gaze and Keating, 1972; Shatz, 1977*a*.) But what about the Midwestern pattern? Here one must resort to a rather unsatisfying number game, and propose that somehow fewer ganglion cell axons are misrouted in Midwestern cats. Thus, input from adjacent retinal regions to the critical parts of laminae A and A1 might not be sufficiently large to produce a complete reorganization of the geniculocortical projection.

At the next stage, cortical neurons might use the positional information conveyed by geniculocortical afferents to construct further connections. For example, in the formation of callosal connections in common cats it is now known that large numbers of neurons initially send their axons across in the corpus callosum (Innocenti, Fiore, and Caminiti, 1977), and this is likely to be the case for Siamese cats as well. One might imagine that this initially diffuse pattern is refined by allowing the survival of only those connections existing between cortical locations in the two hemispheres that, by virtue of their geniculocortical input, represent similar points in the visual field. This scheme would produce the adult pattern of callosal connections seen in common cats, in which the only visual field positions shared by both hemispheres are the vertical midline and the few degrees to either side of it. In Boston Siamese cats, on the other hand, the misrouting of retinogeniculate fibers gives each hemisphere an identical representation of the visual field extending 20° to either side of the vertical midline. Appropriate callosal connections then would be expected to develop between much more widespread regions of cortex—just what is observed.

It is not known how a particular neuron might receive information concerning the success with which its axon has terminated during development. One possibility is that some sort of chemical recognition occurs between pre- and postsynaptic elements. If so, then neurons involved in the formation of callosal connections must be capable of distinguishing the right from the left visual hemifield, but those supplying the anomalous corticocortical connections must not. An alternative possibility is suggested by the observation that in common cats callosal connections may be modified by abnormal visual experience (Lund, Mitchell, and Henry, 1978; Innocenti and Frost, 1979). Thus, it may be that under normal circumstances callosal connections (but not cortico-cortical) are refined and corrected by visual experience after birth. Whatever the case, it is remarkable that from a study of connections in the Siamese cat's visual system something can be inferred about developmental events likely to operate in normal animals as well.

ACKNOWLEDGMENTS

I wish to thank David Hubel and Torsten Wiesel for providing advice and encouragement throughout these studies, and Simon LeVay for his lively and stimulating interest in this subject. This work was supported by National Institutes of Health grants EY00082, EY00605, and EY00606, and by the Harvard University Society of Fellows.

REFERENCES

Berlucchi, G., M. S. Gazzaniga, and G. Rizzolatti (1967). Microelectrode analysis of transfer of visual information by the corpus callosum, *Arch. Ital. Biol.* **105:**583–596.

Cooper, M. L. and J. D. Pettigrew (1977). Naso-temporal division of the retinothalamic pathway in normal and Siamese cats, *Soc. Neurosci. Abstr.*, Vol. 3, p. 556.

Fisken, R. A., L. J. Garey, and T. P. S. Powell (1975). The intrinsic, association, and commissural connections of area 17 of the visual cortex, *Philos. Trans. R. Soc. London B Biol. Sci.* **272:**487–536.

Garey, L. J. and T. P. S. Powell (1967). The projection of the lateral geniculate nucleus upon the cortex in the cat, *Proc. R. Soc. London B Biol. Sci.* **169:** 107–126.

Garey, L. J. and T. P. S. Powell (1968). The projection of the retina in the cat, *J. Anat.* **102:**189–222.

Gaze, R. M. and M. J. Keating (1972). The visual system and "neuronal specificity," *Nature (London)* **237:**375–378.

Gilbert, C. D. and J. P. Kelly (1975). The projections of cells in different layers of the cat's visual cortex, *J. Comp. Neurol.* **163:**81–105.

Guillery, R. W. (1969). An abnormal retinogeniculate projection in Siamese cats, *Brain Res.* **14:**739–741.

Guillery, R. W. (1970). The laminar distribution of retinal fibers in the dorsal lateral geniculate nucleus of the cat: a new interpretation, *J. Comp. Neurol.* **138:**339–368.

Guillery, R. W. and J. H. Kaas (1971). A study of normal and congenitally abnormal retinogeniculate projections in cats, *J. Comp. Neurol.* **143:**73–100.

Guillery, R. W., M. D. Oberdorfer, and E. H. Murphy (1979). Abnormal retinogeniculate and geniculo-cortical pathways in several genetically distinct color phases of the mink (*Mustela vison*), *J. Comp. Neurol.*, in press.

Hayhow, W. R. (1958). The cytoarchitecture of the lateral geniculate body in relation to the distribution of crossed and uncrossed fibers, *J. Comp. Neurol.* **110:**1–51.

Hickey, T. L. and R. W. Guillery (1974). An autoradiographic study of the retinogeniculate pathways in the cat and fox, *J. Comp. Neurol.* **156:**239–254.

Hubel, D. H. and T. N. Wiesel (1961). Integrative action in the cat's lateral geniculate body, *J. Physiol. (London)* **155:**385–398.

Hubel, D. H. and T. N. Wiesel (1965). Receptive fields and functional architecture in two nonstriate visual areas (18 and 19) of the cat, *J. Neurophysiol.* **28:**229–289.

Hubel, D. H. and T. N. Wiesel (1967). Cortical and callosal connections concerned with the vertical meridian of the visual fields in the cat, *J. Neurophysiol.* **30:**1561–1573.

Hubel, D. H. and T. N. Wiesel (1971). Aberrant visual projections in the Siamese cat, *J. Physiol. (London)* **218:**33–62.

Innocenti, G. M., L. Fiore, and R. Caminiti (1977). Exuberant projection into the corpus callosum from the visual cortex of newborn cats, *Neurosci. Lett.* **4:**237–242.

Innocenti, G. M. and D. O. Frost (1979). Abnormal visual experience stabilizes juvenile patterns of interhemispheric connections, *Nature (London)*, in press.

Kaas, J. H. and R. W. Guillery (1973). The transfer of abnormal visual field representations from the dorsal lateral geniculate nucleus to the visual cortex in Siamese cats, *Brain Res.* **59:**61–95.

Laties, A. M. and J. M. Sprague (1962). The projection of optic fibers to the visual centers in the cat, *J. Comp. Neurol.* **127:**35–70.

LaVail, J. H., R. A. Nixon, and R. L. Sidman (1978). Genetic control of retinal ganglion cell projections, *J. Comp. Neurol.* **182:**399–422.

Lund, R. D., D. E. Mitchell, and G. H. Henry (1978). Squint-induced modification of callosal connections in cats, *Brain Res.* **144:**169–172.

Sanderson, K. J. (1971). The projection of the visual field to the lateral geniculate and medial interlaminar nuclei in the cat, *J. Comp. Neurol.* **143:**101–118.

Sanderson, K. J., R. W. Guillery, and R. M. Shackelford (1974). Congenitally abnormal visual pathways in the mink (*Mustela vison*) with reduced retinal pigment, *J. Comp. Neurol.* **154:**225–245.

Shatz, C. J. (1977*a*). A comparison of visual pathways in Boston and Midwestern Siamese cats, *J. Comp. Neurol.* **171:**205–228.

Shatz, C. J. (1977*b*). Abnormal interhemispheric connections in the visual system of Boston Siamese cats: a physiological study, *J. Comp. Neurol.* **171:**229–246.

Shatz, C. J. (1977*c*). Anatomy of interhemispheric connections in the visual system of Boston Siamese and ordinary cats, *J. Comp. Neurol.* **173:**497–518.

Shatz, C. J. and S. LeVay (1979). Siamese cat: altered connections of visual cortex, *Science* **204:**328–330.

Stone, J. S. (1966). The naso-temporal division of the cat's retina, *J. Comp. Neurol.* **126:**585–600.
Stone, J., J. E. Campion, and J. Leicester (1978). The nasotemporal division of the retina in the Siamese cat, *J. Comp. Neurol.* **180:**783–798.
Stone, J. S. and S. M. Hansen (1966). The projection of the cat's retina on the lateral geniculate nucleus, *J. Comp. Neurol.* **126:**601–624.
Wilson, M. E. (1968). Cortico-cortical connexions of the cat visual areas, *J. Anat.* **102:**375–386.

PLASTICITY OF VISUAL CALLOSAL PROJECTIONS

Raymond D. Lund and Donald E. Mitchell

University of Washington, Seattle, Washington
and
National Vision Research Institute of Australia, Carlton, Victoria, Australia

INTRODUCTION

An important feature of central nervous system organization is the presence of orderly sensory "maps" of the external world represented throughout the brain. The details of such maps have been well documented in the central visual system (e.g., Montero, Rojas, and Torrealba, 1973; Kaas, 1977; Palmer, Rosenquist, and Tusa, 1978; Van Essen and Zeki, 1978). When neurons located in one visual map connect with those in another, they tend to do so in a logical fashion. For example, that part of cortical visual area 17 of cats receiving thalamic relay from one retinal quadrant projects only to that part of area 18 receiving from the same retinal quadrant (Hubel and Wiesel, 1965; Garey, Jones, and Powell, 1968). Some interconnections between central maps are restricted to only part of the map. This is true of the fibers running in the corpus callosum that connect the visual areas of the two cerebral hemispheres. One of these visual callosal projections interconnects the border region between areas 17 and 18. In normal cats, the projection extends into area 17 a small distance, connecting mainly with cells responding to a narrow strip of visual field within a few degrees of the representation of the vertical midline.

Because of their normally limited distribution, the callosal connections of area 17 provide a system in which it is possible to investigate the constraints that direct an association pathway to distribute in a coherent manner with respect to part of the map of the external environment, in this case, the visual fields. Shatz (1977) has shown that in Siamese cats in which there is disorder of the geniculocortical pathway, an associated disorder of the callosal projection within area 17 also occurs, suggesting that the pattern of the thalamic afferents plays a part in defining secondary connections. In our studies, we have investigated the role of function in defining the pattern of callosal projection to area 17 in cats in which no aberration of geniculocortical mapping is suspected (Lund, Mitchell, and Henry, 1978; Lund and Mitchell, 1979*a*, *b*). We will describe first the normal callosal distribution and then how this is generated in development and how it is modified by squint induced surgically during development and by dark-rearing.

NORMAL CALLOSAL DISTRIBUTION WITHIN AREA 17

The normal distribution of callosal axons has been determined by unilateral removal of the lateral gyrus in four regular adult cats and two further animals aged 173 and 208 days (which served to match some of the experimental series). After a survival period of 4 days, the animals were anesthetized and fixed by perfusion with buffered paraformaldehyde. The brains were sectioned transversely and one series was stained by the Fink-Heimer method for degenerating axons and another with cresyl violet to define cytoarchitectonic boundaries. The border between areas 17 and 18 was defined using criteria outlined by others (Otsuka and Hassler, 1962; Garey, 1971). Particular attention was given to the disposition of pyramidal cells of layer III at the border and to the thinning of layer IV and the widening of layer V as one moves from area 17 to 18.

The degenerating callosal axons are distributed as shown in Figure 1, with the majority of degeneration occurring close to the border between areas 17 and 18. In area 17, most degeneration is found within the region to which projects the central 2° of representation of the visual field. Very occasional callosal axons are encountered on the medial bank of the lateral gyrus of some cats. From studies in which horseradish peroxidase was injected into one visual cortex, it appears that the principal source of the callosal pathway is the border region between areas 17 and 18, with very occasional cells being identified on the medial bank of the lateral gyrus. These results are in accord with previous observations (see Fisken, Garey, and Powell, 1975; Shatz, 1977).

DEVELOPMENT OF CALLOSAL PROJECTION

The development of the visual callosal projection has been studied from shortly after the time the eyelids open at 1 week postnatal to 6

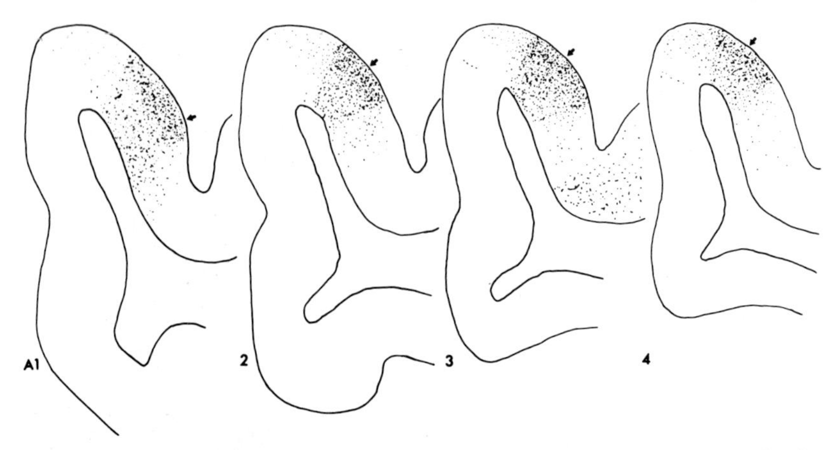

FIGURE 1. Transverse sections of lateral gyrus of normal cat, showing distribution of degeneration produced by removal of contralateral lateral gyrus. The approximate position of the 17/18 border, as judged from Nissl-stained sections, is marked by an arrow. Sections 1–4 are taken at 5 mm, 6.5 mm, 8 mm, and 9.5 mm, respectively, from posterior border of cortex.

weeks postnatal. Most of the work has been done on kittens in which a mixture of tritiated proline and leucine was injected in several penetrations into the lateral gyrus. After 24 hours, the animals were fixed as before. The brains were sectioned transversely and prepared for autoradiography. In two animals HRP was injected with the amino acids in order to label cells of the opposite cortex by retrograde transport. In these two animals we also studied the patterns of degeneration stainable by the Fink-Heimer method, resulting from the damage caused by the injection pipette.

In accord with previous studies (Anker and Cragg, 1974; Stein and Edwards, 1979), degeneration and autoradiography showed substantial corticotectal and corticogeniculate pathways in all animals. The HRP injections indicated a substantial thalamocortical pathway by day 12, as would be expected from previous work (Anker and Cragg, 1974). The callosal projection on day 12 was sparse, particularly in the more superficial layers. Grain counts made at different places around the lateral gyrus revealed that on the medial bank of the lateral gyrus they were at background levels, while around the position of the 17/18 border they averaged four times background, with deeper layers of the cortex having up to eight times background levels. This confirms the impression from visual inspection that the callosal projection at this age is distributed similarly to that in a normal adult animal. It would not be expected that single fibers on the medial bank could be detected above background, however. Degeneration material confirms the findings from autoradiography, although the sensitivity of the method is limited by the presence of some spontaneous degeneration. The HRP material shows most callosally projecting cells to be located around the 17/18 border, with, as in the adult, very occasional labeled cells on the medial bank of the lateral gyrus.

At older ages, the autoradiographic grain becomes heavier in the more superficial layers and the overall density of grain increases markedly. It appears from comparison of kittens injected at 4 and 6 weeks of age that, while the one injected at 6 weeks shows a projection more comparable to adults, the projection at 4 weeks is somewhat reduced in density.

In summary, it appears that a callosal pathway has already invaded the cortex by the time the eyelids first open. It arises from cells spatially distributed as in adult animals and terminates in the same area as in adults. With increased time, the projection becomes denser.

EFFECTS OF INDUCED STRABISMUS UPON DISTRIBUTION OF CALLOSAL FIBERS

Strabismus was induced in kittens between 8 and 112 days of age. After a further survival of 6 months to 4 years, one lateral gyrus was removed and the animals were processed in the same manner as the normal adult cats. In each of the cats used in the preliminary study (Lund et al., 1978), we found callosal degeneration to be heavier than normal at the 17/18 border and to extend in significant amount onto the medial bank of the lateral gyrus as far as the suprasplenial sulcus (Figure 2). Recording from one of these animals showed that the degeneration must extend in area 17 up to 10° of visual field representation from the vertical midline. The effect was present with both convergent and divergent strabismus. In animals with more extreme squint, the amount of callosal degeneration on the medial bank of the lateral gyrus was greater, but the area covered was the same as after a less pronounced squint. There was no diminution in the amount of projection or area covered in an animal in which eye muscle surgery was performed at 112 days, compared with animals operated at 10 days postnatal. One noticeable feature of the abnormal projection was that it was composed of coarse axons, and even in the area of normal distribution there was a noticeable increase in the amount of coarse axon degeneration.

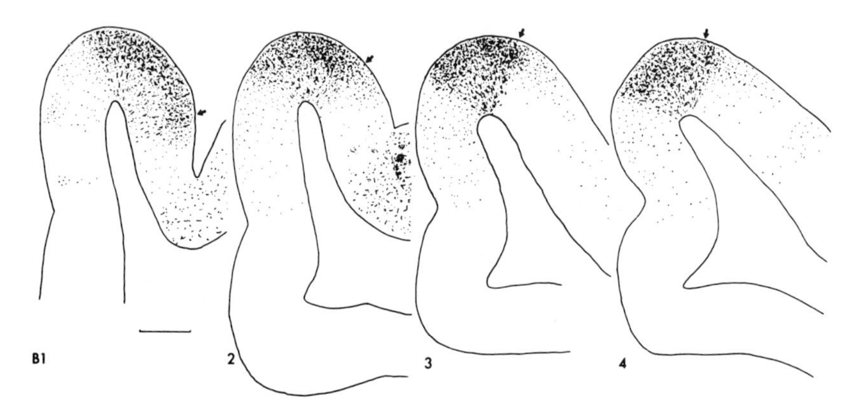

FIGURE 2. Corresponding set of sections from cat in which divergent strabismus was induced on day 112 postnatal. Scale signifies 1 mm.

Subsequent studies (Lund and Mitchell, 1979*b*) have shown that in animals in which squint was induced by unilateral eye surgery, there is an asymmetry in the pattern of expansion of callosal projections. This is illustrated in Figure 3. In cats with a divergent strabismus of the left eye, the callosal pathway from the left to the right cortex is abnormally broadly spread, while the projection in the opposite direction (from right to left hemisphere) is normal (Figure 3*A*). However, with convergent strabismus, the converse is true (Figure 3*B*).

In summary, these findings show that eye alignment plays a part in determining the distribution of callosal axons. This leads to the question of what happens in the absence of a visual input throughout development.

EFFECTS OF DARK-REARING

Kittens were reared in the dark for 127, 208 (two animals), and 256 days from birth. On exposure to light at the end of this period, all but the 127-day-old showed a marked divergent strabismus, and all showed severe visuomotor impairment in accord with previous studies (Lund and Mitchell, 1979*a*).

After unilateral removal of the lateral gyrus and 4 days' survival, we found that the callosal projections tested by degeneration techniques appeared severely reduced in three cats compared with age-matched normal controls (Figure 4). In the fourth cat the callosal pathway was only slightly reduced compared with normal (Figure 5). Unlike results in cats with induced squint, degeneration was absent from the medial bank of the lateral gyrus. The reason for variation between animals

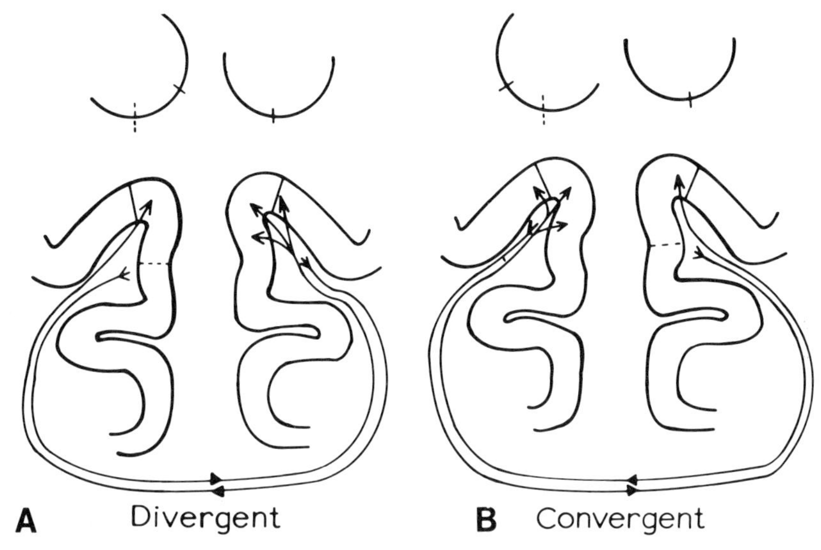

FIGURE 3. Diagrams of eyes and visual cortices showing asymmetry of expansion of callosal projections in animals with (*A*) divergent strabismus and (*B*) convergent strabismus. See text for further description.

could not be easily accounted for by technical factors, since the two brains illustrated were from littermates reared together and were processed together. Both animals showed eye misalignment.

It appears, therefore, that dark-rearing leads to a variable but some-

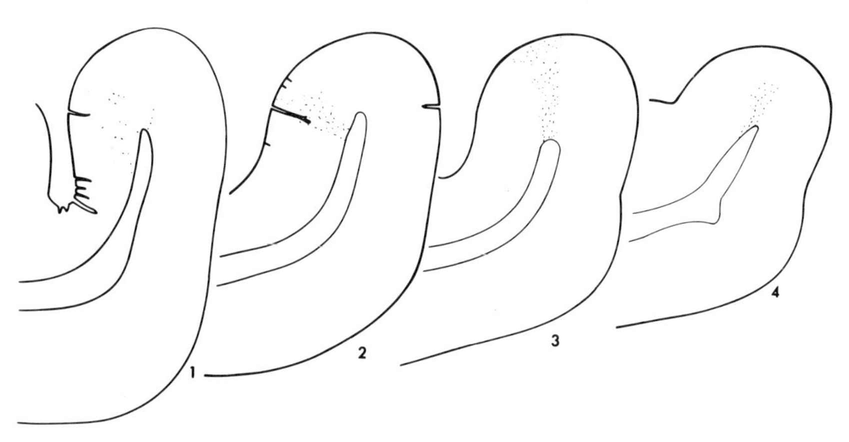

FIGURE 4. Series of sections corresponding to those shown in Figure 1, indicating the pattern of callosal degeneration in a dark-reared cat.

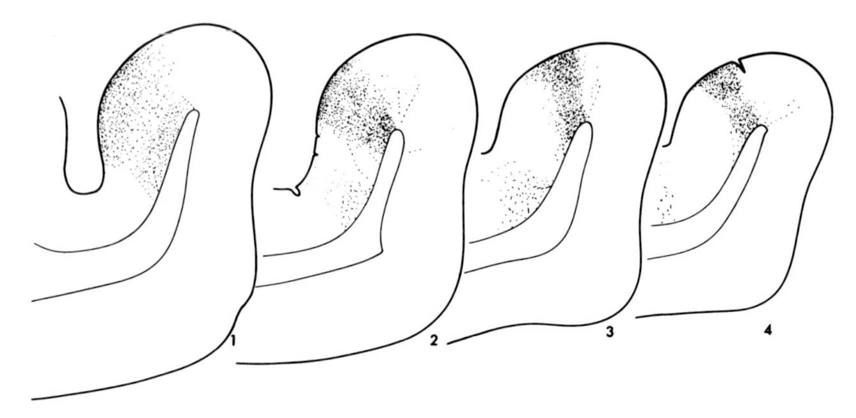

FIGURE 5. Similar set of sections of degeneration in a dark-reared littermate of the cat illustrated in Figure 4.

times substantial reduction of the callosal projection that is independent of whether or not the animal has a misalignment of the eyes.

COMPARABLE EXPERIMENTS IN OTHER ANIMALS

Two studies have continued the investigation of the effects of afferents on callosal pathways. In one, we (J. Lund, R. Lund, and P. Kramar, unpublished) induced convergent and divergent squints, respectively, in one eye of two monkeys (*M. nemestrina*) 1 week old. At 6 months postnatal, the corpus callosum was cut and the degeneration patterns were examined in comparison with a control 6-month-old animal. No significant difference could be detected between experimental and control animals in the amount of callosal projection extending into area 17.

A second study, part of a larger investigation of callosal plasticity in rats (Cusick and Lund, 1977 and unpublished; Cusick, 1978), has shown that optic tract lesions at birth lead to substantial diminution of the callosal pathway distributing at the 17/18 border on the side ipsilateral to the tract, while leaving other callosal projections unaffected. This contrasts with experiments in which the thalamus is damaged, depriving the visual cortex of geniculocortical afferents. In these animals, there is expansion of callosal axons into area 17.

DISCUSSION

It is apparent that the callosal pathways interconnecting the primary visual cortical areas are in an early stage of development at the time the eyes open. By day 12, callosal axons have already entered the cortex at the 17/18 border, although at that age the projection is light compared with adults, and is mostly concentrated on the deeper layers. Over the next few weeks the projection increases significantly in density to attain approximately normal proportions by 6 weeks of age. The early appearance of the callosal pathway, while not detected in a degeneration study (Anker and Cragg, 1974), has support from two other autoradiographic studies (Stein and Edwards, 1979; LeVay and Shatz, unpublished) in which a callosal projection was identified in the cortex shortly after birth. By 12 days, the cells of origin of the callosal pathway are already disposed as in normal adults, with most of them at the 17/18 border and occasional ones on the medial bank of the lateral gyrus.

This means that the anatomical aberrations detected after squint or visual deprivation are likely to develop from a normally disposed but low-density pathway, rather than from a broadly projecting, diffuse system. The results obtained from the dark-rearing experiments might signify that the pathway fails to develop beyond the initial low-density projection in most animals, but it is equally possible that it could represent a state of degeneration from a more mature projection.

The broad callosal projection in cats with strabismus may have several origins: (1) It may arise from cells located at the 17/18 border. However, the HRP studies of Innocenti and Frost (1978) show that even if this is the case, there is, nevertheless, a major contribution from cells located on the medial bank of the lateral gyrus. (2) The pathway may sprout from the occasional cells on the medial bank shown to contribute callosal axons normally. This would seem unlikely in view of the large numbers of cells on the medial bank shown by Innocenti and Frost (1978) as giving rise to callosal axons. (3) It is possible that cells throughout area 17 are supplying callosal axons over a staggered but prolonged period of development and that most axons terminating elsewhere than the 17/18 border region degenerate or their cells of origin die. The concept of degeneration either of cells or of axons, if appropriate connections are not made, is one for which there is considerable evidence in the peripheral nervous system (see Bennett and Pettigrew, 1976; Hendry, 1976; Landmesser and Pilar, 1976; Cowan, 1978; Lund, 1978) and for which there appears to be a parallel for neurons distributing within the central nervous system (Clarke and Cowan, 1976). It is uncertain whether the retraction is the result of a competition for a limited number of synaptic sites or a limited amount of a trophic substance like nerve growth factor, or whether some other mechanism is operative.

In the present experiments, the source of the variation in callosal connections between regular and strabismic cats is difficult to explain, especially when taking into account the asymmetry that relates to the direction of eye deviation. It is presumed that in some way visual function plays a part in leading to the abnormal patterns found in the strabismic cats. In normal cats, a line seen directly in front of the animal will project to the area centralis and to the 17/18 border of both hemispheres. This is true of the normal eye of cats with a unilaterally induced squint, as illustrated in Figure 3. For a laterally deviated eye, the line projects upon the temporal retina and ipsilateral cortex only (Figure 3*A*, dotted line), while for a medially deviated eye, the line projects upon the nasal retina and contralateral cortex (Figure 3*B*, dotted line). Such an asymmetry may be the source of the asymmetry in the callosal abnormality, and as such emphasizes the importance of representation of the vertical midline in determining the pattern of callosal connections.

The failure to find comparable effects in monkeys may be due to a number of factors. The development of the cortex appears to be more advanced at birth than in the young kittens studied here; the 17/18 border is a much more precise landmark than in cat; unlike cat, there is no indication of cells within area 17 contributing callosal axons in normal animals (Lund, Lund, Hendrickson, Bunt, and Fuchs, 1975).

The studies on rat, by contrast, although not exactly comparable in design to the cat experiments, emphasize considerable plasticity of visual cortical connections and suggest that the principles derived from the cat may have parallels in other animals born at an immature stage of development.

In summary, it appears that the basic plan of callosal connections is formed independent of functional considerations. The studies on Siamese cats (Shatz, 1977) and on rats with thalamic lesions (Cusick, 1978) emphasize that the organization of the geniculocortical pathway is an important determinant of adult callosal patterns. However, our results show that this basic plan can be substantially modified if the functional patterns carried by the geniculocortical pathway are abnormal.

ACKNOWLEDGMENTS

We wish to thank Doris Ringer for secretarial help and Renée Wise for technical assistance. Supported by grants from the United States Public Health Service (EY00596) to R.D.L., and from the National Research Council of Canada (AP7660) to D.E.M.

REFERENCES

Anker, R. L. and B. G. Cragg (1974). Development of the extrinsic connections of the visual cortex in the cat, *J. Comp. Neurol.* **154**:29–42.

Bennett, M. R. and A. G. Pettigrew (1976). The formation of neuromuscular synapses, *Cold Spring Harbor Symp. Quant. Biol.* **40**:409–424.

Clarke, P. G. H. and W. M. Cowan (1976). The development of the isthmo-optic tract in the chick, with special reference to the occurrence and correction of developmental errors in location and connections of isthmo-optic neurons, *J. Comp. Neurol.* **167**:143–164.

Cowan, W. M. (1978). Aspects of neural development, *Int. Rev. Physiol.* **17**: 149–191.

Cusick, C. G. (1978). Normal and abnormal callosal connections in the occipital cortex of albino rats, *Soc. Neurosci. Abstr.*, Vol. 4, p. 470.

Cusick, C. and R. D. Lund (1977). Plasticity and specificity of the corpus callosum in the albino rat following neonatal lesions, *Soc. Neurosci. Abstr.*, Vol. 3, p. 424.

Fisken, R. A., L. J. Garey, and T. P. S. Powell (1975). The intrinsic association and commissural connections of area 17 of the visual cortex, *Philos. Trans. R. Soc. London B Biol. Sci.* **272**:487–536.

Garey, L. J. (1971). A light and electron microscope study of the visual cortex of the cat and monkey, *Proc. R. Soc. London B Biol. Sci.* **179**:21–40.

Garey, L. J., E. G. Jones, and T. P. S. Powell (1968). Interrelations of striate and extrastriate cortex with the primary relay sites of the visual pathway, *J. Neurol. Neurosurg. Psychiatry* **31**:135–157.

Hendry, I. A. (1976). Controls in the development of the vertebrate sympathetic nervous system, *Rev. Neurosci.* **2**:149–194.

Hubel, D. H. and T. N. Wiesel (1965). Receptive fields and functional architecture in two nonstriate visual areas (18 and 19) of the cat, *J. Neurophysiol.* **28**:229–289.

Innocenti, G. M. and D. O. Frost (1978). Visual experience and the development of the efferent system to the corpus callosum, *Soc. Neurosci. Abstr.*, Vol. 4, p. 475.

Kaas, J. H. (1977). Sensory representations in mammals, pp. 65–80 in *Function and Formation of Neural Systems*, Stent, G. S., ed. Dahlem Konferenzen, Berlin.

Landmesser, L. and G. Pilar (1976). Fate of ganglionic synapses and ganglion cell axons during normal and induced cell death, *J. Cell Biol.* **68**:357–374.

Lund, R. D. (1978). Transneuronal influences in development, pp. 183–201 in *Development and Plasticity of the Brain.* Oxford University Press, New York.

Lund, J. S., R. D. Lund, A. E. Hendrickson, A. H. Bunt, and A. F. Fuchs (1975). The origin of efferent pathways from the primary visual cortex, area 17, of the macaque monkey as shown by retrograde transport of horseradish peroxidase, *J. Comp. Neurol.* **164**:287–304.

Lund, R. D. and D. E. Mitchell (1979*a*). The effects of dark rearing on visual callosal connections of cats, *Brain Res.* **167**:172–175.

Lund, R. D. and D. E. Mitchell (1979*b*). Asymmetry in the visual callosal connections of strabismic cats, *Brain Res.* **167**:176–179.
Lund, R. D., D. E. Mitchell, and G. H. Henry (1978). Squint-induced modification of callosal connections in cats, *Brain Res.* **144**:169–172.
Montero, V. M., A. Rojas, and F. Torrealba (1973). Retinotopic organization of striate and peristriate visual cortex in the albino rat, *Brain Res.* **53**:197–201.
Otsuka, R. and R. Hassler (1962). Über Aufbau und Gliederung der corticalen Sehsphäre bei der Katze, *Arch. Psychiatr. Nervenkr.* **203**:212–234.
Palmer, L. A., A. C. Rosenquist, and R. J. Tusa (1978). The retinotopic organization of lateral suprasylvian visual areas in the cat, *J. Comp. Neurol.* **177**:237–256.
Shatz, C. J. (1977). Anatomy of interhemispheric connections in the visual system of Boston Siamese and ordinary cats, *J. Comp. Neurol.* **173**:497–518.
Stein, B. E. and S. B. Edwards (1979). Corticotectal and other corticofugal projections in neonatal cat, *Brain Res.* **161**:399–409.
Van Essen, D. C. and S. M. Zeki (1978). The topographic organization of rhesus monkey prestriate area, *J. Physiol. (London)* **277**:193–226.

GROWTH AND DIFFERENTIATION FACTORS IN NEURONAL DEVELOPMENT

CELL CULTURE STUDIES ON THE MECHANISM OF ACTION OF NERVE GROWTH FACTOR

Lloyd A. Greene, David E. Burstein, Jeffrey C. McGuire, and Mark M. Black

Harvard Medical School and Children's Hospital Medical Center, Boston, Massachusetts

The biological effects of nerve growth factor (NGF) on its target cells are familiar to most neuroscientists and have been well described in a number of recent reviews (Levi-Montalcini, 1976; Mobley, Server, Ishii, Riopelle, and Shooter, 1977; Bradshaw, 1978). These effects include: increase in somatic volume, greatly stimulated proliferation of neurites, maintenance of survival, and increased levels of several enzymes involved in synthesis of neurotransmitters. On the other hand, the mechanism of action of NGF remains largely unclear. The purpose of this paper is to critically review selected aspects of the state of the art with regard to NGF's mechanism of action. Although we cannot yet provide a complete description of the means whereby NGF produces its effects, we hope to provide at least a conceptual framework upon which future work can be based.

EXPERIMENTAL SYSTEMS

Like many scientific problems, the study of the mechanism of action of NGF is dependent on the availability of suitable experimental models or "preparations." Since the actions of NGF may be demonstrated in vitro (Levi-Montalcini and Angeletti, 1968), many studies on the factor have employed cultures of whole or dissociated sympathetic or dorsal root ganglia. Among the well-known advantages of such in vitro systems is the increased ability to control both the chemical environment and cellular composition of the cultures. Cultures of ganglia also have disadvantages for certain types of studies. These include: (1) the heterogeneity of the tissue, particularly in cultures of whole ganglia; (2) the limited amount of material that can be obtained; (3) the dependence of neurons on NGF for their survival, and thus the difficulty of obtaining NGF-untreated controls; (4) the likelihood that the neurons have already been exposed and have responded to NGF in vivo before being placed in culture, thus raising the possibility that one is studying the *continuation* rather than the *initiation* of the effects of NGF and that, with respect to fiber outgrowth, one is studying the *regeneration* rather than the *generation* of neurites.

Another system that has recently been introduced to study the mechanism of action of NGF is a clonal cell line (PC12) derived (Greene and Tischler, 1976) from a transplantable rat pheochromocytoma (Warren and Chute, 1972). When grown in serum-containing medium without exogenous NGF, PC12 cells possess many of the differentiated neurotransmitter and ultrastructural features associated with pheochromocytomas and with their nonneoplastic counterparts, adrenal medullary chromaffin cells (Greene and Tischler, 1976; Greene and Rein, 1977*a,b*; Greene and Rein, 1978; Tischler and Greene, 1978). Significantly, several days after PC12 cells are exposed to physiological levels of NGF (ca. 10 ng/ml), they cease mitosis (Greene and Tischler, 1976) and acquire a number of properties characteristic of sympathetic neurons. These responses include increased somatic volume; outgrowth of neurites (see Figure 1) (Greene and Tischler, 1976); increased electrical excitability and responsiveness to acetylcholine (Dichter, Tischler, and Greene, 1977); and the presence of synaptic-like vesicles (Greene and Tischler, 1976; Tischler and Greene, 1978). Such effects are reversible in that withdrawal of NGF results in disintegration of neurites and recommencement of cell division (Greene and Tischler, 1976). Another feature of the PC12 line is that the cells can also undergo rapid "regeneration" of neurites. That is, when PC12 cells that have been preexposed to NGF for at least 1 week are divested of their neurites by mechanical means and then replated, NGF-dependent regrowth of long neurites (100–200 μm in length) occurs within 24 hours (Greene, 1977; Burstein and Greene, 1978). In contrast, NGF-dependent de novo generation of neurites by PC12 cells previously unexposed to NGF occurs with a lag of at least 24 hours and is characterized by an elongation rate of about 30 μm per day. Such findings suggest that the PC12 line has certain advantages for studying the mechanism(s) of action of NGF,

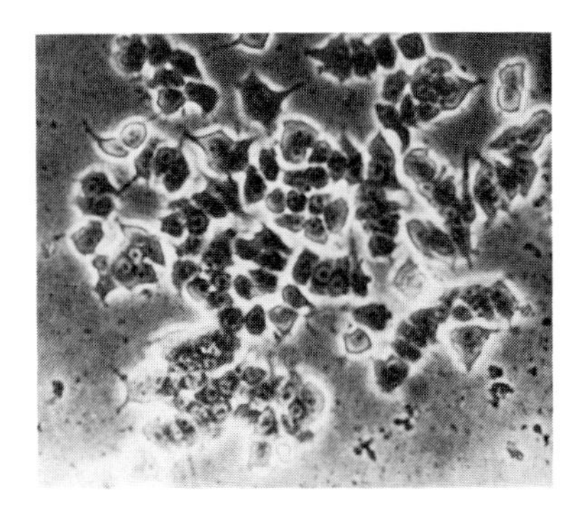

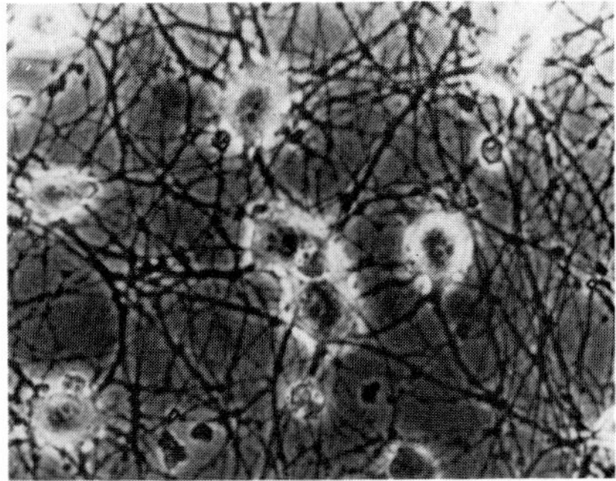

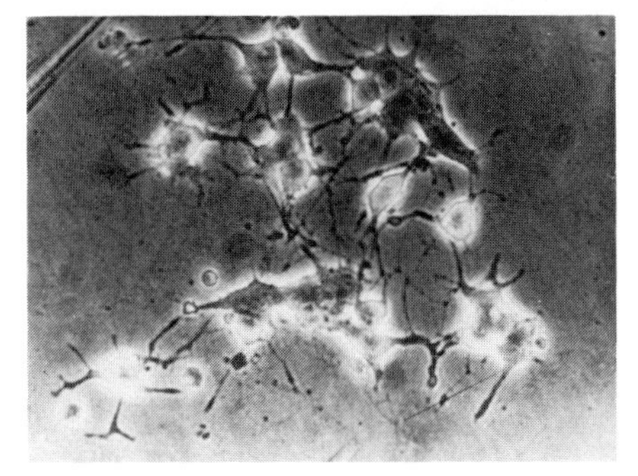

FIGURE 1. Phase contrast photomicrographs of PC12 cells untreated with NGF (upper left), treated with NGF (50 ng/ml) for 13 days (upper right), or treated with Db cAMP for 13 days (lower center).

including: (1) the availability of large numbers of homogenous cells, (2) the ability of PC12 cells to respond to NGF, without requiring it for survival (thus providing viable NGF-untreated cells as controls), (3) the opportunity to study the *initial* effects of NGF on previously unexposed target cells, and (4) the opportunity to study the reversibility of the effects of NGF. The PC12 system also has certain potential disadvantages. For example, in contrast to normal sympathetic neurons in vivo or in vitro, NGF treatment of PC12 cells does not cause an increase in the specific activity of tyrosine hydroxylase (Greene and Tischler, 1976). Also, since PC12 cells are derived from a tumor, there is always the possibility that some or all of their responses to NGF are different from those of normal tissues. Thus, the final description of the mechanism(s) of action of NGF will, in all likelihood, require data derived from work with both normal and neoplastic tissues.

NGF RECEPTORS

Although a detailed description of studies on NGF receptors is beyond the scope of this article, a summary of work in this area is clearly relevant. It is generally assumed that, like other peptide effectors, NGF interacts with its target cells primarily via specific receptors. However, work over the last several years has indicated multiple means whereby NGF can interact with its target cells.

Surface membrane receptors. Binding studies employing [^{125}I]NGF and either membrane preparations or intact cell dissociates of sympathetic or sensory ganglia have revealed the existence of specific NGF receptors on the plasma membrane (Banerjee, Snyder, Cuatrecasas, and Greene, 1973; Herrup and Shooter, 1973; Frazier, Boyd, and Bradshaw, 1974). Kinetic evidence has been presented that suggests that binding of NGF to cell membranes may show several apparent dissociation constants (Frazier et al., 1974; Mobley et al., 1977). It has not yet been resolved whether this is due to negative cooperativity or to multiple binding sites.

Internalization of NGF and retrograde axonal transport. Work from several laboratories indicates that NGF can also be internalized by target cells (Burnham and Varon, 1973; Hendry, Stöckel, Thoenen, and Iversen, 1974; Andres, Jeng, and Bradshaw, 1977; Yankner and Shooter, 1979; Calissano and Shelanski, personal communication). Moreover, it has been well established that NGF internalized at nerve endings can be retrogradely transported to the cell body (Hendry et al., 1974; Paravicini, Stöckel, and Thoenen, 1975). NGF transported by this means is biologically active in inducing tyrosine hydroxylase (Paravicini et al., 1975) and in supporting cell viability and neurite outgrowth (Campenot, 1977).

Nuclear receptors. Recent evidence suggests that at least a portion of the NGF that is internalized reaches the nucleus, where it binds to specific receptors (Andres et al., 1977; Yankner and Shooter, 1979). The site of these receptors has been identified as being the chromatin (Andres et al., 1977) and/or the nuclear membrane (Yankner and Shooter, 1979).

The above findings clearly suggest that NGF may interact with its target cells at several sites (plasma membrane, cytoplasm, nucleus). Such findings raise the presently unresolved issue as to the possible relevance of each site of interaction to NGF's mechanism of action. Several possibilities can be envisioned. One is that the action of NGF requires only interaction with a plasma membrane receptor. The observation that NGF bound to Sepharose beads appears to stimulate neurite outgrowth from cultured sensory ganglia was interpreted to support this view (Frazier, Boyd, and Bradshaw, 1973*a*). However, it has been pointed out (Bradshaw, 1978) that such experiments did not exclude the possibility that NGF (or an active fragment thereof) was cleaved from the beads and internalized. Another possibility is that the site of action of NGF is within the cell, either in the cytoplasm or nucleus, and that the plasma membrane receptors are in essence only "carriers" that serve in the uptake and internalization of the factor. This alternative would be consistent with the biological effects of retrogradely transported NGF. However, it has not been ruled out that transported NGF somehow reattains the cell surface or that it promulgates its effects by interaction with a "plasma membrane" receptor that has been internalized and trapped within the cell. A third possibility is that the action of NGF involves multiple sites of interaction. For instance, promotion of NGF's effects may require that the factor interact with target cells at all levels. Alternatively, different types of receptors (e.g., nuclear vs. plasma membrane) could regulate different aspects of NGF's mechanism of action. As will be presented below, NGF appears to have separable pathways of action on its target cells. It is attractive to consider (see also Bradshaw, 1978) the possibility that different actions of NGF are mediated by different sites at which it interacts with its target cells. Finally, it is noteworthy that evidence exists for the interaction of several other polypeptide hormones (such as epidermal growth factor [EGF] [Carpenter and Cohen, 1976] and insulin [Kahn and Baird, 1978]) with the cell membrane and for their internalization.

RNA TRANSCRIPTION AND THE MECHANISM OF ACTION OF NGF

A crucial aspect of the mechanism of action of NGF is the role of RNA synthesis. That is, does the interaction of NGF with its target cells lead to specific effects on RNA transcription that are in turn required for expression of NGF's other actions? In reviewing the actions of NGF, Levi-Montalcini and Angeletti (1968) hypothesized that NGF might work via stimulation of synthesis of specific mRNAs that in turn would lead to enhanced synthesis of specific proteins. Among the major bases for this suggestion were experiments that appeared to show that NGF stimulated RNA and protein synthesis in cultured sensory ganglia (Angeletti, Gandini-Attardi, Toschi, Salvi, and Levi-

Montalcini, 1965). However, it has been pointed out more recently that apparent enhancements in synthesis were due to decreased synthesis by the control (NGF-untreated) cells and to the effect of NGF on the rate of uptake (rather than the incorporation) of precursor molecules (Horii and Varon, 1977). In a more rigorous test of the transcription hypothesis, Partlow and Larrabee (1971) found that NGF stimulated neurite outgrowth from freshly explanted chick sympathetic ganglia even when RNA synthesis was blocked with actinomycin D. This led to the alternative suggestion that the effects of NGF (at least on neuronal survival and on neurite outgrowth) do not require transcription.

Because of their unique properties, PC12 cells have been used for further studies on the role of RNA synthesis in the mechanism of action of NGF (Burstein and Greene, 1978). These studies employed three compounds (actinomycin D, cordycepin, and camptothecin), each of which blocks RNA synthesis by a different mechanism of action. Significantly, NGF-stimulated generation of neurite outgrowth was blocked by each drug, and at levels that appeared to have little effect on cell viability. Confirmation of cell viability and vitality was obtained with camptothecin-treated cultures. The effects of this inhibitor on RNA synthesis are completely and rapidly reversible (Horwitz, 1975). Removal of camptothecin from NGF-treated cultures (even after a week of exposure to the inhibitor) was followed by initiation of neurite outgrowth. Such experiments strongly suggest that de novo synthesis of RNA is required for NGF-stimulated initiation or generation of neurite outgrowth. Another interesting feature of the camptothecin removal experiment was that when this inhibitor was withdrawn from NGF-treated cultures, the rate of appearance of neurites was accelerated by about 18 hours, in comparison with the rate that occurs in cultures after NGF is first added. This suggests that part of the sequence of actions of NGF can take place in the presence of camptothecin. An additional aspect of interest in the inhibitor experiments was that for camptothecin and actinomycin, neurite outgrowth was blocked at drug levels considerably lower than those required to cause complete inhibition of RNA synthesis. This suggests that neurite outgrowth may involve particularly drug-sensitive species of RNA.

The discovery that initiation of NGF-dependent neurite outgrowth from PC12 cells requires RNA synthesis would appear to be in conflict with the results that Partlow and Larrabee (1971) found with sympathetic ganglia. However, as pointed out above, such ganglion cells probably have had prior exposure to NGF in vivo and thus might be considered to have undergone NGF-stimulated regeneration (as opposed to generation) of neurites in vitro. To test the possibility that generation and regeneration of neurites might have different requirements for RNA synthesis, analogous experiments were carried out with PC12 cells (Burstein and Greene, 1978). That is, PC12 cells were treated with NGF for several weeks, divested of their neurites by mechanical means, and replated with either no NGF, NGF, or NGF and concentrations of actinomycin, cordycepin, or camptothecin sufficient to block >95% of cellular RNA synthesis. Significantly, while the rapid (i.e., occurring within 24 hours) regrowth of neurites required the presence of NGF, it was not blocked by the inhibitors of RNA synthesis. Such results are consistent with those of Partlow and Larrabee (1971); that is, regeneration of neurites by cells previously exposed to NGF requires the physical presence of the factor, but does not require transcription of RNA.

The above findings have led to the following model for the mechanism of action of NGF in stimulating neurite outgrowth (Burstein and Greene, 1978). (1) NGF-stimulated neurite outgrowth requires the physical presence of NGF for a nontranscriptional event. (2) Neurite outgrowth also requires the presence of material whose synthesis is regulated by NGF and requires transcription. (3) During the initial generation of neurites, the pools of such materials are low, and thus neurite outgrowth is transcriptionally dependent. However, after the cells have been exposed to NGF for a sufficient length of time to accumulate pools of these materials, the pools can sustain rapid neurite regeneration for 24 hours, even when transcription is blocked.

One crucial test of the above model is that withdrawal of NGF should result in cessation of transcriptional effects and gradual diminution of the pool of material(s) through turnover. In agreement with this prediction are the findings (Burstein and Greene, 1978) that when NGF-pretreated PC12 cells are divested of their neurites and then subcultured for various lengths of time before being reexposed to NGF, they undergo a progressive loss (half-time about 16 hours) of the capacity for rapid, transcription-independent neurite regeneration.

The process whereby responsive cells acquire the ability for rapid, transcription-independent neurite outgrowth in the presence of NGF has been termed "priming." Several additional aspects of priming are of interest here. First, priming of PC12 cells does not appear to be merely a consequence of the effect of NGF on cell division. Thus, blockade of cell division in PC12 cultures with mitotic inhibitors or by lowering the serum concentration causes neither neurite outgrowth nor priming. Also, growth-inhibited PC12 cells show a time-lag for initiation of neurite outgrowth, as well as a time-dependent loss of priming upon withdrawal of NGF. Another aspect of priming is that this process requires neither attachment to a substrate nor process outgrowth. Our recent experiments indicate that PC12 cells can become primed by NGF in spinner-suspension culture and when neurite outgrowth of attached cells is blocked with inhibitors of tubulin assembly.

PROTEIN SYNTHESIS

The fact that generation of neurite outgrowth by NGF requires RNA transcription raises the question as to the role of such RNA. One possibility is that the RNA itself plays a direct role in neurite outgrowth. For example, there is evidence (Heidemann, Sander, and Kirschner, 1977) that RNA could play a structural role in organization of microtubules. A second possibility is that neurite outgrowth requires specific proteins that are the translational products of RNAs whose synthesis is regulated by NGF. Consistent with this are our findings that both generation and regeneration of neurites by PC12 cells are blocked by inhibitors of protein synthesis.

To test for effects of NGF on protein synthesis, one- and two-dimensional electrophoretic analyses have been carried out on the polypeptides synthesized by PC12 cells before and after treatment with the factor (McGuire, Greene, and Furano, 1978). Somewhat surprisingly, it was found that NGF treatment results in no qualitative, and only very few quantitative, changes in the relative proportions of the 1,000-or-so polypeptides resolved by such methods. This indicates that promotion of neurite outgrowth by NGF takes place in large part via rearrangement of the types of abundant polypeptides the cell already makes prior to treatment with the factor. This again emphasizes the presence of a nontranscriptional role for NGF. Although the overall pattern of synthesis of abundant proteins by PC12 cells shows few alterations, striking quantitative changes have been found in synthesis of a small number of minor species. One of these is a polypeptide of apparent molecular weight 80,000 (McGuire and Greene, in preparation). Increased synthesis of this species is detectable within 24 hours of NGF treatment and reaches maximal levels by 48 hours of treatment. Withdrawal of NGF causes return to control levels of synthesis within 24–48 hours. Significantly, the effect of NGF on synthesis of this species is selectively blocked by camptothecin at concentrations that also block neurite outgrowth. A second component whose relative synthesis also appears to be greatly increased by NGF treatment is a glycoprotein (detectable by labeling with fucose or glucosamine) of apparent MW 230,000 (McGuire et al., 1978). This glycoprotein (designated NILE glycoprotein) is exposed on the cell surface, and is present on the surface of normal sympathetic neurons (Lee, Greene, and Shelanski, in preparation). Moreover, the effect of NGF on NILE synthesis is also selectively blocked by low levels of camptothecin. In addition to the two polypeptides mentioned above, it is also possible that NGF treatment affects the synthesis of components that are among the class of nonabundant proteins not easily detectable by present electrophoretic methods. Such proteins of minor abundance could play an important regulatory or structural role in the mechanism of action of NGF.

NGF AND CELL SURVIVAL

In contrast to findings on neurite outgrowth, data concerning the role of RNA synthesis in the effect of NGF on cell survival are less clear. It has been well documented that NGF is required for the in vitro survival of embryonic sympathetic and responsive sensory neurons (Levi-Montalcini and Angeletti, 1963). Thus, it follows from the ability of NGF-treated neurons to extend neurites in the presence of actinomycin D that at least some of the effects of NGF on survival are nontranscriptional. However, for such experiments, one cannot exclude the possibility that the effect of NGF on survival also requires a transcriptionally dependent step and that (as with neurite outgrowth) the neurons are already primed by their prior in vivo exposure to the factor.

Experiments with PC12 cells (Greene, 1978) have begun to offer some insight on the matter of transcription and NGF-dependent cell survival. PC12 cells do not require NGF for survival in medium containing horse or rat serum. In serum-free medium, however, the cells die unless NGF is present. Under these conditions, NGF also stimulates neurite outgrowth. Although total blockade of RNA synthesis in serum-free PC12 culture results in cell death even in the presence of NGF, it is not clear whether death is due to specific interference with NGF-dependent transcription or to toxicity resulting from nonspecific inhibition of total RNA synthesis. Significantly, however, exposure of NGF-treated serum-free cultures to levels of camptothecin that reversibly block initiation of neurite outgrowth does not interfere with the ability of NGF to maintain cell viability. While these findings do not altogether rule out a possible transcription-dependent component of NGF's effect on cell survival, they do suggest that if such a component exists, it is different from that required for NGF-stimulated initiation of neurite outgrowth.

MODELS FOR MECHANISM OF ACTION OF NGF

The above discussion has indicated that the mechanism of action of NGF includes interaction of the factor with one or more specific receptors and activation of at least two separable pathways, one of which requires RNA synthesis and the other of which does not. There still remain, however, a number of unknown steps linking the interaction of NGF with its target cells to the final biological response to the factor. In this section, we shall briefly review and attempt to evaluate several major models proposed for other steps in the mechanism of action of NGF.

Direct Effects on Assembly of Tubulin and Actin

Calissano and his associates have carried out in vitro experiments that indicate that NGF can interact with tubulin and actin in a stoichiometric fashion and can favor their polymerization (Calissano and Cozzari, 1974; Levi, Cimino, Mercanti, Chen, and Calissano, 1975; Calissano, Monaco, Castellani, Mercanti, and Levi, 1978). This has led to the suggestion that NGF could directly interact with such molecules in vivo and catalyze their assembly into filamentous macromolecular structures (composed of microtubules and microfilaments) that in turn play a necessary role in neurite outgrowth and elongation. While this model would be consistent with the nontranscriptional role of NGF in neurite outgrowth, it has certain aspects that remain to be clarified. First is the question of the specificity of the interaction between NGF and actin or tubulin. For instance, the mouse NGF used for the experiments has an isoelectric point of about 9.2, and it has been shown that a number of basic proteins as well as polycations can promote assembly of tubulin (Lee, Tweedy, and Timasheff, 1978). Second, it is not clear whether the macromolecular assemblies whose formation is catalyzed in the presence of NGF are products of true polymerization or are due to a less specific aggregation. Third, it has not been demonstrated that NGF can interact with such molecules within living cells.

Ornithine Decarboxylase

Rapid induction of ornithine decarboxylase (ODC) has been found to be associated with the responses of a number of cell types to various growth- and differentiation-promoting hormones (Russell, Byus, and Manen, 1976). It has been suggested that this enzyme and the polyamines whose synthesis it catalyzes may play a causal role in mediating subsequent cellular response to such hormones (Russell et al., 1976). The demonstration that NGF can also cause rapid induction of ODC in brain (Roger, Schanberg, and Fellows, 1974; Lewis, Lakshmanan, Nagaiah, MacDonnell, and Guroff, 1978) and sympathetic ganglia (MacDonnell, Nagaiah, Lakshmanan, and Guroff, 1977) has thus led to the hypothesis that ODC may also play a causal role in the mechanism of action of NGF (MacDonnell et al., 1977). This possibility has recently been evaluated with PC12 cells (Greene and McGuire, 1978). Treatment of PC12 cultures with ng/ml levels of NGF results in a rapid (maximal within 5–8 hours) 20- to 40-fold transcription-dependent increase in ODC activity. However, when either the enzymatic activity or the induction of this enzyme is blocked by specific inhibitors, there is no detectable inhibition of the effect of NGF on neurite outgrowth, increase in cell volume, or maintenance in serum-free medium. Moreover, specific blockade of ODC synthesis or of ODC activity in cultures of chick sympathetic ganglia has no apparent effect on neurite outgrowth, cell size, or cell viability. These findings suggest that ODC induction is not required (and therefore plays no mechanistic role) in at least a number of the actions of NGF.

Uptake of Small Molecules

Early in vitro work on the mechanism of action of NGF was interpreted to show that NGF stimulated synthesis of protein and RNA (Angeletti et al., 1965). However, more recent work (Horii and Varon, 1977) has indicated that these results could instead be explained by effects of NGF on the uptake of the radiolabeled precursor molecules used for the experiments. In the presence of NGF, uptake by chick embryo sensory neurons was found to be maintained at a constant level, while in NGF-untreated "control" cultures, uptake progressively declined with time. In addition, Horii and Varon (1977) found that withdrawal of NGF from cultured sensory neurons would result in a decline in uptake and that if NGF was reintroduced to the cells within 6 hours, uptake was rapidly (within 10 minutes) restored to original levels. This has led to the suggestion (Horii and Varon, 1977) that the primary action of NGF on its target cells may be to regulate the uptake of small molecules and that the effects on uptake may in turn trigger further events in the sequence of action of the factor. Recent experiments with PC12 cells (McGuire and Greene, 1979) may be relevant to this proposal. Addition of NGF to PC12 cultures results in a rapidly on-setting (detectable with 15 minutes) increase in the uptake of amino acids. However, similar increases can also be induced by exposure to elevated levels of serum. Since serum treatment did not induce neurite outgrowth, it would appear that stimulation (or maintenance) of uptake of small molecules is not *sufficient* to trigger all of the effects of NGF. Nevertheless, such considerations do not rule out the possibility that effects on small molecule uptake may play a *necessary* role in one or several actions of NGF. One particularly inviting possibility is that the mechanism by which NGF supports survival of responsive neurons (or of PC12 cells in serum-free medium) is to maintain the uptake of required small molecules.

Cyclic AMP

A number of hormones have been demonstrated to increase intracellular levels of cyclic AMP (cAMP), which in turn serves as a "second messenger" to trigger subsequent responses. However, as will be discussed below, the evidence for the role of cAMP in the mechanism of action of NGF is far from being conclusive.

Support for a cAMP role in NGF's mechanism of action has come principally from two types of evidence. One is that NGF causes increases in cAMP levels of its target cells. Nikodijevic, Nikodijevic, Yu, Pollard, and Guroff (1975) have reported an increase (approximately twofold) in cAMP levels in organ cultures of rat superior cervical ganglia in response to NGF, while Schubert and Whitlock (1977) and Schubert, LaCorbiere, Whitlock, and Stallcup (1978) reported a 70% increase in cultures of PC12 cells following addition of NGF. In each case, the increase was transitory and peaked within 5 minutes of addition of the factor. The second type of evidence for a cAMP role is the finding that exposure to cAMP or its analogues (such as dibutyryl cAMP [Db cAMP]) can mimic the effects of NGF. A number of groups have reported that such compounds elicit fiber outgrowth from cultured embryonic dorsal root ganglia (Roisen, Murphy, and Braden, 1972; Hier,

Arnason, and Young, 1973; Frazier, Ohlendorf, Boyd, Aloe, Johnson, Ferrendelli, and Bradshaw, 1973*b*). Also, Schubert and Whitlock (1977) and Schubert et al. (1978) have claimed to elicit neurite outgrowth from PC12 cells with Db cAMP.

Nevertheless, there are a number of pieces of evidence that do not support a role for cAMP in the mechanism of action of NGF. (1) Frazier et al. (1973*b*) reported that NGF had no effect on cAMP levels in chick embryo sensory ganglia. These workers also found that NGF did not stimulate adenyl cyclase in broken cell preparations of the ganglia. In addition, Otten, Hatanaka, and Thoenen (1978) found that NGF treatment caused no statistically significant change in either cAMP or cGMP levels of rat superior cervical ganglia in vivo or in vitro. Furthermore, Hatanaka, Otten, and Thoenen (1978) found no affect of NGF on cAMP or cGMP levels in PC12 cultures. In each of the above studies, the times examined were comparable to those in which positive effects of NGF on cAMP were reported. (2) Frazier et al. (1973*b*) reported that while Db cAMP caused neurite outgrowth from cultured embryonic chick sensory ganglia, it did not stimulate outgrowth from cultured chick embryo sympathetic ganglia. Also, Db cAMP (in a wide range of tested concentrations) fails to support the survival of neurons in dissociated cell cultures of chick sympathetic ganglia (Greene, unpublished). Furthermore, as shown in Figure 1, we find that while treatment of PC12 cells with Db cAMP does elicit morphological changes, such responses are not comparable with those caused by NGF. (3) Hier et al. (1973) found that chick embryo sensory ganglia treated with Db cAMP did not accumulate colchicine-binding material (presumably tubulin) during 24 hours of culture. This led to the suggestion that the outgrowth caused by Db cAMP was the result of direct effects on polymerization of preformed tubulin. In contrast, NGF-treated ganglia were found to accumulate colchicine-binding material during this time period. This led to the suggestion that NGF and cAMP elicit neurite outgrowth by different mechanisms. (4) It must be considered that even if data concerning cAMP increases and the effects of Db cAMP were reproducible, they would still not conclusively prove a cAMP mechanism for NGF, since increased cAMP, like increased ODC, could be without further causal consequences. Also, it is conceivable that Db cAMP could indirectly trigger an NGF-like response even though cAMP normally plays no physiological role in the response. In summary, it appears that the arguments supporting a role for cAMP in NGF's mechanism of action are far from being conclusive. However, even though cAMP does not appear to be sufficient to moderate *all* of NGF's actions, a role for cAMP in *some* of these actions has not been thoroughly ruled out.

Ion Flux

It has been suggested that the mechanism of action of NGF could include regulation of ion flux in responsive cells (Shelanski, 1973). Recently, Schubert et al. (1978) have proposed that NGF treatment leads to the mobilization of Ca^{2+} ions, which in turn leads to neurite outgrowth. Their evidence for this was that NGF treatment enhanced Ca^{2+} efflux from PC12 cells and that exposure to conditions that caused increased Ca^{2+} influx (such as depolarization with elevated K^+) promoted neurite outgrowth. However, Cohen and Shooter (personal communication), also working with PC12 cells, found no effect of NGF on Ca^{2+} flux. Also, in our laboratory and in another (U. Otten, personal communication), exposure of PC12 cells to elevated K^+ did not cause initiation of neurite outgrowth. Furthermore, elevated K^+ fails to support neurons in dissociated cell cultures of chick sympathetic ganglia (Greene, unpublished). Thus, evidence for a role for Ca^{2+} mobilization in the mechanism of action of NGF has not been confirmed and appears, on the contrary, to be negative.

Adhesion to Substrate

Varon (1975) has proposed that a primary action of NGF, particularly with respect to neurite outgrowth, might be to alter membrane adhesiveness via effects on the plasma membrane. In this vein, it was recently reported (Schubert and Whitlock, 1977; Schubert et al., 1978) that NGF treatment caused rapid (detectable within 10 minutes of exposure) changes in the adhesion of PC12 cells to one another and to plastic tissue culture dishes, as well as increased lectin agglutinability. It was theorized that such changes in turn induce neurite outgrowth.

Since neurite outgrowth does not, of course, occur in the absence of attachment to a substrate, these proposals are difficult to evaluate. One would like to know whether NGF causes specific changes in adhesion of its target cells, whether the changes are required for neurite outgrowth, and whether the changes are sufficient for neurite outgrowth. Nevertheless, certain critical points can be raised. First, although Schubert and Whitlock (1977) have described increases in adhesion of PC12 cells to tissue culture plastic occurring within 10 minutes of NGF exposure, we find that by 24 hours of NGF treatment, the PC12 cells show very poor attachment to tissue culture plastic and tend to detach. Normal sympathetic neurons (grown in presence of NGF) also show poor adhesion to tissue culture plastic (Varon and Raiborn, 1972). (Such findings account for the use of collagen-coated substrates to culture sympathetic neurons and NGF-treated PC12 cells.) Second, plating of PC12 cells in the absence of NGF on substrates to which they have high adhesion, such as concentrated collagen or polylysine, does not induce neurite outgrowth. Third, despite the rapidly occurring changes in adhesion, induction of neurite outgrowth by NGF does not commence before a lag of at least 24 hours (Greene and Tischler, 1976; Burstein and Greene, 1978). Thus, altered adhesion to substrate, while an attractive possibility for one aspect of NGF's mechanism of action, has yet to be convincingly demonstrated or critically tested.

CONCLUSIONS

(1) There is evidence for multiple means of interaction of NGF with its target cells. These include plasma membrane receptors, internalization, retrograde transport, and nuclear receptors. The role of each of these in the mechanism(s) of action of NGF remains to be determined. However, given that NGF has multiple actions and has both transcriptional and nontranscriptional effects, it is highly plausible that triggering of any given response to NGF requires interaction of the factor with several different types of sites, and/or that different responses to NGF may each be mediated via different sites of interaction. (2) The mechanism of action of NGF requires two separable pathways; one of these requires RNA synthesis, the other does not. (3) The effect of NGF on survival has a nontranscriptional component. A transcriptional requirement for survival is not yet clear, but can at least be separated from that required for neurite outgrowth. (4) NGF induces neurite outgrowth without bringing about major changes in protein composition. Thus, the effects of NGF on outgrowth of neurites take place in large part via rearrangement of the types of molecules already synthesized prior to treatment with the factor. (5) NGF treatment does cause several quantitative changes in the relative levels of several cell polypeptides. These or other polypeptides of low abundance may play required regulatory or structural roles in neurite outgrowth. (6) A number of proposed models for steps in the mechanism of action of NGF have been evaluated. While several of these are attractive, none has yet been conclusively demonstrated.

ACKNOWLEDGMENTS

This work was supported by United States Public Health Service grants NS11557 and NS12200, the National Foundation-March of Dimes, and the Sloan Foundation.

REFERENCES

Andres, R. Y., I. Jeng, and R. A. Bradshaw (1977). Nerve growth factor receptors: identification of distinct classes in plasma membranes and nuclei of embryonic dorsal root neurons, *Proc. Natl. Acad. Sci. USA* **74**:2672–2676.

Angeletti, P. U., D. Gandini-Attardi, G. Toschi, M. I. Salvi, and R. Levi-Montalcini (1965). Metabolic aspects of the effect of nerve growth factor on sympathetic and sensory ganglia: protein and ribonucleic acid synthesis, *Biochim. Biophys. Acta* **95**:111–120.

Banerjee, S. P., S. H. Snyder, P. Cuatrecasas, and L. A. Greene (1973). Nerve growth factor receptor binding in sympathetic ganglia, *Proc. Natl. Acad. Sci. USA* **70**:2519–2523.

Bradshaw, R. A. (1978). Nerve growth factor, *Annu. Rev. Biochem.* **47**:191–216.

Burnham, P. and S. Varon (1973). In vitro uptake of active nerve growth factor by dorsal root ganglia of embryonic chick, *Neurobiology (Copenh.)* **3**:232–245.

Burstein, D. E. and L. A. Greene (1978). Evidence for both RNA-synthesis-dependent and -independent pathways in stimulation of neurite outgrowth by nerve growth factor, *Proc. Natl. Acad. Sci. USA* **75**:6059–6063.

Calissano, P. and C. Cozzari (1974). Interaction of nerve growth factor with mouse-brain neurotubule protein(s), *Proc. Natl. Acad. Sci. USA* **71**:2131–2135.

Calissano, P., G. Monaco, L. Castellani, D. Mercanti, and A. Levi (1978). The nerve growth factor potentiates actomyosin ATPase, *Proc. Natl. Acad. Sci. USA* **75**:2210–2214.

Campenot, R. B. (1977). Local control of neurite development by nerve growth factor, *Proc. Natl. Acad. Sci. USA* **74**:4516–4519.

Carpenter, G. and S. Cohen (1976). ^{125}I-Labeled human epidermal growth factor: binding, internalization and degradation in human fibroblasts, *J. Cell Biol.* **71**:159–171.

Dichter, M. A., A. S. Tischler, and L. A. Greene (1977). Nerve growth factor-induced change in electrical excitability and acetylcholine sensitivity of a rat pheochromocytoma cell line, *Nature (London)* **268**:501–504.

Frazier, W. A., L. F. Boyd, and R. A. Bradshaw (1973*a*). Interaction of nerve growth factor with surface membranes: biological competence of insolubilized nerve growth factor, *Proc. Natl. Acad. Sci. USA* **70**:2931–2935.

Frazier, W. A., L. F. Boyd, and R. A. Bradshaw (1974). Properties of specific binding of ^{125}I-nerve growth factor to responsive peripheral nerves, *J. Biol. Chem.* **249**:5513–5519.

Frazier, W. A., C. E. Ohlendorf, L. F. Boyd, L. Aloe, E. M. Johnson, J. A. Ferrendelli, and R. A. Bradshaw (1973*b*). Mechanism of action of nerve growth factor and cyclic AMP on neurite outgrowth in chick embryonic sensory ganglia: demonstration of independent pathways of stimulation, *Proc. Natl. Acad. Sci. USA* **70:**2448–2452.

Greene, L. A. (1977). A quantitative bioassay for nerve growth factor (NGF) activity employing a clonal pheochromocytoma cell line, *Brain Res.* **133:** 350–353.

Greene, L. A. (1978). Nerve growth factor prevents the death and stimulates neuronal differentiation of clonal PC12 pheochromocytoma cells in serum-free medium, *J. Cell Biol.* **78:**747–755.

Greene, L. A. and J. C. McGuire (1978). Induction of ornithine decarboxylase by nerve growth factor dissociated from effects on survival and neurite outgrowth, *Nature (London)* **276:**191–194.

Greene, L. A. and G. Rein (1977*a*). Release, storage and uptake of catecholamines by a clonal cell line of nerve growth factor (NGF) responsive pheochromocytoma cells, *Brain Res.* **129:**247–263.

Greene, L. A. and G. Rein (1977*b*). Release of ^{3}H-norepinephrine from a clonal line of pheochromocytoma cells (PC12) by nicotinic cholinergic stimulation, *Brain Res.* **138:**521–528.

Greene, L. A. and G. Rein (1978). Short-term regulation of catecholamine synthesis in an NGF-responsive clonal line of rat pheochromocytoma cells, *J. Neurochem.* **30:**549–555.

Greene, L. A. and A. S. Tischler (1976). Establishment of a noradrenergic clonal line of rat adrenal pheochromocytoma cells which respond to nerve growth factor, *Proc. Natl. Acad. Sci. USA* **73:**2424–2428.

Hatanaka, U., U. Otten, and H. Thoenen (1978). Nerve growth factor-mediated selective induction of ornithine decarboxylase in rat pheochromocytoma: a cyclic AMP-independent process, *FEBS Lett.* **92:**313–316.

Heidemann, S. R., G. Sander, and M. W. Kirschner (1977). Evidence for a functional role of RNA in centrioles, *Cell* **10:**337–350.

Hendry, I. A., K. Stöckel, H. Thoenen, and L. L. Iversen (1974). The retrograde axonal transport of nerve growth factor, *Brain Res.* **68:**103–121.

Herrup, K. and E. M. Shooter (1973). Properties of the β nerve growth factor receptor of avian dorsal root ganglia, *Proc. Natl. Acad. Sci. USA* **70:**3884–3888.

Hier, D. B., B. G. Arnason, and M. Young (1973). Nerve growth factor: relationship to the cyclic AMP system of sensory ganglia, *Science* **182:**79–81.

Horii, Z. I. and S. Varon (1977). Nerve growth factor on membrane permeation to exogenous substrates in dorsal root ganglionic dissociates from the chick embryo, *Brain Res.* **124:**121–123.

Horwitz, S. B. (1975). Camptothecin, pp. 649–656 in *Handbook of Experimental Pharmacology,* Vol. 28, Part 2, Sartorelli, A. C. and D. G. Johns, eds. Springer-Verlag, Berlin.

Kahn, C. R. and K. Baird (1978). The fate of insulin bound to adipocytes, *J. Biol. Chem.* **253:**4900–4906.

Lee, J. C., N. Tweedy, and S. Timasheff (1978). In vitro reconstitution of calf brain microtubules: effects of macromolecules, *Biochemistry* **17:**2783–2790.

Levi, A., M. Cimino, D. Mercanti, J. S. Chen, and P. Calissano (1975). Interaction of nerve growth factor with tubulin. Studies on binding and induced polymerization, *Biochim. Biophys. Acta* **399:**50–60.

Levi-Montalcini, R. (1976). The nerve growth factor: its role in growth, differentiation and function of the sympathetic adrenergic neuron, *Prog. Brain Res.* **45**:235–258.

Levi-Montalcini, R. and P. U. Angeletti (1963). Essential role of the nerve growth factor in the survival and maintenance of dissociated sensory and sympathetic embryonic nerve cells in vitro, *Dev. Biol.* **7**:653–659.

Levi-Montalcini, R. and P. U. Angeletti (1968). Nerve growth factor, *Physiol. Rev.* **48**:534–569.

Lewis, M. E., J. Lakshmanan, K. Nagaiah, P. C. MacDonnell, and G. Guroff (1978). Nerve growth factor increases activity of ornithine decarboxylase in rat brain, *Proc. Natl. Acad. Sci. USA* **75**:1021–1023.

MacDonnell, P. C., K. Nagaiah, J. Lakshmanan, and G. Guroff (1977). Nerve growth factor increases activity of ornithine decarboxylase in superior cervical ganglion of young rats, *Proc. Natl. Acad. Sci. USA* **74**:4681–4684.

McGuire, J. C. and L. A. Greene (1979). Rapid stimulation by nerve growth factor of amino acid uptake by clonal PC12 pheochromocytoma cells, *J. Biol. Chem.*, in press.

McGuire, J. C., L. A. Greene, and A. V. Furano (1978). NGF stimulates incorporation of fucose or glucosamine into an external glycoprotein in cultured rat PC12 pheochromocytoma cells, *Cell* **15**:357–365.

Mobley, W. C., A. C. Server, D. N. Ishii, R. J. Riopelle, and E. M. Shooter (1977). Nerve growth factor, *N. Engl. J. Med.* **297**:1096–1104, 1149–1158, 1211–1217.

Nikodijevic, B., O. Nikodijevic, W. M.-Y. Yu, H. Pollard, and G. Guroff (1975). The effect of nerve growth factor on cyclic AMP levels in superior cervical ganglia of the rat, *Proc. Natl. Acad. Sci. USA* **72**:4769–4771.

Otten, U., H. Hatanaka, and H. Thoenen (1978). Role of cyclic nucleotides in NGF-mediated induction of tyrosine hydroxylase in rat sympathetic ganglia and adrenal medulla, *Brain Res.* **140**:385–389.

Paravicini, U., K. Stöckel, and H. Thoenen (1975). Biological importance of retrograde axonal transport of nerve growth factor in adrenergic neurons, *Brain Res.* **84**:279–291.

Partlow, L. M. and M. E. Larrabee (1971). Effects of a nerve growth factor, embryo age and metabolic inhibitors on growth of fibers and on synthesis of ribonucleic acid and protein in embryonic sympathetic ganglia, *J. Neurochem.* **18**:2101–2118.

Roger, L. J., S. M. Schanberg, and R. F. Fellows (1974). Growth and lactogenic hormone stimulation of ornithine decarboxylase in fetal rat brain, *Endocrinology* **95**:904–911.

Roisen, F. J., R. A. Murphy, and W. G. Braden (1972). Neurite development in vitro. I. The effects of adenosine 3′,5′-monophosphate (cyclic AMP), *J. Neurobiol.* **3**:347–368.

Russell, D. H., C. V. Byus, and C.-A. Manen (1976). Proposed model of major sequential biochemical events of a trophic response, *Life Sci.* **19**:1297–1306.

Schubert, D., M. LaCorbiere, C. Whitlock, and W. Stallcup (1978). Alterations in the surface properties of cells responsive to nerve growth factor, *Nature (London)* **273**:718–723.

Schubert, D. and C. Whitlock (1977). Alteration of cellular adhesion by nerve growth factor, *Proc. Natl. Acad. Sci. USA* **74**:4055–4058.
Shelanski, M. (1973). Chemistry of the filaments and tubules of brain, *J. Histochem. Cytochem.* **21**:529–539.
Tischler, A. S. and L. A. Greene (1978). Morphological and cytochemical properties of a clonal line of rat adrenal pheochromocytoma cells which respond to nerve growth factor, *Lab. Invest.* **39**:77–89.
Varon, S. (1975). Nerve growth factor and its mode of action, *Exp. Neurol.* **48**(2):75–92.
Varon, S. and C. Raiborn (1972). Dissociation, fractionation and culture of chick embryo sympathetic ganglionic cells, *J. Neurocytol.* **1**:211–221.
Warren, S. and R. Chute (1972). Pheochromocytoma, *Cancer* **29**:327–331.
Yankner, B. A. and E. M. Shooter (1979). Nerve growth factor in the nucleus: interaction with receptors on the nuclear membrane, *Proc. Natl. Acad. Sci. USA* **76**:1269–1273.

ENVIRONMENTAL DETERMINATION OF NEUROTRANSMITTER FUNCTIONS

Paul H. Patterson

Harvard Medical School, Boston, Massachusetts

It is clear that many steps in neuronal development can be influenced by interactions between neurons and their embryonic environment. One such step is the neuronal decision as to which type of neurotransmitter to produce and which type of synapse to form. This article will briefly discuss recent work on the role of cellular interactions in the selection of transmitter type in a number of neuronal systems, both in vitro and in vivo. For more detailed reviews of this subject see Patterson, 1978, and Bunge, Johnson, and Ross, 1978.

An interesting system to study in terms of control of the choice of phenotype is the neural crest. The crest is a transient embryonic structure consisting of an apparently homogeneous group of cells lying on the dorsal part of the neural tube. The crest cells migrate to a great many different sites in the embryo and give rise to a variety of neuronal as well as nonneuronal progeny (Weston, 1970; Johnston, Bhakdinaronk, and Reid, 1974). A number of techniques have been used to show that certain derivatives arise from distinct axial levels of the crest. Of particular relevance in the present context is the observation that the adrenergic (catecholamine-containing) cells of the sympathetic ganglia and adrenal medulla arise primarily from the lumbar region of the crest, whereas the cholinergic (acetylcholine-containing) neurons of the gut arise from the more rostral and more caudal levels (Le Douarin, 1977). These findings raise the question whether the crest cells are irreversibly predetermined at each axial level as to their direction of migration and final differentiated phenotype or whether they are a homogeneous group of naive cells dependent on the embryonic structures at each axial level to direct their migratory and developmental fates.

That the crest population at each axial level is not, in fact, irreversibly predetermined has been shown by recent work of Le Douarin and colleagues. To demonstrate this, they grafted neural tube plus the associated crest from one axial level of a quail embryo to a different axial level of a chick embryo (Le Douarin, 1969). Individual cells in the graft can be recognized at later stages because the nucleoli of the grafted quail cells can be distinguished from the nucleoli of the surrounding chick cells in histological sections. One type of experiment involved transplanting the more rostral crest cells (which would have given rise, among other things, to cholinergic neurons of the gut) from a quail embryo into a chick embryo at a lumbar axial level (which normally gives rise to adrenergic progeny). When this chimeric embryo developed further, grafted rostral crest cells were found to populate sympathetic ganglia and the adrenal medulla. Furthermore, the cells from the graft were found to contain catecholamines (CA), as determined by the formaldehyde-induced fluorescence technique (Le Douarin and Teillet, 1974). Thus some cells in the presumptive cholinergic population, when placed at a new axial level, migrated to sites and developed a phenotype typical of that axial level. The reverse type of experiment gave the same result: presumptive adrenergic crest cells from the lumbar region, when transplanted to a more rostral level, migrated to the gut and displayed apparent cholinergic properties (Le Douarin, Renaud, Teillet, and Le Douarin, 1975). Again, the choice of adrenergic vs. cholinergic phenotype corresponded to the environment in which the crest cells were placed rather than to that from which they arose. While these experiments show that the transmitter choice of a population as a whole is not completely and irreversibly predetermined before the cells have migrated from the crest, it remains possible that a crest population contains subgroups of cells, each capable of expressing only one developmental fate. Such subgroups could then be selected for survival by the environment.

What is the origin of the environmental factors that influence the differentiated fate of neural crest cells? Early experiments of Cohen (1972), which were confirmed and extended by Norr (1973), suggested that the ventral structures past which crest cells migrate can promote adrenergic expression. For instance, when crest was cultured with somitic mesenchyme and ventral neural tube, the number of cells showing catecholamine fluorescence was increased over controls of crest with either neural tube, somite, or nonsomitic mesenchyme alone. Teillet, Cochard, and Le Douarin (1978) have suggested that the notochord may also be a source of this adrenergic signal during migration. In apparent contradiction of these findings, Cohen (1977) and Greenberg and Schrier (1977) have reported that neural crest cells can express adrenergic or cholinergic functions in cell culture without the continued presence of somite or neural tube. It will be of interest to see if lumbar and rostral crest can each give rise to cholinergic and adrenergic progeny under these low-density culture conditions and whether neural tube, mesenchyme, etc., can influence the outcome. It will also be important to determine if the serum in the culture medium contains differentiation signals like those suggested to be produced by neural tube-somite combination. Purification and characterization of these putative "factors" is obviously an important, if very difficult, goal.

If there is an early adrenergic influence operating on migrating crest cells, then how do cholinergic parasympathetic cells develop? Two possibilities are: (1) The environment through which the presumptive cholinergic population migrates is different; it does not exert the early adrenergic influence. For instance, precursors of the cholinergic neurons of the ciliary ganglion do not migrate past the notochord. Or perhaps the somites in the rostral region are different from those at the lumbar level. (2) Alternatively, all autonomic precursor cells may receive an early adrenergic cue that is reversed when the cells arrive in a cholinergic target tissue such as the gut. Consistent with this notion is the recent finding of adrenergic neurons in ganglia in the very early gut (Cochard, Goldstein, and Black, 1978). The fluorescent cells disappear during further development; whether these neurons are converted to cholinergic neurons by their local environment or whether they simply die remains to be determined.

That a population of crest derivatives can still reverse its phenotypic choice after it has arrived at its normal destination has been shown in vivo by Le Douarin, Teillet, Ziller, and Smith (1978). They have taken early cholinergic ganglia (ciliary and Remak's), after expression of cholinergic properties has begun, and "back-transplanted" them into the lumbar region of the crest. The cells in these ganglia disaggregate and undergo a second migration. Some of the transplanted cells reach sympathetic ganglia and the adrenal medulla and display CA fluorescence. It is not clear, however, whether any of these adrenergic cells previously produced choline acetyltransferase (CAT), the cholinergic marker enzyme. That is, did the environment determine the developmental fate of the population by instructing single cells which transmitter to produce, or did it act by selecting a predetermined cell type for survival, while those with the inappropriate transmitter simply died?

Studies under the more defined conditions of cell culture have shown that the fluid and cellular environment surrounding sympathetic neurons can indeed determine the type of transmitter produced and type of synapse formed by these neural crest derivatives. It is consistent with the idea of an early adrenergic signal that all of these neurons taken from the newborn rat superior cervical ganglion display adrenergic properties in the first few days in culture (Johnson, Ross, Meyers, Rees, Bunge, Wakshull, and Burton, 1976; Mains and Patterson, 1973*c*). Furthermore, virtually all of the neurons in the ganglion show CA fluorescence at birth (Eränkö, 1972). In fact, sympathetic neuroblasts can produce CA even before ceasing mitosis (Cohen, 1974; Rothman, Gershon, and Holtzer, 1978). Under appropriate culture conditions, the sympathetic neurons will continue to develop adrenergically along a time course qualitatively similar to that seen in vivo (Mains and Patterson, 1973*c*; Patterson and Chun, 1977*b*). The neurons can synthesize, store, release, and take up CA, and can form synapses with each other of adrenergic morphology; they produce little acetylcholine (ACh) (Mains and Patterson, 1973*a,b*; Rees and Bunge, 1974; Burton and Bunge, 1975; Patterson, Reichardt, and Chun, 1975; Landis, MacLeish, Potter, Furshpan, and Patterson, 1976; Patterson and Chun, 1977*a*).

However, if the neurons are cocultured with appropriate types of nonneuronal cells or are grown in culture medium that has been incubated on nonneuronal cells in a separate dish (conditioned medium, CM), the adrenergic development ceases and cholinergic properties appear (Johnson et al., 1976; Patterson and Chun, 1977*b*). These neurons produce little CA, but considerable ACh; they form functional cholinergic synapses with each other or with other targets such as skeletal or heart muscle (O'Lague, Obata, Claude, Furshpan, and Potter, 1974; Patterson and Chun, 1974; Patterson et al., 1975; Nurse and O'Lague, 1975; O'Lague, MacLeish, Nurse, Claude, Furshpan, and Potter, 1975; Furshpan, MacLeish, O'Lague, and Potter, 1976; Landis, 1976; Johnson et al., 1976; Ko, Burton, Johnson, and Bunge, 1976; Patterson and Chun, 1977*a,b*). The diffusable cholinergic signal in CM can cause these dramatic changes in neuronal development without altering neuronal survival or growth (Patterson and Chun, 1977*a*). This is especially striking because single-cell experiments have shown that the transmitter decision of most if not all of the neurons is influenced in this manner by the culture environment. It is possible to grow single neurons in microcultures containing various concentrations of CM or nonneuronal cells (Reichardt, Patterson, and Chun, 1976). As many as 80–90% of individual neurons grown on heart or skeletal muscle are cholinergic after 4 or more weeks in culture, while as few as 0% are cholinergic under "control" conditions (Reichardt and Patterson, 1977; Nurse, 1977). At intermediate CM concentrations, there are similar numbers of adrenergic and cholinergic neurons (Reichardt and Patterson, 1977). The observation that the population can be shifted from all (or virtually all) adrenergic to virtually all cholinergic, coupled with the finding that CM does not alter neuronal survival, argues that individual sympathetic neurons, even after aggregating as a ganglion and beginning adrenergic development, still have the capacity to become functional cholinergic neurons. Thus the early adrenergic signal is indeed reversible.

The biochemical studies on mature single neurons further showed that virtually all of the neurons "chose" one transmitter or the other in an apparent "flip-flop" manner; very few cells produced neither of the two transmitters or detectable quantities of both (Reichardt and Patterson, 1977). However, electrophysiological (Furshpan et al., 1976) and electron microscopic (Landis, 1976) studies on immature single neurons have described three classes of neurons grown on heart cells: (1) neurons that formed hexamethonium-sensitive cholinergic synapses (with no small granular vesicles) on their own somas and elicited atropine-sensitive cholinergic responses in the heart cells; (2) neurons that elicited propanolol-sensitive adrenergic responses in the heart cells and had neuronal varicosities containing high proportions of small granular vesicles; and (3) neurons that were "dual-function," i.e., that elicited first a cholinergic response and then an adrenergic response in beating myocytes, and that had varicosities containing only occasional small granular vesicles. The difference between the electrophysiological and biochemical results with respect to dual-function neurons may reflect an inability of the biochemical methods to detect a low level of one of the two transmitters in cells that were actually bifunctional. Alternatively, the young dual-function neurons detected electrophysiologically may have been passing through a transient stage that leads to one of the two differentiated states observed with biochemical methods in the older cells. A priori, it should not be surprising to find such a stage, since, as previously described, virtually all of the neurons are adrenergic to begin with and virtually all can be made to become cholinergic.

Given the importance of nerve growth factor (NGF) in the development of sympathetic neurons, it is relevant to inquire whether the CM factor can be distinguished from NGF. The known physical properties of the CM factor appear to be different from NGF. The active entity appears to be a basic macromolecule with an apparent molecular weight on Sephadex chromatography of about 50,000. It is highly sensitive to periodate treatment, suggesting that carbohydrate residues may be important for the activity (Weber and Patterson, unpublished). It is also possible to distinguish clearly these two factors on functional grounds. NGF is necessary for sympathetic neuron survival and stimulates growth and adrenergic differentiation (e.g., Chun and Patterson, 1977*a,b*). Perhaps surprisingly, NGF also stimulates cholinergic differentiation in sympathetic neurons grown on heart cells or in CM (Chun and Patterson, 1977*c*; see also Hill and Hendry, 1977). In fact, NGF stimulates the production of both transmitters to the same extent. That is, the ratio of ACh to CA, which is a reflection of the relative rates of synthesis of these transmitters, remains constant over a wide range of NGF concentrations for both adrenergic (ACh/CA = 0.02) and cholinergic (ACh/CA = 200) cultures (Chun and Patterson, 1977*c*). Thus, with respect to transmitter production, NGF is permissive rather than instructive, in that it is necessary for survival and stimulates growth and differentiation but does not tell the neuron which transmitter to produce. Therefore NGF and CM are qualitatively different developmental signals: as previously described, CM does not affect neuronal survival or growth, but does tell the neurons which type of transmitter and synapse to produce. These findings also argue against NGF being the early signal for adrenergic determination, although NGF is obviously necessary for further adrenergic development (see also Black, Coughlin, and Cochard, this volume).

Is this environmental influence on transmitter determination confined to the autonomic nervous system, or are there examples in other areas as well? Progress with cultures of spinal cord and dorsal root ganglia is promising in this respect. Mouse spinal cord cells display some CAT activity in monolayer cultures, but this can be significantly increased by either coculture with skeletal muscle cells or by growth in CM from various nonneuronal cultures (Giller, Schrier, Shainberg, Fisk, and Nelson, 1973; Giller, Neale, Bullock, Schrier, and Nelson, 1977; Godfrey, Schrier, and Nelson, unpublished). These findings raise interesting questions. For instance, is there selection of cholinergic neurons for survival? The nonneuronal cells and CM increase overall neuronal survival, but survival in the cholinergic subpopulation is really what needs to be determined. Does CM stimulate the growth and differentiation of the cholinergic population, as NGF does with sympathetic neurons, or does CM affect the decision of spinal cord neurons as to which transmitter to produce, as in the case of cultured sympathetic neurons?

These considerations raise the further question of the similarity of the CM factors in the spinal cord and sympathetic systems. The only physical comparison that can now be made is that both agents appear to be macromolecules. However, there is a physiological similarity in terms of types of nonneuronal cells that can produce the effect. In both systems, skeletal and heart muscle cells and fibroblasts are effective, whereas liver is relatively ineffective. This observation suggests a role of target organs in transmitter determination, since heart and skeletal muscle receive cholinergic innervation, whereas liver cells receive only adrenergic innervation. However, this simple idea does not fit easily with the observation that ganglionic nonneuronal cells, gliomas, and fibroblasts can all condition the medium effectively (Patterson and Chun, 1974, 1977*a*). Another similarity between the sympathetic and spinal cord systems is that both show species specificity: CM produced by chick cells is not effective on rat or mouse neurons (Patterson and Chun, 1977*a*; Giller et al., 1977; Godfrey, Schrier, and Nelson, unpublished).

Another aspect of the specificity of the agents concerns the other types of neurons affected. Godfrey, Schrier, and Nelson (unpublished) have found that muscle CM is effective in raising CAT activity in cultures of spinal cord and medulla but does not raise CAT in cultures of midbrain and cortex, even though all four brain areas display a basal CAT activity in culture. These observations may correspond to the work of Obata (1977), who found that while spinal cord and medulla cultures formed cholinergic synapses with muscle, cerebellar and cerebral neurons did not. Thus the neurons that can form synapses with muscle may be the ones that can respond to CM.

Diffusible factors from nonneuronal cells may also play a role in the choice of peptides produced by sensory neurons in culture. Neuron-alone cultures from chick dorsal root ganglia produce considerably more substance P than somatostatin, whereas neurons cocultured with ganglionic nonneuronal cells produce much more somatostatin and less substance P (Mudge, unpublished). Thus the nonneuronal cells can cause more than a 50-fold reciprocal change in peptide production. This change is not accompanied by changes in overall neuronal survival and can be mimicked by CM addition to neuron-alone cultures. It is clearly of interest to know whether single sensory neurons can produce either peptide (or both), depending on the environment in which they develop. It would also be interesting to investigate other areas where more than one transmitter is present in the same ganglion or layer, such as the myenteric plexus and the retina.

Thus far the discussion has centered on the potential role of the migratory pathway and the nonneuronal cells in ganglia or target tissue in transmitter determination. What about the role of neuronal activity imposed by synaptic connections made onto a neuron? As discussed by Black et al. in this volume, synaptic input to sympathetic neurons in the superior cervical ganglion is necessary for full adrenergic maturation. Is neuronal activity important not only for the full expression of transmitter differentiation once the choice has been made, but also in making the choice of transmitters in the first place? The role of activity in the adrenergic-cholinergic decision by sympathetic neurons in culture has recently been examined (Walicke, Campenot, and Patterson, 1977). Chronic depolarization of these neurons by treatment with elevated K^+ or veratridine lowers ACh/CA ratios as much as 300-fold after several weeks in culture. These results with chronic depolarization have also been duplicated by stimulating the neurons electrically at physiological frequencies (1 action potential/sec). The primary effect of depolarization is an inhibition of cholinergic development rather than an enhancement of adrenergic development (over normal adrenergic controls). Thus the initial adrenergic choice made by these neurons (see above) appears to be stabilized by activity and the neurons are less susceptible to the cholinergic signal, CM. It may be that activity could also have stabilized an initial cholinergic choice, if one had been made. In this regard, it will be of interest to study the influence of activity in other transmitter choices as well. The influence of depolarization on the adrenergic-cholinergic decision appears to involve Ca^{2+}, and cyclic AMP may also play a role (Walicke et al., 1977; Walicke and Patterson, 1978).

Thus synaptic input on a neuron could be one of the factors regulating the choice of transmitter. This is particularly interesting in view of the fact that these neurons normally make and receive specific synaptic connections with other cells. Such specific circuitry not only determines the function of the nervous system, but may also specify during development the chemical nature of the elements in that pattern. If neuronal cell death represents a mechanism for ensuring proper connectivity in the nervous system, then the influences of nonneuronal cells and synaptic input described here may provide mechanisms to ensure that the connected elements form the chemically appropriate types of synapses.

REFERENCES

Black, I. B., M. D. Coughlin, and P. Cochard (1979). Factors regulating neuronal differentiation, *Soc. Neurosci. Symp.* **4:**184–207.

Bunge, R., M. Johnson, and C. D. Ross (1978). Nature and nurture in development of the autonomic neuron, *Science* **199:**1409–1416.

Burton, H. and R. P. Bunge (1975). A comparison of the uptake and release of [^{3}H]norepinephrine in rat autonomic and sensory ganglia in tissue culture, *Brain Res.* **97:**157–162.

Chun, L. L. Y. and P. H. Patterson (1977*a*). The role of nerve growth factor in the development of rat sympathetic neurons in vitro. I. Survival, growth and differentiation of catecholamine production, *J. Cell Biol.* **75:**694–704.

Chun, L. L. Y. and P. H. Patterson (1977*b*). The role of nerve growth factor in the development of rat sympathetic neurons in vitro. II. Developmental studies, *J. Cell Biol.* **75:**705–711.

Chun, L. L. Y. and P. H. Patterson (1977*c*). The role of nerve growth factor in the development of rat sympathetic neurons in vitro. III. Effect on acetylcholine production, *J. Cell Biol.* **75:**712–718.

Cochard, P., M. Goldstein, and I. Black (1978). Ontogenetic appearance and disappearance of tyrosine hydroxylase and catecholamines in the rat embryo, *Proc. Natl. Acad. Sci. USA* **75:**2986–2990.

Cohen, A. M. (1972). Factors directing the expression of sympathetic nerve traits in cells of neural crest origin, *J. Exp. Zool.* **179:**167–182.

Cohen, A. M. (1974). DNA synthesis and cell division in differentiating avain adrenergic neuroblasts, pp. 359–370 in *Dynamics of Degeneration and Growth in Neurons,* Fuxe, K., L. Olson, and Y. Zotterman, eds. Pergamon Press, New York.

Cohen, A. M. (1977). Independent expression of the adrenergic phenotype by neural crest cells in vitro, *Proc. Natl. Acad. Sci. USA* **74:**2899–2903.

Eränkö, L. (1972). Postnatal development of histochemically demonstrable catecholamines in the superior cervical ganglion of the rat, *Histochem. J.* **4:**225–236.

Furshpan, E. J., P. R. MacLeish, P. H. O'Lague, and D. D. Potter (1976). Chemical transmission between rat sympathetic neurons and cardiac myocytes developing in microcultures: evidence for cholinergic, adrenergic, and dual-function neurons, *Proc. Natl. Acad. Sci. USA* **73:**4225–4229.

Giller, E. L., J. H. Neale, P. N. Bullock, B. Schrier, and P. G. Nelson (1977). Choline acetyltransferase activity of spinal cord cell cultures increased by co-culture with muscle and by muscle-conditioned medium, *J. Cell Biol.* **74:**16–29.

Giller, E. L., B. K. Schrier, A. Shainberg, H. R. Fisk, and P. G. Nelson (1973). Choline acetyltransferase activity is increased in combined cultures of spinal cord and muscle cells from mice, *Science* **182**:588–589.

Greenberg, J. and B. Schrier (1977). Development of choline acetyltransferase activity in chick cranial neural crest cells in culture, *Dev. Biol.* **61**:86–93.

Hill, C. E. and I. A. Hendry (1977). Development of neurons synthesizing noradrenaline and acetylcholine in the superior cervical ganglion of the rat in vivo and in vitro, *Neuroscience* **2**:741–749.

Johnson, M., D. Ross, M. Meyers, R. Rees, R. Bunge, E. Wakshull, and H. Burton (1976). Synaptic vesicle cytochemistry changes when cultured sympathetic neurons develop cholinergic interactions, *Nature (London)* **262**: 308–310.

Johnston, M. C., A. Bhakdinaronk, and Y. C. Reid (1974). An expanded role of the neural crest in oral and pharyngeal development, pp. 37–52 in *4th Symposium on Oral Sensation and Perception,* Bosma, J. F., ed. Fogarty International Center Proceedings No. 21. U.S. Government Printing Office, Washington, D.C.

Ko, C.-P., H. Burton, M. I. Johnson, and R. P. Bunge (1976). Synaptic transmission between rat superior cervical ganglion neurons in dissociated cell cultures, *Brain Res.* **117**:461–485.

Landis, S. C. (1976). Rat sympathetic neurons and cardiac myocytes developing in microcultures: correlation of the fine structure of endings with neurotransmitter function in single neurons, *Proc. Natl. Acad. Sci. USA* **73**:4220–4224.

Landis, S. C., P. R. MacLeish, D. D. Potter, E. J. Furshpan, and P. H. Patterson (1976). Synapses formed between dissociated neurons: the influence of conditioned medium. *Soc. Neurosci. Abstr.*, Vol. 2, p. 197.

Le Douarin, N. (1969). Particularités du noyau interphasique chez la Caille japonaise (*Coturnix coturnix* japonica). Utilisation de ces particularités comme "marquage biologique" dans des recherches sur les interactions tissulaires et les migrations cellulaires au cours de l'ontogènese, *Bull. Biol. Fr. Belg.* **103**:435–452.

Le Douarin, N. M. (1977). The differentiation of the ganglioblasts of the autonomic nervous system studied in chimeric avian embryos, pp. 171–190 in *Cell Interactions in Differentiation,* Karkinen-Jääskelëinen, M., L. Saxén, and L. Weiss, eds. Academic Press, New York.

Le Douarin, N. M., D. Renaud, M. Teillet, and G. H. Le Douarin (1975). Cholinergic differentiation of presumptive adrenergic neuroblasts in interspecific chimeras after heterotopic transplantations, *Proc. Natl. Acad. Sci. USA* **72**:728–732.

Le Douarin, N. M. and M.-A. M. Teillet (1974). Experimental analysis of the migration and differentiation of neuroblasts of the autonomic nervous system and of neuroectodermal mesenchymal derivatives, using a biological cell marking technique, *Dev. Biol.* **41**:162–184.

Le Douarin, N. M., M. Teillet, C. Ziller, and J. Smith (1978). Adrenergic differentiation of the cholinergic ciliary and Remak ganglia in avian embryo after in vivo transplantation, *Proc. Natl. Acad. Sci. USA* **75**:2030–2034.

Mains, R. E. and P. H. Patterson (1973*a*). Primary cultures of dissociated sympathetic neurons. I. Establishment of long-term growth in culture and studies of differentiated properties, *J. Cell Biol.* **59**:329–345.

Mains, R. E. and P. H. Patterson (1973*b*). Primary cultures of dissociated sympathetic neurons. II. Initial studies on catecholamine metabolism, *J. Cell Biol.* **59**:346–360.

Mains, R. E. and P. H. Patterson (1973*c*). Primary cultures of dissociated sympathetic neurons. III. Changes in metabolism with age in culture, *J. Cell Biol.* **59**:361–366.

Norr, S. (1973). In vitro analysis of sympathetic neuron differentiation from chick neural crest cells, *Dev. Biol.* **34**:16–38.

Nurse, C. A. (1977). The formation of cholinergic synapses between dissociated rat sympathetic neurons and skeletal myotubes in cell culture. Ph.D. dissertation, Harvard University, Cambridge.

Nurse, C. A. and P. H. O'Lague (1975). Formation of cholinergic synapses between dissociated sympathetic neurons and skeletal myotubes of the rat in cell culture, *Proc. Natl. Acad. Sci. USA* **72**:1955–1959.

Obata, K. (1977). Development of neuromuscular transmission in culture with a variety of neurons and in the presence of cholinergic substances and tetrodotoxin, *Brain Res.* **118**:141–153.

O'Lague, P. H., P. R. MacLeish, C. A. Nurse, P. Claude, E. J. Furshpan, and D. D. Potter (1975). Physiological and morphological studies on developing sympathetic neurons in dissociated cell culture, *Cold Spring Harbor Symp. Quant. Biol.* **40**:399–407.

O'Lague, P. H., K. Obata, P. Claude, E. J. Furshpan, and D. D. Potter (1974). Evidence for cholinergic synapses between dissociated rat sympathetic neurons in cell culture, *Proc. Natl. Acad. Sci. USA* **71**:3602–3606.

Patterson, P. H. (1978). Environmental determination of autonomic neurotransmitter functions, *Annu. Rev. Neurosci.* **1**:1–18.

Patterson, P. H. and L. L. Y. Chun (1974). The influence of nonneuronal cells on catecholamine and acetylcholine synthesis and accumulation in cultures of dissociated sympathetic neurons, *Proc. Natl. Acad. Sci. USA* **71**:3607–3610.

Patterson, P. H. and L. L. Y. Chun (1977*a*). The induction of acetylcholine synthesis in primary cultures of dissociated rat sympathetic neurons. I. Effects of conditioned medium, *Dev. Biol.* **56**:263–280.

Patterson, P. H. and L. L. Y. Chun (1977*b*). The induction of acetylcholine synthesis in primary cultures of dissociated rat sympathetic neurons. II. Developmental aspects, *Dev. Biol.* **60**:473–481.

Patterson, P. H., L. F. Reichardt, and L. L. Y. Chun (1975). Biochemical studies on the development of primary sympathetic neurons in cell culture, *Cold Spring Harbor Symp. Quant. Biol.* **40**:389–397.

Rees, R. and R. P. Bunge (1974). Morphological and cytochemical studies of synapses formed in culture between isolated rat superior cervical ganglion neurons, *J. Comp. Neurol.* **157**:1–11.

Reichardt, L. F. and P. H. Patterson (1977). Neurotransmitter synthesis and uptake by individual rat sympathetic neurons developing in microcultures, *Nature (London)* **270**:147–151.

Reichardt, L. F., P. H. Patterson, and L. L. Y. Chun (1976). Norepinephrine and acetylcholine synthesis by individual sympathetic neurons under various culture conditions, *Soc. Neurosci. Abstr.,* Vol. 2, p. 225.

Rothman, T. P., M. D. Gershon, and H. Holtzer (1978). Cell division and the acquisition of adrenergic characteristics by developing sympathetic ganglion cells, *Dev. Biol.* **65**:322–341.

Teillet, M., P. Cochard, and N. Le Douarin (1978). Relative roles of the mesenchymal tissues and the complex neural tube-notochord on the expression of adrenergic metabolism in neural crest cells, *Differentiation,* in press.

Walicke, P. A., R. B. Campenot, and P. H. Patterson (1977). Determination of transmitter function by neuronal activity, *Proc. Natl. Acad. Sci. USA* **74**:5767–5771.

Walicke, P. A. and P. H. Patterson (1978). Effects of neuronal activity and cyclic nucleotide derivatives on the development of transmitter function in cultured sympathetic neurons, *Soc. Neurosci. Abstr.,* Vol. 4, p. 129.

Weston, J. A. (1970). The migration and differentiation of neural crest cells, *Adv. Morphog.* **8**:41–114.

FACTORS REGULATING NEURONAL DIFFERENTIATION

Ira B. Black, Michael D. Coughlin, and Philippe Cochard

Cornell University Medical College, New York, New York

INTRODUCTION

Normal neuronal ontogeny involves a number of distinct but often contemporaneous processes, including embryonic induction of the neuronal primordium, migration of the neuroblast to its definitive site, terminal mitosis, transmitter phenotypic expression, axon outgrowth with target innervation, afferent innervation, and initiation of neurotransmission (Black, 1978). To successfully undertake and complete these tasks, the neuron must *survive*. The neuron, however, is not without help and guidance in effecting these differentiative changes, and, in fact, *cannot* survive and grow alone. It is now apparent that cellular interactions at multiple levels of the neuraxis influence and, indeed, govern development of the neuron. In turn, many if not all of these interactions are mediated by specific regulatory molecules. For the purposes of this brief review these molecules will be termed growth factors, although it is recognized that many subserve other functions as well and that, in some cases, influences on development arise from these other actions.

Although the regulatory cellular interactions are as yet poorly characterized, a number of generalizations may be warranted. It is clear, for example, that some growth factors mediate proximate interactions over angstrom distances, while others act as virtual hormones, mediating remote cellular interactions. As might be expected, different factors are specific for different classes of neurons and, excluding hormones such as thyroxine, no known molecule affects all neurons. The factors are capable of eliciting long-term changes in receptive neurons, and at least one molecule, nerve growth factor (NGF), is necessary for *survival* of certain neurons. Several lines of evidence suggest that cellular interactions may influence phenotypic transmitter expression in neurons. Lastly, cellular interactions are not restricted to a single phase of ontogeny, but are critical throughout development. During embryonic life, the microenvironment of the neuroblast governs migration and expression of transmitter characters. In the fetus, target innervation begins, initiating a new spectrum of cellular interactions; at this stage, target factors critically influence neuronal differentiation and, in fact, survival. After birth, with the establishment of synaptic connections, transsynaptic interactions assume increasing importance. Throughout many of the foregoing stages, long-range humoral factors and short-range interactions between neurons and surrounding glia continue to mold neuronal growth and differentiation. In summary, the selective expression of information encoded within the primitive neuron is dependent on a series of cellular interactions, many of which are already known to be mediated by specific effector molecules.

The present discussion focuses on three classes of developmental cellular interactions that illustrate regulation by very different molecules at different stages of ontogeny. Progressing from later to earlier stages of ontogeny, we consider (1) interneuronal transsynaptic regulation, (2) target-neuron interactions, and (3) embryonic microenvironment-neuron relations. Discussion of these particular examples serves to illustrate that certain interactions have already been well characterized, whereas others are poorly understood. (We have omitted discussion of interactions between fixed molecules on cell surfaces, since this represents an entirely different area, in which underlying mechanisms have not yet been defined.) Lastly, we have attempted to indicate that the study of cellular interactions and growth factors involves a number of methodologic approaches and is critically dependent on parallel in vivo and in vitro experimentation.

This review concentrates on the autonomic nervous system, which has been employed extensively to study neuronal growth factors. Initially, the basic relevant structure and biochemistry of this well-defined system is described to illustrate potential loci of growth factor action. Orthograde transsynaptic regulation of development by the neurotransmitter is discussed in some detail. Subsequently, target regulation of neuron development through retrograde transsynaptic interactions is considered, since this represents an entirely different type of influence in normal ontogeny. Finally, regulation of neuronal transmitter phenotypic expression by the embryonic microenvironment is considered.

Although our knowledge is only rudimentary in this area, discussion may help to enunciate outstanding issues and questions that are presently under study.

TRANSSYNAPTIC REGULATION OF DEVELOPMENT IN THE AUTONOMIC NERVOUS SYSTEM

The cholinergic fibers that innervate autonomic ganglia arise from perikarya lying in the intermediolateral columns of the spinal cord. These fibers make synaptic contact with noradrenergic neurons in sympathetic ganglia (Figure 1). The cholinergic neurons themselves receive afferent innervation from suprasegmental levels, including the hypothalamus (Hancock, 1976; Saper, Loewy, Swanson, and Cowan, 1976). In turn, postsynaptic noradrenergic neurons of the ganglia innervate a variety of target structures (Figure 1). Choline acetyltransferase (CAT), the enzyme that synthesizes acetylcholine, is highly localized to presynaptic terminals in sympathetic ganglia and may be used to monitor maturation of these cells (Black, Hendry, and Iversen, 1971*a*). On the other hand, tyrosine hydroxylase (T-OH), the rate-limiting enzyme in catecholamine biosynthesis (Levitt, Spector, Sjoerdsma, and Udenfriend, 1965), is highly localized to postsynaptic neurons in ganglia and may be used as an index of maturation of these elements (Black, Hendry, and Iversen, 1971*b*). A series of studies has indicated that these and other developmental characters are regulated by a series of transsynaptic interactions during ontogeny.

Orthograde and retrograde transsynaptic interactions at multiple levels of the autonomic neuraxis regulate the development of sympathetic neurons. In summary, descending influences within the spinal cord (1 in Figure 1) regulate the development of presynaptic cholinergic neurons and second-order noradrenergic neurons in sympathetic ganglia. In turn, presynaptic cholinergic neurons regulate postsynaptic adrenergic neurons (2 in Figure 1) through transsynaptic mechanisms. Conversely, target organs regulate development of innervating noradrenergic neurons through retrograde transsynaptic factors (4 in Figure 1) and also regulate the development of presynaptic cholinergic neurons. Postsynaptic adrenergic neurons regulate development of presynaptic cholinergic neurons through a retrograde transsynaptic process (3 in Figure 1). Consequently, complex reciprocal interactions at the synapse are critical in the regulation of normal neuronal development in the sympathetic system. We may now examine orthograde

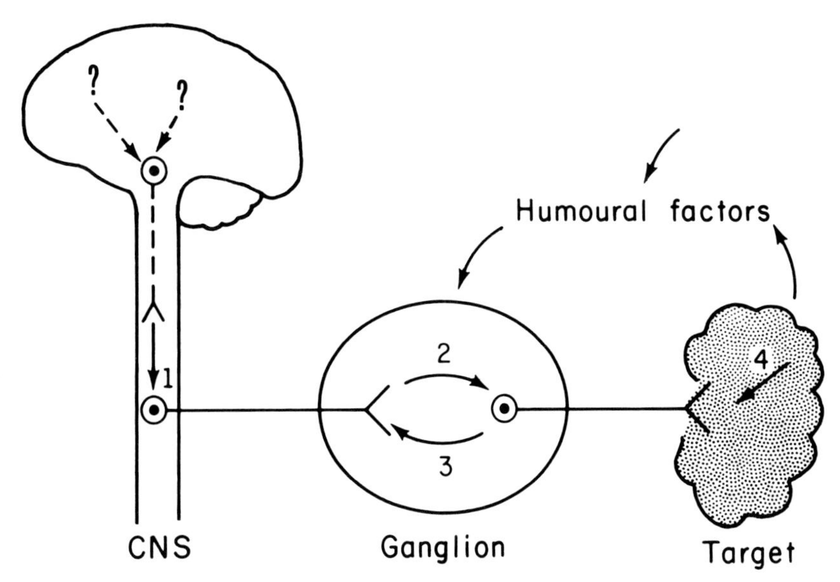

FIGURE 1. Schematic representation of intercellular interactions regulating autonomic development. Numbers refer to sites of regulation: 1, spinal transsynaptic; 2, intraganglionic orthograde transsynaptic; 3, intraganglionic retrograde transsynaptic; 4, target-neuron retrograde. Note that targets may also release humoral factors and that the sites of suprasegmental regulation in the CNS are undefined.

transsynaptic regulation within sympathetic ganglia as an example of this developmental interaction.

Initial studies, performed in mice in vivo, indicated that a major portion of sympathetic development transpires postnatally (Black et al., 1971*a*). During perinatal ontogeny, there is a marked increase in synapse numbers within the sympathetic ganglion, and the developmental increase in synapse numbers immediately precedes the 6- to 10-fold rise in postsynaptic T-OH activity (Black et al., 1971*a*). These observations suggested that the development of T-OH activity might be dependent on formation of intraganglionic synaptic contacts. To examine this possibility, ganglia were unilaterally decentralized in neonatal animals. The presynaptic cholinergic trunk innervating the superior cervical sympathetic ganglion (SCG) was transected in 3- to 4-day-old mice and rats, and in each animal the contralateral intact ganglion served as a control. Decentralization of the ganglion prevented the normal developmental increase in postsynaptic T-OH activity (Black et al., 1971*a*), suggesting that presynaptic cholinergic neurons regulate the development of postsynaptic T-OH activity through a transsynaptic process (Figure 2). In addition, the other postsynaptic catecholamine biosynthetic enzymes, dopa decarboxylase and dopamine-β-hydroxylase, also failed to develop normally in decentralized ganglia.

What is, or what are the transsynaptic messages involved in this regulation? To approach this question, neonatal animals were treated with long-acting ganglionic blocking agents. These drugs prevent postsynaptic depolarization by competing with acetylcholine for postsynaptic receptor sites. In fact, treatment of neonates with chlorisondamine or pempidine prevented the normal development of postsynaptic T-OH activity, reproducing the effects of ganglion decentralization (Black and Geen, 1973, 1974). Moreover, as was the case with decentralization, postsynaptic dopa decarboxylase and dopamine-β-hydroxylase also failed to develop normally. These effects appeared to be specific for nicotinic receptor blockade, since treatment of animals with atropine, the classic muscarinic antagonist, had no effect on development (Black and Geen, 1973, 1974). Consequently, presynaptic neurons regulate postsynaptic development through the mediation of acetylcholine and its interaction with nicotinic receptors on the postsynaptic membrane. On this basis, it is not necessary to postulate the existence of some as yet unidentified presynaptic growth factor, since the normal presynaptic transmitter also subserves this growth function.

The presynaptic neurotransmitter may regulate postsynaptic development through the normal process of depolarization or, alternatively, may operate through unique processes. Recent work performed by Patterson and colleagues (1975), and reported in detail elsewhere in this volume (1979), suggests that the former is the case. Briefly, sympathetic neurons in culture tend to increase the ratio of catecholamine to acetylcholine synthesis when subjected to depolarizing stimuli. Consequently, the data derived from in vivo and in vitro experimentation suggest that presynaptic cholinergic nerves may influence postsynaptic development of neurotransmitter characters through the depolarizing effect of acetylcholine. Hence, acetylcholine may serve as a "growth factor" by interacting with postsynaptic nicotinic receptors and effecting normal postsynaptic depolarization. These observations and contentions imply that for normal development the sympathetic neuron must undergo normal depolarization, suggesting that normal neuronal activity is a requisite for normal ontogeny.

How do acetylcholine and postsynaptic depolarization regulate the

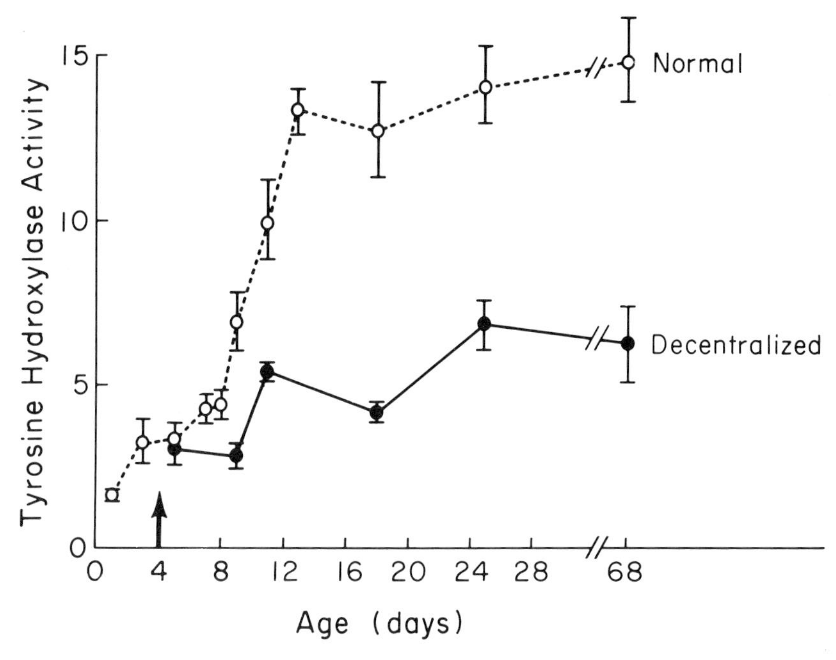

FIGURE 2. Effect of surgical decentralization on day 4 on development of tyrosine hydroxylase activity in mouse superior cervical ganglion. Groups of six mice were killed at various times postoperatively and tyrosine hydroxylase activity (pmol/ganglion·h) was measured in decentralized and contralateral control ganglia. The value obtained 1 day after surgery does not differ significantly from control; all other values from decentralized ganglia are significantly lower than control values ($P < 0.01$).

development of transmitter enzyme activities? Acetylcholine could increase enzyme activity by activating preexistent enzyme molecules or, alternatively, by increasing the number of enzyme molecules within the postsynaptic neuron. In fact, a series of immunotitration experiments using specific antibodies to tyrosine hydroxylase indicated that acetylcholine regulates the accumulation of postsynaptic T-OH molecules (Black, Joh, and Reis, 1974). Decentralization or treatment of neonatal animals with ganglionic blocking agents prevents the development of postsynaptic T-OH activity by blocking the normal increase in the number of T-OH enzyme molecules (Black et al., 1974). Consequently, acetylcholine as a growth factor appears to regulate development by governing the number of specific transmitter enzyme molecules that accumulate within the postsynaptic noradrenergic neuron.

Presynaptic cholinergic neurons could operate autonomously in this transsynaptic regulation, or, in turn, could be dependent on information descending within the spinal cord. These possibilities were examined by transecting spinal cords in neonatal rats. Spinal transection at the fifth thoracic segment prevented the normal development of the caudal sixth lumbar (L-6) ganglion, whereas the rostral superior cervical ganglion developed normally (Black, Bloom, and Hamill, 1976; Hamill, Bloom, and Black, 1977). Transection prevented the normal development of presynaptic CAT activity and postsynaptic T-OH activity within the L-6 ganglion. This effect was long lasting, persisting for at least 1 year. These observations suggest that descending information within the spinal cord, presumably operating through transsynaptic mechanisms, is important in the development of proximate cholinergic neurons as well as second-order noradrenergic neurons in the periphery. At present, the locus of the spinal interactions is unknown and the transmitters involved are undefined.

TARGET REGULATION OF DEVELOPMENT

Increasing evidence suggests that retrograde transsynaptic factors, originating from targets, regulate development of innervating noradrenergic neurons. Target extirpation in neonatal mice or rats prevents the normal survival of those noradrenergic neurons destined to innervate the targets, and this effect may be prevented by treatment with nerve growth factor (Hendry and Iversen, 1973; Dibner and Black, 1976; Dibner, Mytilineou, and Black, 1977). In addition to regulating survival of afferent neurons, targets also determine specificity of the morphologic pattern of innervation. Olson and Malmfors (1970) demonstrated that the characteristic pattern of innervation of a target is determined by the target itself and not by the innervating neuron. For example, in the iris the adrenergic innervation consists of a dense plexus in which thick preterminal axons undergo extensive branching to form a dense network of varicose terminals. This pattern differs radically from that observed in other peripheral organs and from central terminal areas such as cerebral and cerebellar cortices. Clearly, it would be difficult to account for organotypic specificity in the pattern of innervation by invoking a single molecular species such as NGF. In fact, increasing evidence suggests that targets may elaborate a number of growth factors, one of which is NGF.

To evaluate the role of targets in noradrenergic growth in detail, we evolved a tissue culture system to allow precise manipulation and perturbation (Coughlin, Boyer, and Black, 1977). We chose the 14-gestational day (E14) fetal mouse SCG, since target innervation begins prenatally and since we wanted to study target-neuron relations from early to late phases of development. In characterizing this system, we found that the E14 ganglion did not require NGF for survival and differentiation in culture, in contrast to all other sympathetic systems that had previously been described. In fact, the E14 ganglion elaborated neurites and exhibited a 6- to 8-fold rise in T-OH activity over 48 hours in culture, even in the presence of antiserum to NGF (anti-NGF) (Coughlin et al., 1977). These facts greatly simplified matters, allowing us to study target-ganglion interactions without the complicating presence of NGF added to the culture medium.

To examine the interaction of target and sympathetic neurons, the SCG was cultured alone or in contact with the submaxillary salivary gland, one of its targets (Coughlin, Dibner, Boyer, and Black, 1978). Ganglion T-OH activity was examined after various periods of incubation, and histochemical staining for acetylcholinesterase was used to visualize axon outgrowth.

After 2 days of incubation, ganglia cultured alone exhibited a fourfold increase in T-OH activity compared to zero-time controls. After 5 days in culture, the isolated ganglia had begun to degenerate and T-OH activity fell significantly to zero-time levels. The presence of the target submandibular gland radically altered the development of T-OH activity (Figure 3). After 2 days in culture in the presence of the target, ganglion enzyme activity increased over 10-fold compared to zero-time controls and was twice that of the ganglion cultured alone. By 5 days in culture, the ganglion grown with target exhibited T-OH activity 7-fold higher than that in ganglia grown alone, describing more than 11-fold increase compared to the zero-time control (Figure 3). Consequently, the target exerted both a stimulatory and maintenance effect, elevating T-OH activity at 48 hours and causing a continual increase to at least 5 days, when the ganglion normally degenerates in the absence of growth factors.

The specificity of the target effect was tested by examining the influence of nontarget tissues in ganglion development (Coughlin et al., 1978). Ganglia cultured for 5 days with salivary gland exhibited an increase in T-OH activity approximately twice that of ganglia grown with lung or metanephric mesenchyme, nontarget tissue (Figure 4). In the presence of lung and metanephric mesenchyme, T-OH activity was, nevertheless, above the 5-day control level.

Ganglia grown with target salivary gland exhibited greater elaboration of neurites and directionality of fiber outgrowth than ganglia grown

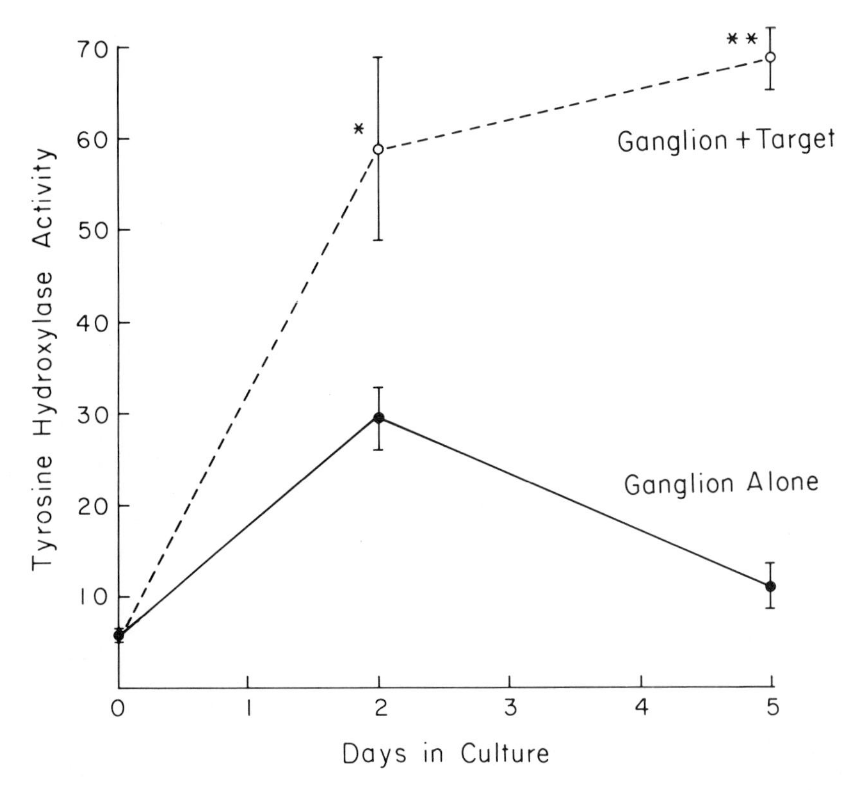

FIGURE 3. Effect of coculture with target tissue on ganglion tyrosine hydroxylase activity. Ganglia were removed from 14-day embryonic mice and cocultured with 14-day embryonic salivary rudiments. Enzyme activity was determined in 5–7 ganglion pairs 2 and 5 days after culture and is expressed as pmol/ganglion·h ± SEM (vertical bars).

* Differs from ganglion alone at $P < 0.05$.

** Differs from ganglion alone at $P < 0.001$.

(Reprinted with permission from Coughlin et al., 1978.)

alone or with nontarget tissues (Coughlin et al., 1978). Ganglia cultured alone in basal medium elaborated neurites that extended radially in all directions at 2 and 3 days of incubation. By 5 days there was a slight decrease in the peripheral fiber network, although the basic pattern was well maintained (Figure 5). In contrast, the salivary gland elicited a much greater production of neurites from the side of the ganglion contacting this target: large, dense bundles of fibers were directed around and into the gland (Figure 5). Ganglia grown with nontarget

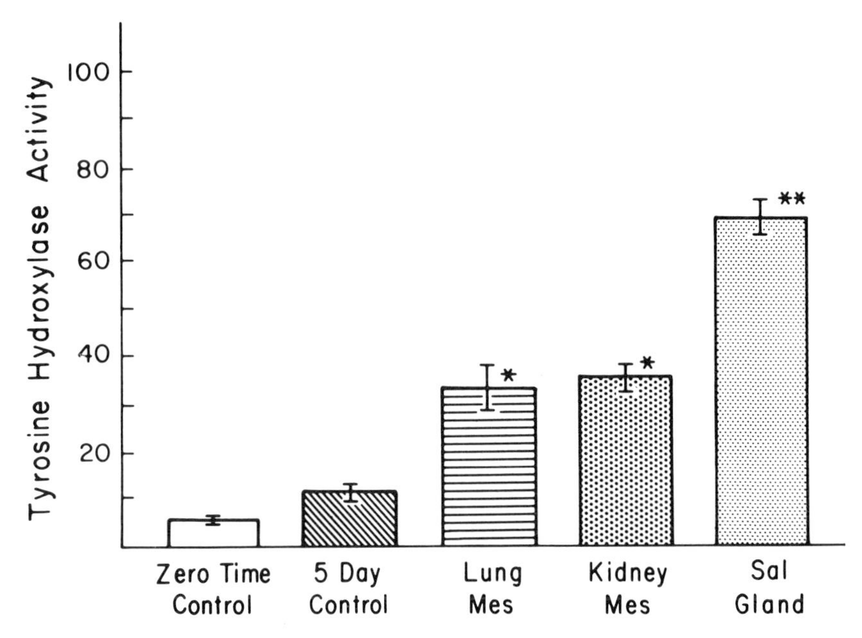

FIGURE 4. Effect of various tissues on ganglionic tyrosine hydroxylase activity. Ganglia from 14-day embryos were grown alone (5 Day Control) or were cocultured with 12-day embryonic lung mesenchyme (Lung Mes), 12-day embryonic metanephric mesenchyme (Kidney Mes), or 14-day embryonic salivary gland rudiments (Sal Gland). After 5 days of culture, five to nine ganglion pairs were assayed for T-OH activity. Activity is expressed as mean pmol/ganglion·h ± SEM (vertical bars).

* Differs from zero-time control or 5-day control at $P < 0.05$ by one-way analysis of variance and Newman-Keuls test.

** Differs from 5-day control, lung mesenchyme, and metanephric mesenchyme at $P < 0.05$ by one-way analysis of variance and Newman-Keuls test.

(Reprinted with permission from Coughlin et al., 1978.)

tissues maintained the peripheral fiber network on the side distant from the tissue, while few fibers grew into such tissues.

To determine whether the target salivary gland regulated ganglion development through the mediation of NGF, cultures were grown in the presence or absence of anti-NGF (Coughlin et al., 1978). Ganglion explants cultured alone for 5 days in medium containing anti-NGF exhibited neurite outgrowth equivalent to that of explants grown in basal medium (Figure 5): fibers extended radially in all directions. Ganglia cultured with salivary glands in the presence of anti-NGF exhibited a marked increase in neurite elaboration compared to ganglia grown

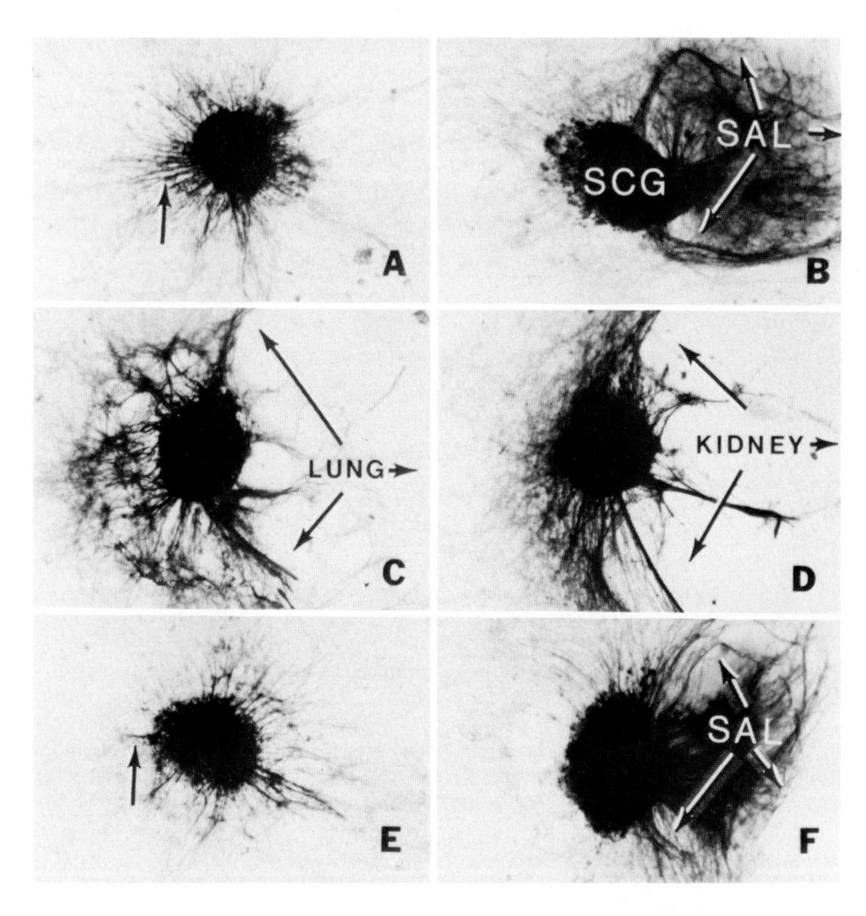

FIGURE 5. Effect of coculture with various tissues and with anti-NGF on fiber outgrowth from embryonic ganglion explants. Embryonic ganglia from mice aged 14 days of gestation were cultured for 5 days under various conditions. *A*: Ganglion grown alone in basal medium. Nerve fibers (arrow) are maintained and extend radially in all directions. *B*: Ganglion (SCG) cocultured with salivary gland (SAL) in basal medium. Extensive fiber outgrowth is most marked from the side of the ganglion in contact with the gland. Thick fiber bundles are directed into the target. In some areas (top of figure) fiber outgrowth appears to be redirected toward the salivary gland. *C*: Ganglion cocultured with 12-day embryonic lung mesenchyme. The mesenchymal tissue (LUNG) has spread over the culture dish surface and covers the area to the right of the SCG. Few fibers grow into the tissues. Most fiber outgrowth is on the side opposite the mesenchyme, with a peripheral network of neurite fibers maintained more fully than in *A*. *D*: Ganglion cocultured with metanephric mesenchyme (KIDNEY). Tissue spreading is similar to that in *C*, and fiber growth into tissue is sparse. *E*: Ganglion grown alone in medium containing anti-

alone. Moreover, the obvious directional outgrowth of large fiber bundles into the salivary gland was not altered by the presence of anti-NGF. Thus, anti-NGF did not alter target stimulation of neurite outgrowth.

In agreement with the morphological observations, anti-NGF did not alter the capacity of the target to elevate ganglion T-OH activity (Coughlin et al., 1978). Ganglia grown alone in basal medium for 5 days exhibited a fivefold rise in T-OH activity, and this increase occurred even in the presence of anti-NGF. Coculture of target and ganglion for 5 days in basal medium resulted in a 10-fold rise in T-OH activity compared to the zero-time control; activity was twice that of ganglia grown alone. Anti-NGF did not alter the target-stimulated elevation of T-OH activity.

Attempts to isolate and characterize the target factor(s) have begun. Initially, we sought to determine whether the factor was diffusible, and cultured a number of sympathetic targets as monolayers; the conditioned medium (CM) derived from such cultures was tested for growth-promoting activity in cultures of dissociated sympathetic neurons from fetal mouse SCG. Measurement of survival and neurite extension in dissociated neurons in culture after 20 hours proved to be the most sensitive index for detection of growth-promoting activity.

To date, the most consistent results have been obtained with CM from fetal heart monolayer cultures (Coughlin and Black, 1979). In basal medium, approximately 20% of the plated neurons survive and extend neurites after 20 hours. Addition of NGF itself doubled neuronal survival, and addition of anti-NGF + NGF reduced survival to control levels (Figure 6). CM also doubled neuronal survival, but addition of anti-NGF to the CM did not reduce survival (Figure 6). Apparently then, the CM factor is antigenically distinct from NGF and most probably represents a new growth-promoting factor. Data obtained in initial characterization studies support the contention that this factor is separate and distinct from NGF.

The active CM factor(s) have been partially characterized (Coughlin and Black, 1979). Activity was retained in CM dialyzed for 18 hours against 0.02 M PO_4, pH 7.4, and was destroyed by heating for 30 min-

---

NGF. Nerve fiber outgrowth (arrow) is essentially the same as in *A*. *F*: Ganglion grown with target in medium containing anti-NGF. Numerous thick bundles of fibers extend into the salivary tissue and show profuse ramification. Fiber outgrowth on the side of the ganglion opposite the target is markedly diminished. ×45.
(Reprinted with permission from Coughlin et al., 1978.)

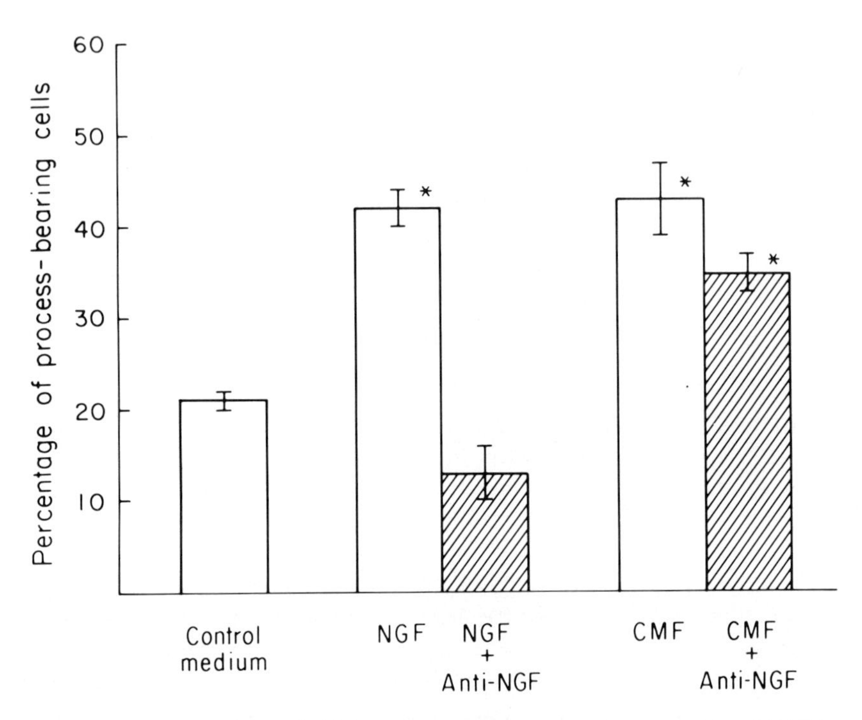

FIGURE 6. Comparison of NGF and mouse heart conditioned medium (CMF) in promoting neurite extension from mouse embryo sympathetic neurons. Histograms represent the percentage of initially plated cells (routinely 20,000–50,000 per 35-mm dish) that survived and bore processes after 20 h in culture. Each value represents the mean ± SEM of 9–17 culture dishes. Phase-bright cells with processes longer than the diameter of the cell body were scored as neurons. Two percent of the surface of each dish was counted. Trypsin-dissociated cells were suspended in Puck's Saline G with 1% horse serum; aliquots were plated into dishes containing medium F-12 with 10% heat-inactivated fetal calf serum (F12FCS10, control medium) and, where indicated, NGF, CMF, and/or anti-NGF. All tissue culture dishes were treated with polyornithine (Helfand et al., 1978). Conditioned medium (CM) was obtained by incubating F12FCS10 for 48–72 h on embryonic mouse heart cells that had previously been grown to confluency. CMF from a partially purified DEAE-cellulose fraction of CM was used at 20 μg/ml medium; in early experiments dishes were pretreated with CM for several hours, after which fresh medium was added. CMF activity remains bound to the dish (Collins, 1978). NGF was used in cultures at 1–10 units/ml. Antiserum to NGF (anti-NGF) was obtained commercially and used at 0.1–1%.

* Differ from control at $P < .001$.

Values for CMF and CMF + anti-NGF are not significantly different.

utes at 60° C. The molecular weight of the factor was apparently greater than 100,000, since activity was retained in the concentrate resulting from pressure filtration through a Millipore PTHK filter with a nominal MW limit of 100,000. Similarly, activity was contained in the excluded volume of a P-150 Bio-Gel column. Partial purification of the active factor on a DEAE-cellulose column yielded a fraction that was maximally active at 10–20 μg protein/ml of medium. Further purification and characterization are in progress.

While it is not established that this putative factor is identical to the submaxillary factor(s) eliciting directed growth and increased T-OH activity, it is apparent that molecules other than NGF are capable of supporting survival and stimulating nerve growth. A growing body of evidence suggests that a number of cell types release factors that support nerve growth. Monard, Solomon, Rentsch, and Gysin (1973) recently reported that C-6 glioma cells release a Millipore-filterable factor into the medium that elicits process formation in a C-1300 neuroblastoma cell line in a concentration-dependent fashion. The morphological differentiation was not associated with changes in intracellular adenosine 3′:5′-monophosphate or in the rate of cell growth. The glial factor was distinct from NGF by a number of criteria: (1) 2.5s NGF itself did not cause neurite formation in the neuroblastoma; (2) the glial factor did not affect chick dorsal root ganglia or rat SCG, both of which respond to NGF; (3) anti-NGF did not inhibit glial factor effect on neuroblastoma; (4) glial factor did not co-migrate with NGF on SDS polyacrylamide gel electrophoresis (Monard, Stöckel, Goodman, and Thoenen, 1975). Consequently, glia are capable of elaborating a number of growth factors. Previously, it had been demonstrated, for example, that glia may support neurons grown in vitro through elaboration of NGF-like molecules (Varon, Raiborn, and Burnham, 1974).

More recently, C-6 glioma cells have been found to release another factor into the medium (Barde, Lindsay, Monard, and Thoenen, 1978). This factor supports survival and neurite formation in dissociated chick embryo dorsal root ganglion neurons in culture. Once again, the effect was anti-NGF insensitive. On the other hand, the neuroblastoma-stimulating factor has no effect on dissociated chick sensory neurons. The new factor was sensitive to pronase and α-chymotrypsin, suggesting that it is a polypeptide or protein.

In summary, glial derivatives are apparently capable of releasing at least three factors with growth-promoting activity: NGF, neuroblastoma-stimulating (glial) factor and sensory-stimulating factor. Isolation and characterization of these factors may help define their structural and functional interrelationships. At present it is unclear whether non-neoplastic glia also elaborate the latter two factors, whether astrocytes, microglia, and oliogdendroglia have similar capacities, and whether a single cell is capable of elaborating all three factors. Conversely, it will be critical to establish the range of neuronal types that are receptive to the different growth factors.

In addition to the glial factors described above, a factor in embryonic chick heart conditioned medium (HCM) has recently been described that supports neurite outgrowth from embryonic chick sympathetic, sensory, and ciliary neurons (Helfand, Riopelle, and Wessells, 1978). The factor is anti-NGF insensitive, does not displace NGF from binding sites, and unlike NGF, is resistant to acid, but labile to alkali. It is inactivated by heating to 90° C and is retained by Amicon P-50 filters. The relationship of this chick heart factor to the fetal mouse heart factor described above awaits characterization of these two (or more) molecules.

EMBRYONIC MICROENVIRONMENT-NEUROBLAST RELATIONS

Increasing evidence indicates that intercellular interactions govern the embryogenesis of the autonomic nervous system and phenotypic expression within the sympathetic neuroblast. It may be useful to enunciate some questions relevant to the role of "growth factors" in embryonic neurogenesis before proceeding to detailed discussion. What factors, if any, govern the conversion of neural plate cells to progenitors of the neural crest, primordium of the autonomic system? What factors influence neural crest migration and cessation of mitosis? What factors mediate the influence(s) of somitic mesenchyme on the migrating cells? What signals guide neuroblasts to their definitive locations and lead to stable association of heterogeneous cellular populations? Although none of these questions have been definitively answered, the present discussion may serve to highlight recent development in this important, emerging area.

The autonomic system arises from the neural crest, a transient embryonic structure that lies in a dorsal midline position on either side of the neural tube (Figure 7). From the neural crest, cells migrate throughout the embryo to give rise to autonomic and sensory neurons, chromatophores, calcitonin-producing cells, nonneuronal cells of the peripheral nervous system, chromaffin cells, and mesenchymal derivatives of the

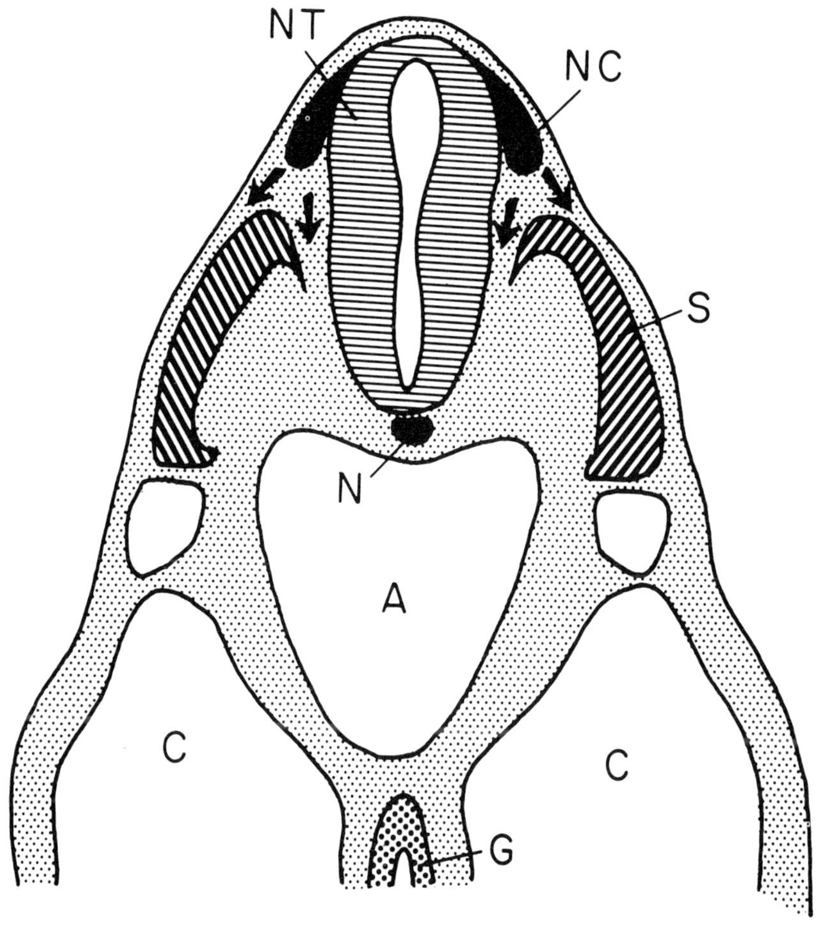

FIGURE 7. Transverse section of the rat embryo. The neural crest (NC) on either side of the neural tube (NT) migrates ventral to the somites (S) and under the ectoderm (arrows). Notation: N, notochord; A, aorta; C, coelom; G, gut.

cephalic region (Coulombre, Johnston, and Weston, 1974). Weston and colleagues demonstrated that grafting of young, newly condensed crest into progressively older hosts results in progressive attenuation of migration, suggesting that the embryonic microenvironment is critical in regulating migration patterns (Weston, 1970).

These cellular interactions appear to play a major role in differentiation of the neural crest. The major pathways of *neuronal* differentiation lead to development of sensory neurons (Tennyson, 1965) and autonomic neurons (Pick, 1963). In turn, autonomic neurons may utilize a variety of transmitters, including acetylcholine and norepinephrine. What factors regulate this critical choice of neurotransmitters? In the chick embryo, appearance of catecholamine histofluorescence in presumptive sympathoblasts is due to interactions with somitic mesenchyme and therefore requires that the crest migrate ventrally (Cohen, 1972, 1974; see Figure 7 for orientation). In addition, ablation of ventral neural tube reduces the quantity of nervous tissue formed, which suggests that neural tube, somitic mesenchyme, and sympathoblasts undergo determinative interactions (Cohen, 1972, 1974). This contention is supported by in vitro studies of chick neural crest that demonstrate that for catecholamine differentiation crest cells must be contiguous with somite (Norr, 1973). In summary, neural tube appears to induce changes in somitic mesenchyme, which then promotes sympathoblast differentiation. In addition, the continued presence of ventral neural tube is apparently required for survival of the differentiating sympathetic neurons. Ventral neural tube may be replaced by NGF in assuring *survival* of the differentiating sympathetic neurons (Norr, 1973).

Le Douarin and colleagues have analyzed migration and differentiation in avian systems (Le Douarin and Teillet, 1973, 1974; Le Douarin, Renaud, Teillet, and Le Douarin, 1975; Le Douarin, Teillet, and Le Lievre, 1977; Smith, Cochard, and Le Douarin, 1977). Normally, the neural crest in the trunk region gives rise to adrenergic sympathetic ganglia and adrenal medullary cells, whereas the "vagal" region (somites 1–7) gives rise to cholinergic enteric ganglion cells. However, if trunk (sympathoadrenal) crest is grafted to the "vagal" region, the crest cells colonize the gut to form cholinergic enteric ganglion cells. If, on the other hand, cephalic crest is transplanted to the "adrenomedullary level" in the trunk region, it becomes adrenergic adrenomedullary chromaffin tissue. It may be concluded that preferential migratory pathways are located at precise levels of the crest in the embryo and lead cells to their definitive sites. Moreover, the expression of a given phenotype may be regulated by the environment of the definitive site and by cells encountered en route.

To define the nature of the intercellular interactions that normally govern phenotypic expression in mammals, we have pursued a strategy of documenting the initial expression of individual noradrenergic transmitter characters in sympathoblasts in embryonic rats in vivo. Using specific antisera, we defined the initial immunocytochemical appearance of T-OH, dopamine-β-hydroxylase (DBH), which converts dopamine to norepinephrine, and phenylethanolamine-*N*-methyltransferase (PNMT), which converts norepinephrine to epinephrine, as well as the catecholamine (CA) neurotransmitters (Cochard, Goldstein, and Black, 1978, 1979).

T-OH, DBH, and the CA transmitters were undetectable in the neural crest itself or the migrating crest cells. These noradrenergic characters first appeared at 11.5 days of gestation in neuroblasts within the sympathetic primordia (Figure 8); simultaneously these noradrenergic characters transiently appeared in a population of neuroblasts in the gut mesenchyme (Figure 8). This synchrony in the expression of different noradrenergic characters may imply that (1) a single, final, common intracellular process governs the expression of many noradrenergic traits, (2) a single extracellular stimulus regulates expression of these diverse traits, and/or (3) multiple intracellular or extracellular signals, governing different noradrenergic characters, are activated simultaneously. Although our studies do not yet distinguish among these alternatives, the previous work described above suggests that interactions among neural tube, notochord, somite, and migrating crest cells are necessary for appearance of CA. These observations are entirely consistent with our results, in which noradrenergic characters were undetectable in the crest itself or in migrating crest cells until positions were assumed in ganglion anlage or in the gut. Our results are consistent with the contention that the ventral neural tube area, in some manner, induces noradrenergic traits in autonomic precursors migrating past, and that expression occurs subsequently in primitive ganglia or gut. We are presently attempting to isolate the factors that mediate the critical cellular interaction. By analyzing cellular components of the ventral neural tube area in vivo and in vitro, we hope to determine whether diffusible or fixed factors are induced in the initial expression of noradrenergic traits in the autonomic neuroblast.

The transient nature of noradrenergic expression in the gut cells implies an additional level of complexity. It is apparent that initial expression of the noradrenergic phenotype is regulated differently from *persistence*, since T-OH, DBH, and CA are only transiently present in the gut cells. Consequently, early in differentiation, phenotypic expression may not be stable, and continued expression may depend on influences from the definitive site. Other interpretations, however, require examination.

The noradrenergic neuroblasts of the gut may be subject to a number

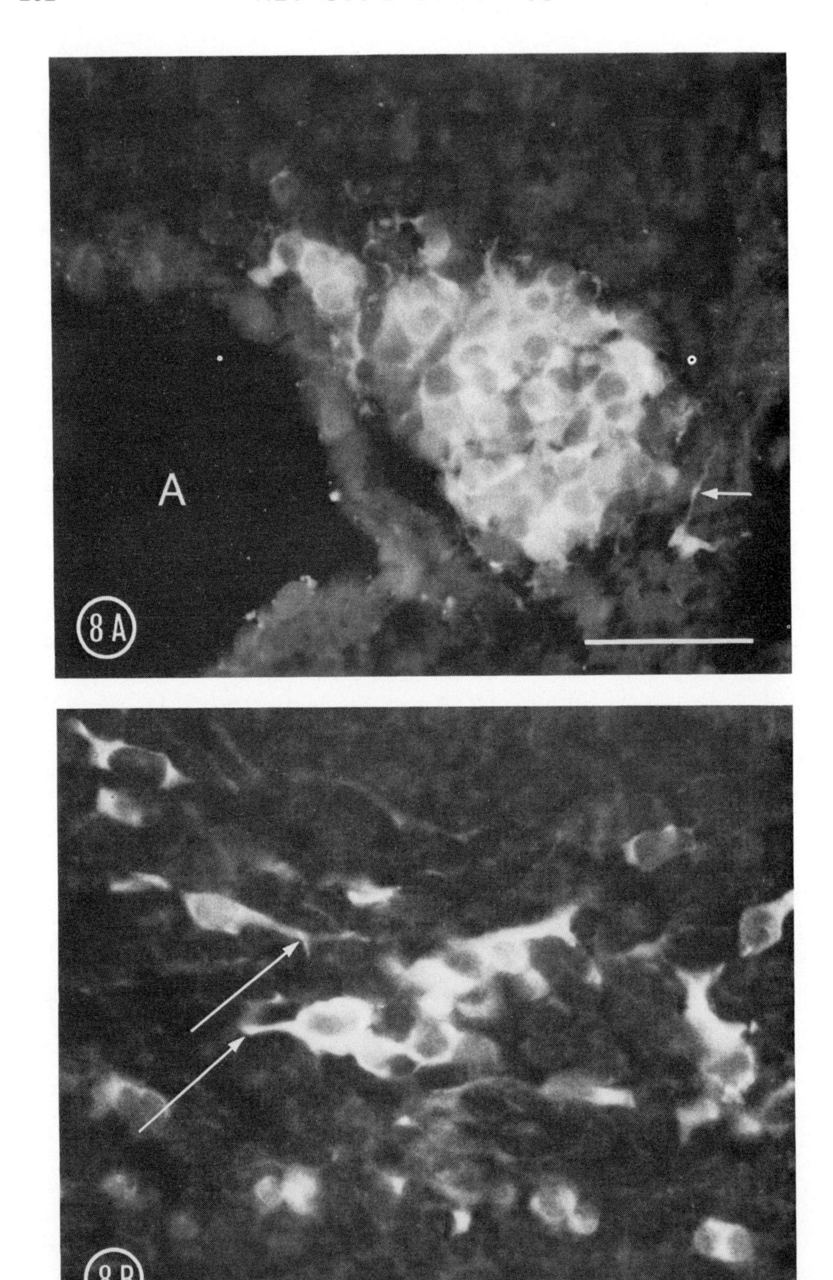
A
8A
8B

of fates, which may provide insights into potential intercellular factors involved. First, the cells may migrate out of the gut. This is unlikely: we have not observed noradrenergic cells in other locations, and the gut cells have already elaborated long cytoplasmic processes, rendering future migration improbable. Second, the gut neuroblasts may be destined to die. This alternative cannot be excluded on the basis of our investigations. Last, the noradrenergic gut neuroblasts may constitute the progenitors of myenteric plexus neurons, which ultimately use other neurotransmitters. Consequently, the disappearance of noradrenergic characters may reflect conversion from norepinephrine to another transmitter. This is a particularly attractive hypothesis for a number of reasons. In our studies, the gut neuroblasts appeared to aggregate to form ganglion-like structures at the time when noradrenergic characters were disappearing. These may be the forerunners of the enteric plexuses. It is well documented that neurons *intrinsic* to the gut use a number of transmitters, including acetylcholine (Dale, 1937) and serotonin (Dreyfus, Bornstein, and Gershon, 1977), as well as norepinephrine (Bennett and Malmfors, 1970; Bennett, Cobb, and Malmfors, 1971; Costa, Furness, and Gabella, 1971). Furthermore, recent experiments in vitro have demonstrated that factors produced by cells of mesenchymal origin may convert noradrenergic sympathetic neurons to cholinergic neurons (Patterson and Chun, 1974; Patterson, Reichardt, and Chun, 1975). Parallel work performed in avian embryos has indicated that gut mesenchyme can induce cholinergic differentiation in crest derivatives (Smith et al., 1977).

On the basis of these considerations, we tentatively conclude that the gut itself either lacks appropriate factors to assure persistence of the noradrenergic phenotype, lacks a survival-promoting factor for adrenergic neuroblasts, and/or elaborates one or more factors that cause conversion to another transmitter phenotype. We are presently attempting to resolve these alternatives and characterize the appropriate factors by using a combination of in vivo and in vitro techniques.

FIGURE 8. Immunocytochemical demonstration of T-OH in 11.5-day embryo. *8A*: Transverse section at the cervical level. Specific T-OH immunofluorescence is located in neuroblasts aggregated lateral to the aorta (A) to form primitive sympathetic ganglia. Note that fluorescence is restricted to the cytoplasm and cytoplasmic processes (arrow) of the sympathoblasts. Bar represents 50 μm. *8B*: Section through the stomach primordium. Numerous neuroblasts located in the gut mesenchyme exhibit intense T-OH-specific immunofluorescence, restricted to the cytoplasm. Note the long T-OH-positive cytoplasmic processes (arrow). Bar represents 50 μm.

CONCLUSION

Increasing evidence indicates that neuronal ontogeny is regulated by a variety of cellular interactions, many of which are mediated by specific effector molecules. In the autonomic nervous system, orthograde and retrograde transsynaptic factors at multiple levels of the neuraxis influence neuronal development. In autonomic ganglia, acetylcholine, the normal presynaptic transmitter, regulates development of postsynaptic noradrenergic neurons. In turn, normal presynaptic cholinergic ontogeny is dependent on influences descending within the spinal cord. Conversely, target organs also regulate development of innervating sympathetic neurons in vivo. Moreover, these effects appear to be mediated by diffusible target "growth factors," since culture medium conditioned by targets or glia promotes neuronal survival and differentiation in vitro.

At earlier stages of ontogeny, embryogenesis of the autonomic system is regulated by cellular interactions, but the nature of these phenomena has yet to be defined. After migration through the ventral neural tube area of the embryo, autonomic neuroblasts of ganglion primordia and primitive gut simultaneously express a number of noradrenergic phenotypic characters. The noradrenergic gut neuroblasts subsequently disappear, whereas in ganglionic cells noradrenergic differentiation continues. These observations have prompted the search for factors in the embryonic microenvironment, including ventral neural tube area, ganglion, and gut, that influence neuronal phenotypic expression.

ACKNOWLEDGMENTS

This work was supported by National Institutes of Health grants NS10259 and HD12108 and was aided by grants from the Dysautonomia Foundation, the National Foundation-March of Dimes, and the National Science Foundation. I.B.B. is a recipient of the Irma T. Hirschl Trust Career Scientist Award. P.C. is a recipient of a fellowship from the Délégation Générale à la Recherche Scientifique et Technique, France. We thank Ms. Dahna Boyer and Ms. Lynn Friedman for excellent technical assistance.

REFERENCES

Barde, Y. A., R. M. Lindsay, D. Monard, and H. Thoenen (1978). New factor released by cultured glioma cells supporting survival and growth of sensory neurones, *Nature (London)* **274:**8181.

Bennett, T., J. L. S. Cobb, and T. Malmfors (1971). Fluorescence histochemical observations on Auerbach's plexus and the problem of inhibitory innervation of the gut, *J. Physiol. (London)* **218:**77P–78P.

Bennett, T. and T. Malmfors (1970). The adrenergic nervous system of the domestic fowl (*Gallus domesticus* (L)), *Z. Zellforsch. Mikrosk. Anat.* **106:** 22–50.

Black, I. B. (1978). Regulation of autonomic development, *Annu. Rev. Neurosci.* **1:**183–214.

Black, I. B., E. M. Bloom, and R. W. Hamill (1976). Central regulation of sympathetic neuron development. *Proc. Natl. Acad. Sci. USA* **73:**3575–3578.

Black, I. B. and S. C. Geen (1973). Trans-synaptic regulation of adrenergic neuron development: inhibition by ganglionic blockade, *Brain Res.* **63:**291–302.

Black, I. B. and S. C. Geen (1974). Inhibition of the biochemical and morphological maturation of adrenergic neurons by nicotinic receptor blockade, *J. Neurochem.* **22:**301–306.

Black, I. B., I. A. Hendry, and L. L. Iversen (1971*a*). Trans-synaptic regulation of growth and development of adrenergic neurons in a mouse sympathetic ganglion, *Brain Res.* **34:**229–240.

Black, I. B., I. A. Hendry, and L. L. Iversen (1971*b*). Differences in the regulation of tyrosine hydroxylase and DOPA-decarboxylase in sympathetic ganglia and adrenal, *Nature (London)* **231:**27–29.

Black, I. B., T. H. Joh, and D. J. Reis (1974). Accumulation of tyrosine hydroxylase molecules during growth and development of the superior cervical ganglion, *Brain Res.* **75:**133–144.

Cochard, P., M. Goldstein, and I. B. Black (1978). Ontogenetic appearance and disappearance of tyrosine hydroxylase and catecholamines in the rat embryo, *Proc. Natl. Acad. Sci. USA* **75:**2986–2990.

Cochard, P., M. Goldstein, and I. B. Black (1979). Initial development of the noradrenergic phenotype in autonomic neuroblasts of the rat embryo *in vivo*, *Dev. Biol.*, in press.

Cohen, A. M. (1972). Factors directing the expression of sympathetic traits in cells of neural crest origin, *J. Exp. Zool.* **179:**167–192.

Cohen, A. M. (1974). DNA synthesis and cell division in differentiating avian adrenergic neuroblasts, pp. 359–370 in *Dynamics of Degeneration and Growth in Neurons*, Fuxe, K., L. Olson, and Y. Zotterman, eds. Pergamon Press, New York.

Collins, F. (1978). Does the same component of conditioned medium induce neurite outgrowth from both sympathetic and parasympathetic neurons? *Soc. Neurosci. Abstr.*, Vol. 4, p. 590.

Costa, M., J. B. Furness, and G. Gabella (1971). Catecholamine-containing nerve cells in the mammalian myenteric plexus, *Histochimie* **25:**103–106.

Coughlin, M. D. and I. B. Black (1979). New factor stimulates growth of mammalian sympathetic neurons in culture, submitted for publication.

Coughlin, M. D., D. M. Boyer, and I. B. Black (1977). Embryologic development of a mouse sympathetic ganglion *in vivo* and *in vitro*, *Proc. Natl. Acad. Sci. USA* **74:**3438–3442.

Coughlin, M. D., M. D. Dibner, D. M. Boyer, and I. B. Black (1978). Factors regulating development of an embryonic mouse sympathetic ganglion, *Dev. Biol.* **66:**513–528.

Coulombre, A. J., M. C. Johnston, and J. A. Weston (1974). Conference on neural crest in normal and abnormal embryogenesis, *Dev. Biol.* **36:**f1–5.

Dale, H. (1937). Acetylcholine as a chemical transmitter of the effects of nerve impulses. I. History of ideas and evidence. Peripheral autonomic actions. Functional nomenclature of nerve fibers, *Mt. Sinai J. Med.* **4:**401–415.

Dibner, M. D. and I. B. Black (1976). The effect of target organ removal on the development of sympathetic neurons, *Brain Res.* **103:**93–102.

Dibner, M. D., C. Mytilineou, and I. B. Black (1977). Target organ regulation of sympathetic neuron development, *Brain Res.* **123:**301–310.

Dreyfus, C. F., M. B. Bornstein, and M. D. Gershon (1977). Synthesis of serotonin by neurons of the myenteric plexus in situ and in organotypic tissue culture, *Brain Res.* **128:**125–139.

Hamill, R. W., E. M. Bloom, and I. B. Black (1977). The effect of spinal cord transection on the development of cholinergic and adrenergic sympathetic neurons, *Brain Res.* **134:**269–278.

Hancock, M. B. (1976). Cells of origin of hypothalamo-spinal projections in the rat, *Neurosci. Lett.* **3:**179–184.

Helfand, S. L., R. J. Riopelle, and N. K. Wessells (1978). Non-equivalence of conditioned medium and nerve growth factor for sympathetic, parasympathetic, and sensory neurons, *Exp. Cell Res.* **113:**39–45.

Hendry, I. A. and L. L. Iversen (1973). Changes in tissue and plasma concentrations of nerve growth factor following removal of the submaxillary glands in adult mice and their effects on the sympathetic nervous system, *Nature (London)* **243:**500–504.

Le Douarin, N. M., D. Renaud, M. A. Teillet, and G. H. Le Douarin (1975). Cholinergic differentiation of presumptive adrenergic neuroblasts in interspecific chimeras after heterotopic transplantations, *Proc. Natl. Acad. Sci. USA* **72:**728–732.

Le Douarin, N. M. and M. A. Teillet (1973). The migration of neural crest cells to the wall of the digestive tract in avian embryo, *J. Embryol. Exp. Morphol.* **30:**31–48.

Le Douarin, N. M. and M. A. Teillet (1974). Experimental analysis of the migration and differentiation of neuroblasts of the autonomic nervous system and of neuroectodermal mesenchymal derivatives, using a biological cell marking technique, *Dev. Biol.* **41:**162–184.

Le Douarin, N. M., M. A. M. Teillet, and C. Le Lievre (1977). Influence of the tissue environment on the differentiation of neural crest cells, pp. 11–27 in *Cell and Tissue Interactions*, Lash, J. and M. Burger, eds. Raven Press, New York.

Levitt, M., S. Spector, A. Sjoerdsma, and S. Udenfriend (1965). Elucidation of the rate-limiting step in norepinephrine biosynthesis in the perfused guinea pig heart, *J. Pharmacol. Exp. Ther.* **148:**1–8.

Monard, D., F. Solomon, M. Rentsch, and R. Gysin (1973). Glia-induced morphological differentiation in neuroblastoma cells, *Proc. Natl. Acad. Sci. USA* **70:**1894–1897.

Monard, D., K. Stöckel, R. Goodman, and H. Thoenen (1975). Distinction between nerve growth factor and glial factor, *Nature (London)* **258**:444–445.
Norr, S. C. (1973). In vitro analysis of sympathetic neuron differentiation with chick neural crest cells, *Dev. Biol.* **34**:16–38.
Olson, L. and T. Malmfors (1970). Growth characteristics of adrenergic nerves in the adult rat, *Acta Physiol. Scand. Suppl.* **348**:1–111.
Patterson, P. H. (1979). Environmental determination of neurotransmitter functions, *Soc. Neurosci. Symp.* **4**:172–183.
Patterson, P. H. and L. L. Y. Chun (1974). The influence of non-neuronal cells on catecholamine and acetylcholine synthesis and accumulation in cultures of dissociated sympathetic neurons, *Proc. Natl. Acad. Sci. USA* **71**: 3607–3610.
Patterson, P. H., L. F. Reichardt, and L. L. Y. Chun (1975). Biochemical studies on the development of primary sympathetic neurons in cell culture, *Cold Spring Harbor Symp. Quant. Biol.* **40**:389–397.
Pick, J. (1963). The submicroscopic organization of the sympathetic ganglion in the frog (*Rana pipiens*), *J. Comp. Neurol.* **120**:409–462.
Saper, C. B., A. D. Loewy, L. W. Swanson, and W. M. Cowan (1976). Direct hypothalamo-autonomic connection, *Brain Res.* **117**:305–312.
Smith, J., P. Cochard, and N. M. Le Douarin (1977). Development of choline acetyltransferase and cholinesterase activities in enteric ganglia derived from presumptive adrenergic and cholinergic levels of the neural crest, *Cell Differ.* **6**:199–216.
Tennyson, V. (1965). Electron microscopic study of the developing neuroblast of the dorsal root ganglion of the rabbit embryo, *J. Comp. Neurol.* **124**: 267–317.
Varon, S., C. Raiborn, and P. A. Burnham (1974). Implication of a nerve growth factor-like antigen in the support derived by ganglionic neurons from their homologous glia in dissociated cultures, *Neurobiology (Copenh.)* **4**:317–327.
Weston, J. A. (1970). The migration and differentiation of neural crest cells, *Adv. Morphog.* **8**:41–114.

NEUROBIOLOGICAL INVESTIGATIONS WITH ANIMAL CHIMERAS AND MOSAICS

MUTANTS AND MOSAICS: TOOLS IN INSECT DEVELOPMENTAL NEUROBIOLOGY

John Palka

University of Washington, Seattle, Washington

In studying the cell-cell interactions that ultimately give rise to the properly assembled nervous system of the adult organism, it is often desirable to interfere with the normal developmental sequence of events. One may wish, for example, to remove set *A* of cells in order to discover whether another set, *B*, will take over the synaptic contacts that the *A* cells would have occupied, or whether the postsynaptic set, *C*, will be affected. Or, one may wish to delay the arrival of axons from set *A* to see whether precise timing is of importance in establishing the pattern of connections of *A* and *B* upon *C*. Or again, one may wish to remove set *C* and study the behavior of *A* and *B* axons when these are deprived of their normal targets. In order to obtain clear results in experiments of this kind, it is essential to be able to manipulate one cell population at a time, and to be sure that the experimental technique does not have complicating side effects.

Surgery has been the most commonly used technique in experimental developmental neurobiology (e.g., Jacobson, 1978) and has yielded many important results. However, there are cases in which it cannot be used because anatomical relations do not permit it. Furthermore, surgical interference is most often accomplished after the establishment of initial neuronal connections during embryogenesis. Consequently, such interference involves the degeneration and regeneration of cells, not just their clean removal or translocation. It is therefore desirable to have alternative methods of manipulating cell populations. One alternative sometimes available is to use mutations that have selective effects on particular classes of cells. Mutants are known that cause particular cells to degenerate (e.g., Simpson and Schneiderman, 1975) or to develop in abnormal locations (e.g., Postlethwait and Schneiderman, 1974), thus paralleling at least some of the manipulations that can be accomplished by surgery. Because mutations affecting a single class of cells intermingled with others in the same region are available (Harris, Stark, and Walker, 1976), "surgery" that could not be accomplished with the finest scalpel can be done for us in the fragile embryo and without any surgical trauma.

However, the use of mutants in the study of neural development brings with it its own set of technical problems. Perhaps the most ubiquitous is the problem of pleiotropy, the common finding that a single gene affects many parts of the organism. We are not entitled to assume that a mutation that has a striking effect upon a group of cells we are interested in has no effect upon any other cells. All cells in the animal, after all, carry the same genotype and are therefore potentially subject to the effects of an alteration in that genotype. The most direct way of dealing with the problem of pleiotropy has been to produce mosaic animals containing a combination of experimental (mutant) and normal (wild-type) cells. Mosaicism can be produced by both genetic and surgical techniques.

The purpose of this paper is to provide some illustrations of how mutants have proved useful in the study of neural development in insects, and how mosaics have enabled investigators to take advantage of the special features of mutants that make them such valuable complements to descriptions of normal development and to surgical manipulation.

Drosophila melanogaster has been the preferred animal in studies using mutants and mosaics. The enormous diversity of available mutations and of genetic techniques has made it an obvious choice, and its small size has been less of a barrier than might have been expected. Many neuroanatomical techniques work well on *Drosophila*, including selective filling methods using cobalt and horseradish peroxidase (HRP), as well as classical reduced silver methods. Small size is actually an advantage in electron microscopy. Electrophysiological recordings are easily made in muscle cells, and the simplicity of the innervation pattern, especially in flight muscles, means that the recordings offer a direct measure of motoneuronal activity.

TYPES OF MOSAICS

In order to illustrate current results and ideas in this field, it seems best to start with a review of types of mosaics that can be produced.

Gynandromorphs

These are individuals containing a mixture of male and female tissues. Such a mosaic will be produced if, usually very early in embryogenesis, a cell of a genetically female zygote (XX in chromosome constitution) loses one of the X-chromosomes. The progeny of that cell will then be XO and phenotypically male. Two features of such mosaics are noteworthy: male and female territories are usually contiguous, presumably because there is but little relative cell displacement or migration in insect development; and the territories are usually very large, because X-chromosome loss occurs very early. In the most frequently employed stocks, loss occurs during the very first nuclear division following fertilization, so half-and-half animals are produced. The orientation of the mitotic spindle in this first division is random, so the precise location of the male-female boundary is also random, and individuals having the boundary in any desired place can be selected. A detailed account of gynandromorphs and other types of genetic mosaics is given by Hall, Gelbart, and Kankel (1976).

Grafts

Grafts are readily accomplished in insects. They can be made between individuals of the same species that may or may not differ in genotype, and also between species and even genera. They can be made at any stage of development, including the adult, and even the tiny *Drosophila* is not beyond the reach of the hands of skillful experimenters. For example, my colleague Margrit Schubiger has adapted the grafting methods developed for much larger flies such as *Musca* by Bhaskaran and Sivasubramanian (1969) to *Drosophila*.

Mitotic Recombination Mosaics

Genetic recombination between chromatids normally occurs during meiosis, when homologous chromosomes pair intimately. In *Drosophila* and a few other animals it can also occur during ordinary mitotic division, perhaps because chromosome pairing also occurs during mitosis. The frequency of such recombination events is normally extremely low, but can be greatly increased by x-irradiation. If a fly is heterozygous for a recessive mutation, recombination can give rise to two sister cells, one of which is homozygous dominant (and usually unimportant or even undetectable) and the other of which is homozygous recessive and therefore shows the recessive trait. Its progeny will constitute a clone of mutant tissue in an otherwise wild-type animal and, as in gynandromorphs, the clone generally stays intact. Since somatic recombination is so rare in the absence of x-rays, the experimenter has control over when in development such a mosaic is produced simply by irradiating animals at different ages (for a technical review, see Becker, 1976).

Temperature-Sensitive Mutants

A number of mutations are known that become expressed only above or below some critical temperature (e.g., Suzuki, Kaufman, Falk, and the U.B.C. Drosophila Research Group, 1976). The experimenter is thus able to create a "mosaic in time," the gene in question being active at one time and inactive at some other time in the same individual animal.

In Table 1 I have ordered these four classes of mosaics according to whether the parameter most under the experimenter's control is space or time or both.

SOME SAMPLE RESULTS

Thesis 1. *The retinal cells associated with a single facet of the compound eye need not have a common clonal origin.*

Behind each facet of the compound eye of insects is a small cluster of receptor cells called retinula cells, collectively forming a repeat unit called an ommatidium (Figure 1). In most species there are eight retinula cells in each ommatidium, and it was believed for many years that these eight cells arise by three successive divisions from a single mother cell, i.e., that they constitute a clone (Bernard, 1937; Kühn, 1965). In the last few years mosaic techniques have been used to test this hypothesis, and the result of all studies to date is a clear *NO*—the cells of an ommatidium need not be clonally related. Whether they are clonally related in some significant proportion of cases is not clear, but they certainly need not be.

TABLE 1. *Various types of mosaics in* Drosophila

Type of mosaic	Domain of effect
Gynandromorph	Space
Graft	Space—Time
Mitotic recombinant	Space—Time
Temperature-sensitive mutant	Time

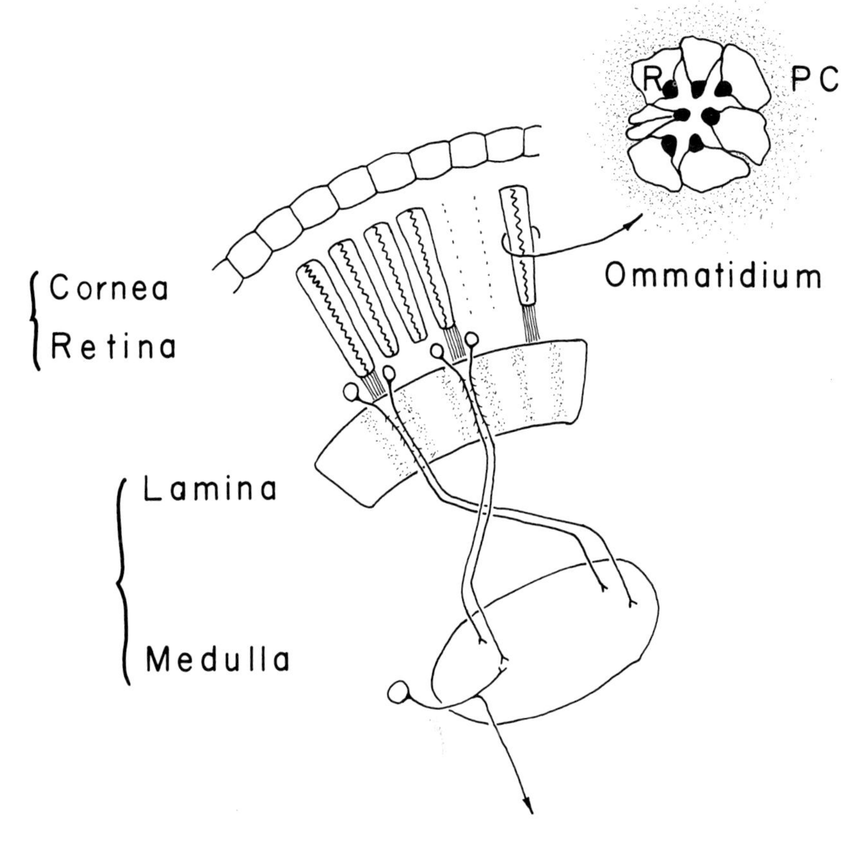

FIGURE 1. Schematic diagram of the insect visual system. The repeat unit of the retina is the ommatidium. Each of its eight retinula cells (R in inset at upper right) contributes a column of microvilli called a rhabdomere, in whose membranes the receptor pigment is lodged. Ommatidia are optically separated from each other by pigment cells (PC). The first and second optic neuropils (lamina and medulla) also have repeating elements called cartridges, as suggested in the drawing. The details of the system vary in different groups of insects. For the experiments of *Thesis 1* it is important to realize that the cornea and retina arise from epidermal cells, whereas the lamina and the more central neuropils arise from neuroblasts. Thus, the lamina can easily be of different genotype from the retina.

The evidence for this conclusion comes from an examination of ommatidia at the borderline between retinal tissues of different genetic composition, differing in a cellular marker that makes it possible to score the genotype of every cell in the ommatidium. If all eight cells form a clone stemming from a single mother cell, then all eight must be of the same genotype; conversely, if ommatidia are found whose component cells differ in genotype, at least two mother cells must have contributed to their formation. By grafting patches of retinal precursor tissue marked with a color mutation (milkweed bugs, Shelton and Lawrence, 1974; cockroaches, Shelton, Anderson, and Eley, 1977) and by producing mosaic eyes by mitotic recombination in *Drosophila* (Hofbauer and Campos-Ortega, 1976; Ready, Hanson, and Benzer, 1976), it has been found that ommatidia are frequently of mixed composition and their cells therefore cannot be clonally related (Figure 2). The results from surgical and genetic mosaics agree entirely, and an older view, based on observational rather than experimental evidence, has been overturned by the use of mosaic methods.

Thesis 2. *The development of the retina is independent of other tissues, but the proper differentiation and even survival of first-order visual interneurons depends on the adjacent retina.*

As long ago as 1922 Kopeć showed that the retina of a moth would survive nicely if transplanted to the abdomen, but the underlying first optic neuropil remaining *in situ* (lamina) became severely reduced and disorganized. Many other studies based on surgical methods and on the analysis of mutants have pointed to the same conclusion (for reviews, see Meinertzhagen, 1973, 1975; Anderson, 1978; Palka, 1979). But perhaps the most elegant and detailed illustration of this one-way trophic relationship has been provided by Meyerowitz and Kankel (1978) using genetic mosaics in *Drosophila* (Figure 3).

Using the recessive mutants *rough* and *glass*, which disrupt retinal organization as well as the regular geometry of the facets of the compound eye, Meyerowitz and Kankel produced mosaics in which part of the eye was mutant and part was normal (wild-type), and the underlying lamina was also wild-type. Both gynandromorph and mitotic recombination techniques were used. Invariably, wild-type optic lobe tissue underlying wild-type retina was normal, but wild-type optic lobe underlying mutant retina was badly disordered (Figure 3*A*). The wild-type lamina was disorganized if and only if the retina was mutant, and causality was inferred.

In a complementary experiment with the dominant mutant *Glued*, which also causes disarray of the facets and the retina, Meyerowitz and Kankel were able to show that mutant optic lobe tissue could be rescued from disorganization by an overlying patch of wild-type retina (Figure 3*B*). In this experiment the whole fly was heterozygous for the dominant mutation *Glued* and both the retina and the lamina were

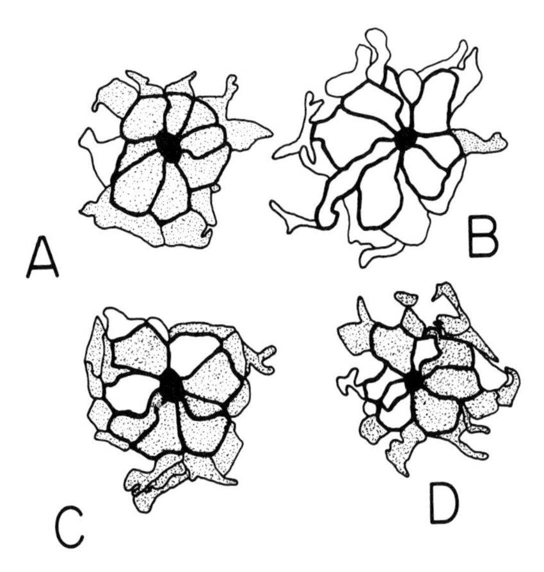

FIGURE 2. Examples of ommatidia with mixed genotypes at a host-graft boundary in a mosaic cockroach eye. In *A*, all the retinula cells are pigmented (stippled), as are all the surrounding pigment cells except one. Note that in the ommatidia of cockroach and many other insect eyes the rhabdomeres contributed by all eight retinula cells are fused into a single central column, the rhabdome, rather than being separated as in the case of the fly eye, illustrated in Figure 1. In *B*, all the cells are white (mutant) except for one pigment cell. In *C* and *D*, various combinations of pigmented and white cells occur. Such results show that neither the eight retinula cells nor the associated pigment cells arise from a single mother cell (*Thesis 1*). (After Shelton et al., 1977.)

badly disorganized. When part of the retina was caused to assume a wild-type genotype and architecture by gynandromorph or mitotic recombination techniques, the underlying lamina also acquired a normal architecture, even though it continued to have a mutant genotype. Not only does this result reemphasize the dependence of the lamina upon the retina, but it also shows that mutant, disorganized lamina does not adversely affect the structure of the retina. This entire study confirms the conclusions reached earlier on the basis of surgical experiments.

Thesis 3. *The developmental programs of different receptor cell types are stable in the face of segmental translocation, and translocated axons interact in orderly fashion with their new substrates.*

Thus far we have encountered mutants for color or other nonessential attributes that are useful as markers of cell origin (*Thesis 1*), and mutants that disrupt selected cell populations and are useful in analyses of trophic relationships (*Thesis 2*). We consider now another class of mutants called homeotic mutants, in which some part of the body is qualitatively transformed into a different part of the body. An excellent example is the mutant *Antennapedia*, in which the more distal parts of the antenna are transformed into leg. The homeotic leg is exceedingly well formed and clearly identifiable as being a middle (mesothoracic) leg rather than a fore- (prothoracic) or hind- (metathoracic) leg, and its neural projections have been studied both anatomically (Stocker, Edwards, Palka, and Schubiger, 1976) and functionally (Deak, 1976; Stocker, 1977). With due caution about possible pleiotropy, homeotic mutants can be regarded as nature's nontraumatic equivalent of surgical translocations.

We have recently been engaged in a more detailed exploration of sensory projections in a different group of homeotic mutants (Palka, Lawrence, and Hart, 1979), those belonging to the so-called *bithorax* series. Flies are members of the order Diptera, so called because its members have only two wings, unlike other insects, which have four. The hind wings of dipterans are modified into halteres, sense organs important in flight orientation in three dimensions. A number of mutations are available that transform halteres into wings and wings into halteres (Lewis, 1963), and some are schematized in Figure 4. Since

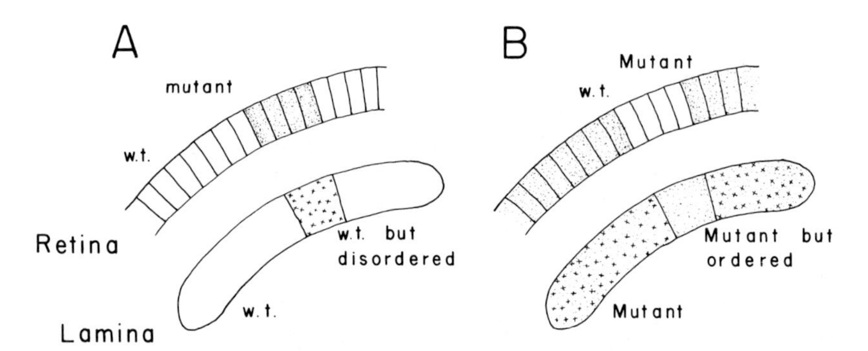

FIGURE 3. Schematic diagram illustrating the experimental basis for *Thesis 2*. Wild-type tissue is shown unshaded and mutant is stippled; disordered lamina is indicated by crosses. In *A*, wild-type (w.t.) but disordered lamina is present under a patch of mutant retina produced by mitotic recombination. In *B*, the lamina is all mutant, but the portion of it under the w.t. retinal patch is rescued from disorder.

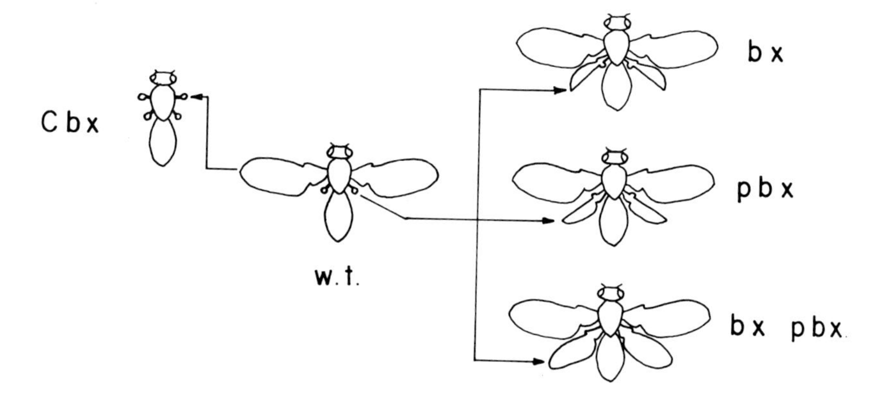

FIGURE 4. Some mutants of the *bithorax* series. *Bithorax (bx)* causes the transformation of anterior metathorax into anterior mesothorax, and thus of anterior haltere into anterior wing. *Postbithorax (pbx)* causes a similar transformation of posterior metathorax into posterior mesothorax. The double mutant has four wings and two sets of mesothoracic legs. *Contrabithorax* (*Cbx*) transforms meso- to metathorax, and in extreme cases flies with four halteres and no wings are found.

the central projections of the sensory cells originating in the two appendages are different, one can ask what will happen when wing axons enter the segment of the CNS that normally receives haltere axons (the metathoracic neuromere, Figure 5), and vice versa.

It is important to emphasize that the cell bodies of virtually all insect sensory neurons are peripherally located in the epidermis, just under the cuticle (Figure 5). Each receptor organ or sensillum is typically formed by four cells believed to be derived by two consecutive divisions from a single mother cell, and thus to be clonally related (Lawrence, 1966). The mother cell starts out as an ordinary epidermal cell, and the differentiative divisions producing the cells of the sensillum occur rather late in development. Thus, the homeotic transformation can be expected to affect the sensory neurons just as much as any other cells of the body surface, and indeed the sensilla constitute the best markers of the source of a piece of cuticle. An expert can identify a small, crumpled patch of cuticle as being leg or thorax or wing primarily on the basis of the sensilla it carries.

We usually think of the wing of an insect as a locomotor organ, the animal's aerofoil and propeller. However, it is also richly endowed with sense organs: in *Drosophila* there are three rows of bristle sensilla along its leading edge, and sensilla campaniformia on the veins of the

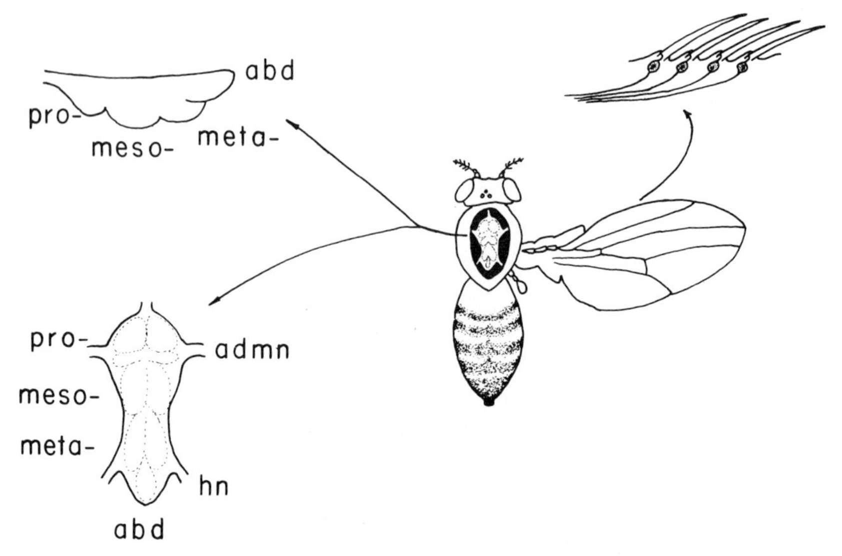

FIGURE 5. The thoracicoabdominal nervous system of *Drosophila*. On the left is shown the CNS in lateral and dorsal view. The three thoracic ganglia, which are separate in most insects and even in some primitive flies, are here fused and called pro-, meso-, and metathoracic neuromeres. The several abdominal ganglia (abd) are also fused together and to the thoracic ones. The sensory fibers arising from peripheral receptor cell bodies in the wing (upper right) enter the CNS through the mixed anterior dorsal mesothoracic nerve (admn). Axons from bristle sensilla on the wing terminate in a rather discrete anterior mesothoracic region called the accessory mesothoracic neuromere. The sensory fibers arising in the haltere enter through the purely sensory haltere nerve (hn).

wing blade (Figure 6). The bristles are both mechanosensory and chemosensory in function. The campaniform sensilla respond to stresses and strains in the cuticle, not unlike the strain gauges an aeronautical engineer might glue to the surface of an airplane wing to monitor its flexions and oscillations. There are two distinct classes of sensilla campaniformia (s.c.): a few large ones distributed on several veins, and about 55 small ones on the principal proximal vein, the radius. Like wings, halteres have small campaniform sensilla near the base. These number several hundred, and are activated as the slender stalk of the haltere is deformed during turning or acceleration. The heavy head of the haltere, functioning primarily as a gyroscopic mass, carries only a few slender bristles called sensilla trichodea.

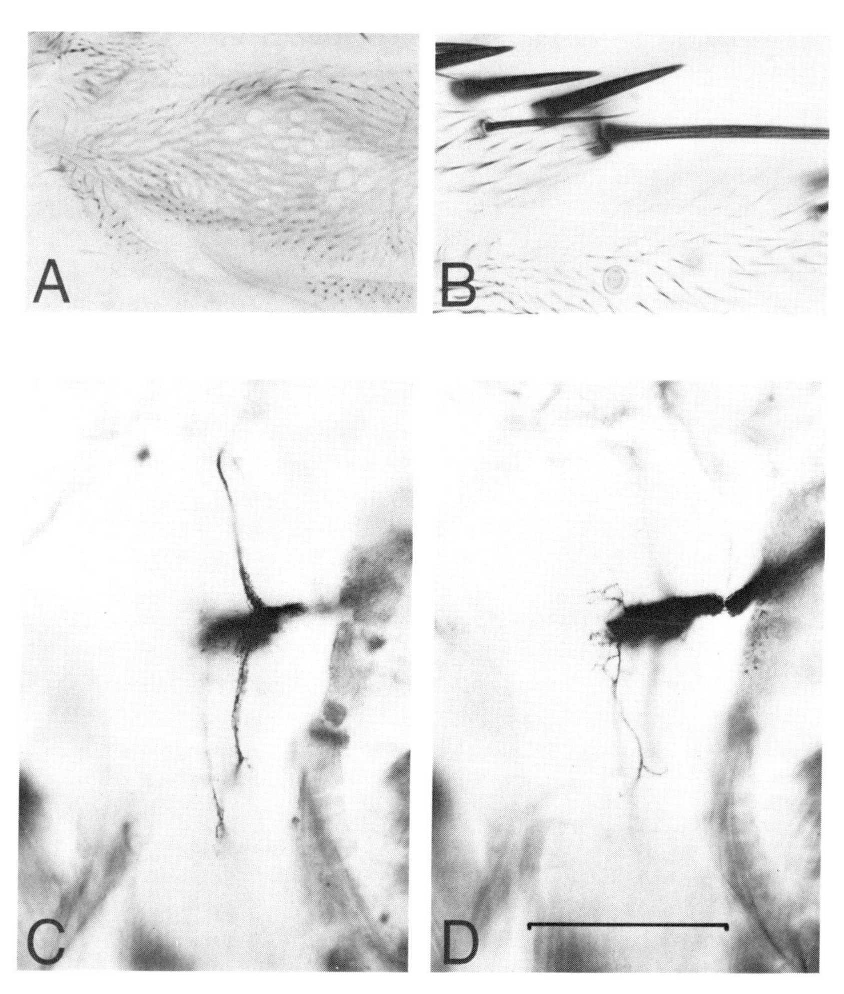

FIGURE 6. Some details of the wing projection in wild-type *Drosophila*. *A*: Small campaniform sensilla on the radial vein of the proximal wing show as light ovals with a faint rim. *B*: Bristle sensilla on the wing margin and two of the large campaniform sensilla—one in sharp focus with a distinct double rim and gray center, the other largely out of focus. *C* and *D*: A cobalt-filled preparation showing the distribution of wing sensory fibers in the CNS in two different focal planes. *C* shows the dorsal component formed by axons from small s.c. such as those shown in *A*; the ovoid is deeper and out of focus. *D* shows the densely filled ovoid formed by axons from bristles and located in the accessory mesothoracic neuromere (see Figure 5), and the widely ramifying large ventral fibers arising from large s.c. such as those shown in *C*. Scale = 50 micrometers.

The complex sensory apparatus of the wing is reflected in a complex central projection, studied recently in our laboratory (Palka, 1977; Palka et al., 1979) using the cobalt backfilling method, and independently by Ghysen (1978) using HRP. Corresponding to the three classes of sensilla on the wing (bristles, large s.c., and small s.c.) are three major groups of central fibers (Figures 6 and 7). The bristles send fine axons into a compact ventral area in the anterior mesothoracic neuromere. The axons of the large s.c. are unusually thick and ramify in all three thoracic neuromeres of the CNS. The small s.c. send their axons into a bifurcating dorsal tract and through this anteriorly into the head and posteriorly into the metathoracic neuromere. The haltere, with a great predominance of small s.c., has only a dorsal projection with characteristic tufts of fibers in the meta- and mesothorax (Figure 7). In cobalt fills, nothing at all is recognizable in the deep ventral levels where the bristles and large s.c. of the wing terminate.

In homeotic mutants that develop a wing in place of a haltere, the homeotic wing fibers enter the CNS at the point normally entered by the haltere nerve. The following features are seen in the central projection (Figure 7): (1) The fine axons from bristles accumulate in the ventral anterior part of the third neuromere; coexisting bristle axons from the normal wing accumulate as usual in the ventral anterior part of the second neuromere. (2) The large axons from the large s.c. also grow ventrally. Unlike the bristle axons, they form a pattern which seems to overlap the distribution of the coexisting large ventral axons from the normal wing in all three neuromeres (see especially Ghysen, 1978). (3) The axons from the small s.c. stay dorsal, as they would in their own segment. Their distribution in the whole ganglion is not distinguishable from the distribution of *haltere* s.c. by any method employed thus far, even though the cuticular components of the s.c. are clearly of wing rather than haltere anatomy in the vast majority of cases.

Why do these three classes of receptors respond differently to the translocation accomplished by the homeotic mutation? One might have expected that all homeotic wing axons would seek native wing axon destinations, but this is only true of some fibers (large s.c.). Or one might have expected that the metathoracic neuromere would dominate all incoming fibers in such a way that they would be steered to haltere destinations regardless of the fact that they were coming from a wing. Again, this happens only to some fibers (small s.c.). And the bristle axons behave in yet another way—they seem to respond to the metathoracic neuromere as if it were a mesothoracic one.

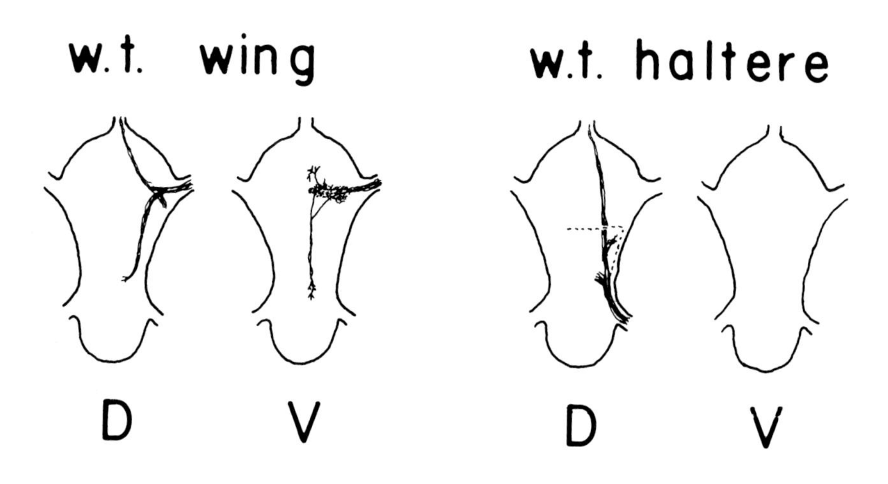

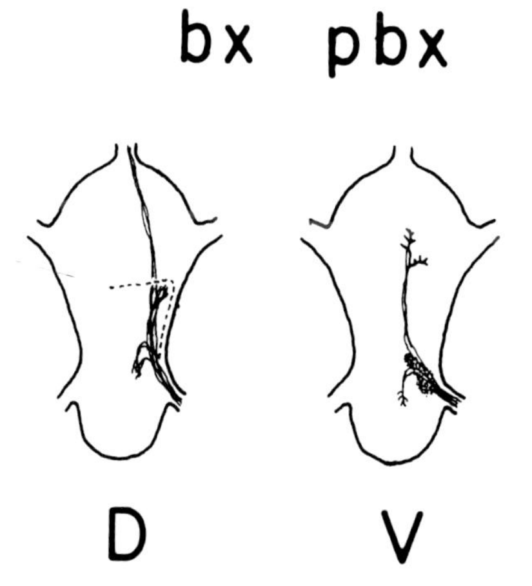

FIGURE 7. Central projections of wild-type and homeotic wings and halteres. In each panel, dorsal (D) and ventral (V) components are shown separately. Details in text. (After Palka et al., 1979.)

We have proposed (Palka et al., 1979) that no single expectation is fulfilled by all classes of fibers because each class carries different developmental instructions. These instructions relate to the nature of the interaction between a growing fiber and the substrate it grows through, namely the ganglion, which is already significantly differentiated at the time of axon ingrowth. We suppose that among the instructions carried by bristle axons is some statement such as "grow ventral and anterior until you reach the neuromere boundary." As long as the metathoracic neuromere is reasonably similar to the mesothoracic one, this instruction will result in an accumulation of bristle axons in its anterior ventral region. But we expect the axons of large s.c. to have a different set of instructions, because their distribution in wild-type animals is different. Suppose their instructions include a search command for particular ventral landmarks distributed in all three neuromeres, which would be consistent with their normal anatomy. Then in the homeotic situation we would not expect them to behave as bristle axons do, but rather to search out those same landmarks.

The case of the homeotic small s.c. whose axons seem to distribute along the same paths and to the same destinations as haltere axons may be slightly different, not in principle but because the normal haltere also bears small s.c. Perhaps the difference between wing and haltere small s.c. is so slight that the cells' pathfinding program, whatever instructions it may contain in addition to "stay dorsal," responds to haltere recognition sites nearly as well as native haltere axons do. We may be seeing a reflection of the fundamental homology between wing and haltere, which is not apparent in the other two cases because the haltere does not normally carry those classes of receptors.

Thus, we would like to think that studying the behavior of sensory axons in homeotic mutants helps us to appreciate the properties of the developmental programs that characterize different types of receptors. At this point, however, we have to face the possibility that meso- and metathoracic neuromeres are normally significantly different, provide different cues to incoming sensory axons, but that the same mutation that transforms the haltere into a wing simultaneously transforms the CNS, which then contains two mesothoracic neuromeres. Mitotic recombination mosaics have allowed us to test this possibility.

Large numbers of flies heterozygous for the *bithorax* (*bx*) mutation were x-irradiated with a dose of 1000 R during two age windows: as late embryos or as first instar larvae. These times were late enough to assure that the cells giving rise to the CNS had already separated from those giving rise to the epidermis, but early enough to produce quite large clones. The dose was sufficiently low to make the probability of two independent recombination events in a single animal very low. We collected 38 animals that had wing tissue (clones of cells homozygous for *bx*) on the haltere but were otherwise wild-type in phenotype. When the mosaic appendages were filled with cobalt, all but one showed a substantial population of axons in the ventral part of the metathoracic neuromere. Remember that normal halteres send no axons to this area, but homeotic appendages do. Clearly, the ability of homeotic sensory axons to reach ventral destinations in the CNS does not require a homeotic transformation of the meta- into a mesothoracic neuromere.

However, there were indications that some transformation of the CNS does indeed occur. Although the general arrangement of this ventral projection in mosaic flies resembled that in mutant flies, a close examination showed a number of subtle differences. For example, the point at which the ventral separated from the dorsal fibers was in a slightly different location, and fibers sometimes appeared in novel locations where we had never seen them either in wild-type or in mutant flies. Some indications of a transformation of neuromeres have also been found in a histological study of the CNS by D. G. King (personal communication). Thus, we conclude that the homeotic mutations we have worked with do affect the CNS, and their effects do have a bearing on the distribution of primary sensory axons. However, many features of that distribution do not require that the CNS be transformed, probably because the serially homologous thoracic neuromeres are so similar to begin with.

This mosaic study has given us evidence of subtle changes in the CNS, changes that are not apparent when only mutant flies are studied. Is this finding of significance beyond the study of an interesting class of mutants? I believe that it reminds us of the basic importance of segmentation in the construction of insect bodies. Segments, after all, are variations on a common theme. Comparing mutants and mosaics may help us to discover what is common to the CNS modules of the various segments, and what the variations consist of. And segmentation is a very widespread characteristic in the animal kingdom. Not only do worms and leeches, bugs, lobsters, and spiders have it, but we do, too.

REVIEW AND PROSPECT

In this brief discussion we have considered three classes of mutants: those that provide markers by which the origin of a given cell can be traced but that do not affect developmental processes pertinent to neural organization; those that cause disruption or elimination of certain cells during development; and those that alter the course of development, switching it from one normal path to another. We have reviewed three specific studies, one using each type of mutant. And finally, we have seen that the construction of mosaics has been essential to the experimental design of each of these studies. It is probably not overstating the matter to assert that in studies of cell-cell interactions in neural development, the intriguing possibilities offered by mutants usually cannot be exploited without the use of mosaics.

Granted that both mutants and mosaics are fascinating to work with, can we point to new findings made possible by their use? *Thesis 1* (nonclonal origin of cells in an ommatidium) contradicts the earlier literature and arises directly from the use of mosaics. Mutants provided the necessary cellular markers. If mitotic recombination mosaics and mosaics produced by grafting had yielded different results, we would have had to consider the possibility of disruptive aftereffects at the surgically produced host-graft borderline. In fact, however, genetic and surgical mosaics both strongly support *Thesis 1*.

Thesis 2 (independence of the retina, dependence of lamina upon retina) was well established by the time the genetic mosaic analysis was done. However, if one carefully reviews each of the previous studies, it becomes clear that no single one of them provides complete evidence for the thesis. The use of suitable mutants and mosaics made it possible to draw many lines of evidence together in a single study.

Thesis 3 (stable expression of developmental programs following "translocation," similarity of serially homologous regions of CNS) is more in the nature of an exploration than a statement of a highly specific proposition. I personally have been especially intrigued with the observation that different classes of receptors behave differently in the homeotic situation—that no single statement such as "homeotic axons seek that region of the CNS appropriate to the nature of the cuticle in which they arise" is an adequate description of the behavior of all cells. Such a description seems to fit the axons of the large s.c. on the wing, but not the small s.c. or the bristles (compare Ghysen, 1978; and Palka et al., 1979) or, for that matter, the axons of the sensilla on the homeotic leg of *Antennapedia*, which terminate primarily in brain areas appropriate to the antenna (Stocker et al., 1976). To me, the lesson of this diversity is that we are seeing the responses of cells with different developmental programs to a spatial perturbation of normal development. It may be possible to use such perturbations to gain further insight both into the kinds of instructions that the receptor axons carry and into the guiding factors within the CNS to which the axons respond, each in its own way.

Likewise, I am intrigued with the subtlety of the changes in the CNS introduced by mutations that alter the surface of the fly so profoundly. Most of the pieces of evidence suggesting changes come from a comparison of mosaic and mutant animals. Individually they are not striking, but in the aggregate I find them convincing. It may be that a closer analysis will show not only which aspects of ganglionic structure are stable in the face of homeotic transformation and which are labile, but also which aspects are important to the incoming axons as they seek to express the developmental programs they carry.

Finally, I want to point out that the various homeotic mutants are a big part of the empirical basis for the theory of compartments (García-Bellido, Ripoll, and Morata, 1973; García-Bellido, 1975), perhaps the most explicitly formulated theory of insect development thus far proposed. This theory seeks to explain how the complex insect body can be built out of a series of building blocks (revealed only in mosaic animals!) that would develop identically except for the action of a small number of specific "selector genes," genes that select which of two alternative routes of development a given, spatially defined "compartment" of the body will follow. The theory is subtle and cannot be reviewed in this short space. (See Crick and Lawrence, 1975; Lawrence and Morata, 1976. Palka, 1979, gives a summary in relation to neurobiology.) But it repays close study, and stands as a challenge to students of neural development to integrate their findings with those of developmental biologists. Surely the notion of building blocks with many similar features underlying their specific differences is an appealing and economical one, well worth testing as new data become available.

In sum, then, mutants and mosaics offer a rich array of tools for probing the steps by which a complex nervous system becomes assembled. They have already offered some important insights, and many more can be expected.

ACKNOWLEDGMENTS

I am grateful to my friends and colleagues, Margrit Schubiger, Hilary Anderson, and John Edwards, for careful reading of the manuscript and for many challenging discussions. Steve Hart shared with me much of the original work presented here. Our research has been supported by United States Public Health Service grant NS-07778.

REFERENCES

Anderson, H. (1978). Postembryonic development of the visual system of the locust, *Schistocerca gregaria*. I. Patterns of growth and developmental interactions in the retina and optic lobe, *J. Embryol. Exp. Morphol.* **45:** 55–83.

Becker, H. J. (1976). Mitotic recombination, pp. 1019–1087 in *The Genetics and Biology of Drosophila*, Vol. 1c, Ashburner, M. and E. Novitski, eds. Academic Press, New York.

Bernard, F. (1937). Recherches sur la morphogénèse des yeux composés d'arthropodes, *Bull. Biol. Fr. Belg. (Suppl.)* **23:**1–162.

Bhaskaran, G. and P. Sivasubramanian (1969). Metamorphosis of imaginal disks of the housefly: evagination of transplanted disks, *J. Exp. Zool.* **171:**385–396.

Crick, F. H. C. and P. A. Lawrence (1975). Compartments and polyclones in insect development, *Science* **189:**340–347.

Deak, I. I. (1976). Demonstration of sensory neurones in the ectopic cuticle of spineless-aristapedia, a homeotic mutant of *Drosophila*, *Nature (London)* **260:**252–254.

García-Bellido, A. (1975). Genetic control of wing disc development in *Drosophila*, pp. 161–178 in *Cell Patterning*, Ciba Foundation Symposium 29 (New Series). Elsevier/North Holland, New York.

García-Bellido, A., P. Ripoll, and G. Morata (1973). Developmental compartmentalisation of the wing disc of *Drosophila*, *Nat. New Biol.* **245:**251–253.

Ghysen, A. (1978). Sensory neurones recognize defined pathways in *Drosophila* central nervous system, *Nature (London)* **274:**869–872.

Hall, J. C., W. M. Gelbart, and D. R. Kankel (1976). Mosaic systems, pp. 265–314 in *The Genetics and Biology of Drosophila*, Vol. 1a, Ashburner, M. and E. Novitski, eds. Academic Press, New York.

Harris, W. A., W. S. Stark, and J. A. Walker (1976). Genetic dissection of the photoreceptor system in the compound eye of *Drosophila melanogaster*, *J. Physiol. (London)* **256:**415–439.

Hofbauer, A. and J. A. Campos-Ortega (1976). Cell clones and pattern formation: genetic eye mosaics in *Drosophila melanogaster*, *Wilhelm Roux's Arch. Dev. Biol.* **179:**275–289.

Jacobson, M. (1978). *Developmental Neurobiology*, 2nd ed. Plenum Press, New York.

Kopeć, S. (1922). Mutual relationship in the development of the brain and eyes of Lepidoptera, *J. Exp. Zool.* **36:**459–467.

Kühn, A. (1965). *Lectures on Developmental Physiology*, 2nd ed., Milkman, R., trans. Springer-Verlag, New York.

Lawrence, P. A. (1966). Development and determination of hairs and bristles in the milkweed bug, *Oncopeltus fasciatus* (Lygaeidae, Hemiptera), *J. Cell Sci.* **1:**475–498.

Lawrence, P. A. and G. Morata (1976). The compartment hypothesis, pp. 132–149 in *Insect Development*, Lawrence, P. A., ed. *Symp. R. Entomol. Soc. London*, No. 8, Blackwell Scientific Publications, Oxford.

Lewis, E. B. (1963). Genes and developmental pathways, *Am. Zool.* **3:**33–56.

Meinertzhagen, I. A. (1973). Development of the compound eye and optic lobes of insects, pp. 51–104 in *Developmental Neurobiology of Arthropods*, Young, D., ed. Cambridge University Press, Cambridge.

Meinertzhagen, I. A. (1975). The development of neuronal connection patterns in the visual systems of insects, pp. 265–288 in *Cell Patterning*, Ciba Foundation Symposium 29 (New Series). Elsevier/North Holland, New York.

Meyerowitz, E. M. and D. R. Kankel (1978). A genetic analysis of visual system development in *Drosophila melanogaster*, *Dev. Biol.* **62**:63–93.

Palka, J. (1977). Neurobiology of homeotic mutants in *Drosophila*, *Soc. Neurosci. Abstr.*, Vol. 3, p. 187.

Palka, J. (1979). Theories of pattern formation in insect neural development, *Adv. Insect Physiol.*, in press.

Palka, J., P. A. Lawrence, and H. S. Hart (1979). Neural projection patterns from homeotic tissue of *Drosophila* studied in *bithorax* mutants and mosaics, *Dev. Biol.* **69**:549–575.

Postlethwait, J. H. and H. A. Schneiderman (1974). Developmental genetics of *Drosophila* imaginal discs, *Annu. Rev. Genet.* **7**:381–433.

Ready, D. F., T. E. Hanson, and S. Benzer (1976). Development of the *Drosophila* retina, a neurocrystalline lattice, *Dev. Biol.* **53**:217–240.

Shelton, P. M. J., H. J. Anderson, and S. Eley (1977). Cell lineage and cell determination in the developing compound eye of the cockroach, *Periplaneta americana*, *J. Embryol. Exp. Morphol.* **39**:235–252.

Shelton, P. M. J. and P. A. Lawrence (1974). Structure and development of ommatidia in *Oncopeltus fasciatus*, *J. Embryol. Exp. Morphol.* **32**:337–353.

Simpson, P. and H. A. Schneiderman (1975). Isolation of temperature sensitive mutations blocking clone development in *Drosophila melanogaster*, and the effects of a temperature sensitive cell lethal mutation on pattern formation in imaginal discs, *Wilhelm Roux's Arch. Dev. Biol.* **178**:247–275.

Stocker, R. F. (1977). Gustatory stimulation of a homeotic mutant appendage, *Antennapedia*, in *Drosophila melanogaster*, *J. Comp. Physiol.* **115**:351–361.

Stocker, R. F., J. S. Edwards, J. Palka, and G. Schubiger (1976). Projection of sensory neurons from a homeotic mutant appendage, *Antennapedia*, in *Drosophila melanogaster*, *Dev. Biol.* **52**:210–220.

Suzuki, D. T., T. Kaufman, D. Falk, and the U. B. C. Drosophila Research Group (1976). Conditionally expressed mutations in *Drosophila melanogaster*, pp. 207–263 in *The Genetics and Biology of Drosophila*, Vol. 1a, Ashburner, M. and E. Novitski, eds. Academic Press, New York.

AMPHIBIAN CHIMERAS AND THE NERVOUS SYSTEM

William A. Harris

Harvard Medical School, Boston, Massachusetts

"The field of causal neurogenesis lies in a kind of biological no-man's-land between neurology and experimental embryology and must draw on one or both of these well grounded disciplines for its substance. It has no true historical background; its geneology is spurious. To put it rather fancifully, causal neurogenesis has been the female partner in a sort of morganatic marriage, the issue of which has been without recognition or title."

Jean Piatt, 1948

INTRODUCTION

The study of amphibian chimeras begins with the fantastic experiments of Born (1894, 1897). In the course of his studies on regeneration, Born discovered that amphibian embryos, after being completely severed in two, might heal together and participate in normal development. Born further found that embryos from different species or even different genera can be united to form composite individuals. In Figure 1 we see two of Born's monsters, the first scientifically created animal chimeras.

Until relatively recently there were just two types of amphibian chimeras: "heteroplastic," composed of tissues from animals of different species within one genus, and "xenoplastic," composed of tissues from animals of different genera or more distant systematic relationships. Because of progress in amphibian genetics, we can now study more subtle chimeras composed of tissues that differ in genotype by only one gene function. Thanks to the pioneering work of the late Rufus Humphrey on axolotl genetics (Humphrey, 1975; Briggs, 1973;

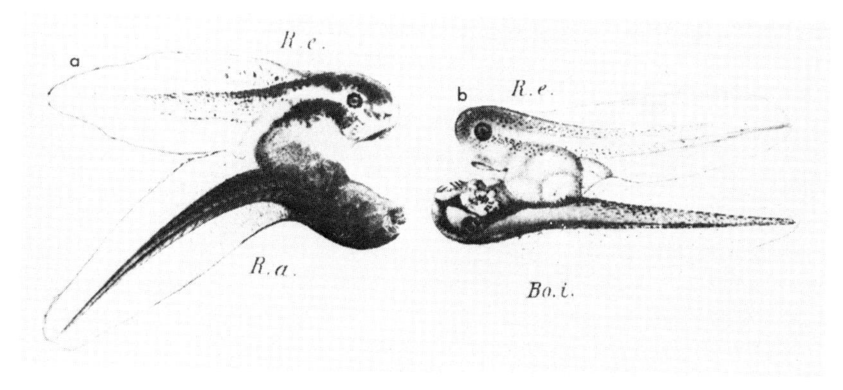

FIGURE 1. Two amphibian chimeras of Born. Left, heteroplastic chimera with two species of frog (*Rana arvalis, R.a.*; and *Rana esculenta, R.e.*). Right, xenoplastic chimera with frog and toad (*Bombinator igneus, Bo.i.*). (From Born, G. [1897]. Über Verwachsungsversuche mit Amphibienlarven, *Wilhelm Roux' Arch. Entwicklungsmech. Org.* **4**:349 ff. With permission of Springer-Verlag.)

Malacinski and Brothers, 1974) and others on the genetics of *Rana* (Browder, 1975) and *Xenopus* (Gurdon and Woodland, 1975), there are now dozens of interesting and useful amphibian mutants.

Nuclear transplantation into amphibian oocytes (Briggs and King, 1952) brought with it a new class of composite animals with nuclear material from one species and cytoplasm from another (Hennen, 1967). More recently still, plasmid clones of single genes from immensely distant species such as *Drosophila* can be injected into amphibian oocyte nuclei, where they are able to function (Gurdon, 1977). Whether these last two types of composite individuals will have any particular significance for neurobiology is a question for the future.

Why talk about chimeras as a unifying theme for a neurobiological symposium? In themselves chimeras do not present any particular set of problems in the nervous system. Rather they represent a biological technique like extracellular electrophysiology, immunohistochemistry, or tissue culture. The chimeric technique is powerful for studying causal neurogenesis. Its power lies in the fact that one can combine tissues from organisms that differ from each other in many ways (e.g., pigmentation, rate of growth and development, ultimate size and morphology, presence or absence of organs, etc.) and yet that show a mutual immunological tolerance when combined in one organism (Harrison, 1934). With such chimeras one can ask which tissues govern cell proliferation, growth rate, morphogenesis, or size, which adult tissues derive from which embryonic ones, what are the potencies of the various embryonic germ layers, and which parts of the organism are able to induce and organize the nearby tissues of another species. One can also learn what developmental chemistry may be common to different species.

Not all grafts between amphibians take. Intolerance increases with age (Eakin and Harris, 1945). Graft rejection has, in fact, recently been used to characterize the genetics of the axolotl immune system (Delanney, 1978).

Many of the studies with heteroplastic chimeras mentioned in this paper are more than 30 years old and were conducted by embryologists who knew none of the recent advances in the understanding of nervous system function through electrophysiology. Some of these studies, while not directly related to the nervous system, have interesting neurophysiological implications. As an example, take Harrison's (1924) paper entitled "Some unexpected results of heteroplastic transplantations of limbs." Forelimb buds from the small salamander *Ambystoma punctatum* were interchanged with those of the large *Ambystoma tigrinum*. The basic result is that the limbs maintain the growth rate and ultimate size characteristic of the donor species even when transplanted. Figure 2 shows a *punctatum* bearing a right limb grafted in the

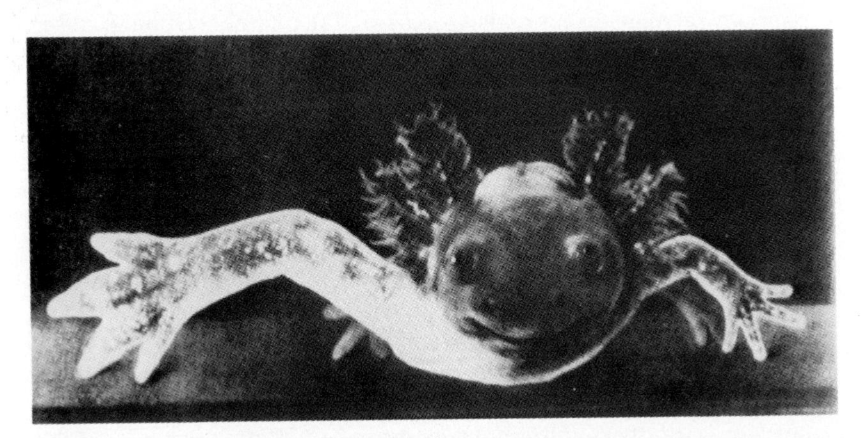

FIGURE 2. Larva of *Ambystoma punctatum* with a right forelimb grafted in the embryonic stage 70 days earlier from the larger-growing *Ambystoma tigrinum*. (From Harrison, R. G. [1934]. Heteroplastic grafting in embryology, *Harvey Lect.* **29:** 116–157. With permission of Academic Press.)

embryonic stage from the larger-growing *tigrinum*. Chimeric studies combining limb mesoderm from one species with ectoderm from another show that the embryonic mesoderm, including the primordial muscular and skeletal systems, is dominant in forming the basic character of the limb (i.e., its size and shape), whereas the ectoderm plays only a minor role in the limb's rate of growth (Rotmann, 1929, 1931, 1933; Harrison, 1931).

Although the *tigrinum* limbs eventually become much bigger than *punctatum* ones, they develop more slowly. Throughout most of embryogenesis the transplanted *tigrinum* limb is relatively underdeveloped with respect to the rest of the animal. It seems as though the central nervous system does not respect the underdeveloped state of the limb, which becomes innervated precociously. Such a limb begins to move sooner than control limbs of the same size and shape. In the reverse transplant, the more developmentally advanced *punctatum* limb remains immobile, waiting for its nervous connections, well past the stage when it would normally begin to function. Both types of transplanted limb eventually develop perfect coordination. These results indicate the rather astonishing fact that the periphery cannot have much accelerating or decelerating effect on the development of the peripheral nervous system. Are motoneurons in the one case synapsing with abnormally immature muscle fibers? In a reciprocal way, these results argue that the rate of development and the morphology of the limb are not a function of the host's invading nerves.

This example demonstrates the way chimeras can be used to pinpoint which components of the developing animal are involved in the growth of a particular organ and which components are not. Much of the nervous system itself has been studied in this way with chimeras.

BASIC METHODS AND RESULTS

Growth Rate and Size Regulation

Many components of the nervous system have been grafted heteroplastically in order to study their growth. The following studies represent some of the more detailed analyses of this kind.

Eye

The initial size of the retina and lens, the size relationships between these and other components of the eye, and the growth of the entire organ relative to the rest of the animal have been studied by heteroplastic transplantation of the eye or its parts. Much of this literature has been reviewed by Twitty (1940, 1955).

The original size of the optic vesicle is a property of the inducing mesoderm and not an intrinsic property of the ectoderm, which will in fact form the eventual eye cup. Rotmann (1939) and Balinsky (1958) transplanted gastrula ectoderm into the future eye region of the neural plate between different species of anurans and urodeles. The eye rudiment that was then induced from the graft was generally of the same size as the normal host eye rudiment. In other words, the inducing mesoderm decided how many cells or how much ectodermal territory of the neural plate should be set aside for making an eye. The same is apparently true for different parts of the forebrain (Balinsky, 1958).

Later in development, however, the foreign optic vesicle, although initially the size prescribed by the host, begins to grow disproportionately for the host at a rate of growth directed by the genetic instructions intrinsic to the eye tissue (Twitty, 1940, 1955). Eyes, like limbs, grafted between two species of urodeles replicate essentially the growth rate of the donor species (Harrison, 1924, 1929; Stone, 1930; Twitty and Schwind, 1931). The size increase of the total organ appears to be relatively unaffected by the growth of its immediate environment. Thus a *tigrinum* eye becomes disproportionately large for its *punctatum* host, and a *punctatum* eye grafted to a *tigrinum* preserves its small size, in spite of the fact that it is nourished by a much larger and more rapidly growing host (Figure 3) (Harrison, 1929; Stone, 1930; Twitty, 1930).

These results do not necessarily indicate a complete absence of the effect of environment on growth rate. Twitty (1930) and Twitty and Elliott (1934) switched eyes of *tigrinum* and *punctatum* larvae when the animals were the same length and therefore not at the same developmental stage. The transplanted eye, if more developmentally advanced than the host, seems to wait without growing until the host has reached an equivalent developmental stage before it starts growing at its own intrinsic rate. Similarly, if the transplanted eye is developmentally immature with respect to the rest of the animal, it grows faster than usual,

FIGURE 3. An adult *Ambystoma punctatum* bearing a transplanted right eye from a *tigrinum*. Transplant performed in the early larval stage of development. (From Stone, L. S. [1930]. Heteroplastic transplantation of eyes between larvae of two species of Amblystoma, *J. Exp. Zool.* **55**:193–261. With permission of Wistar Press.)

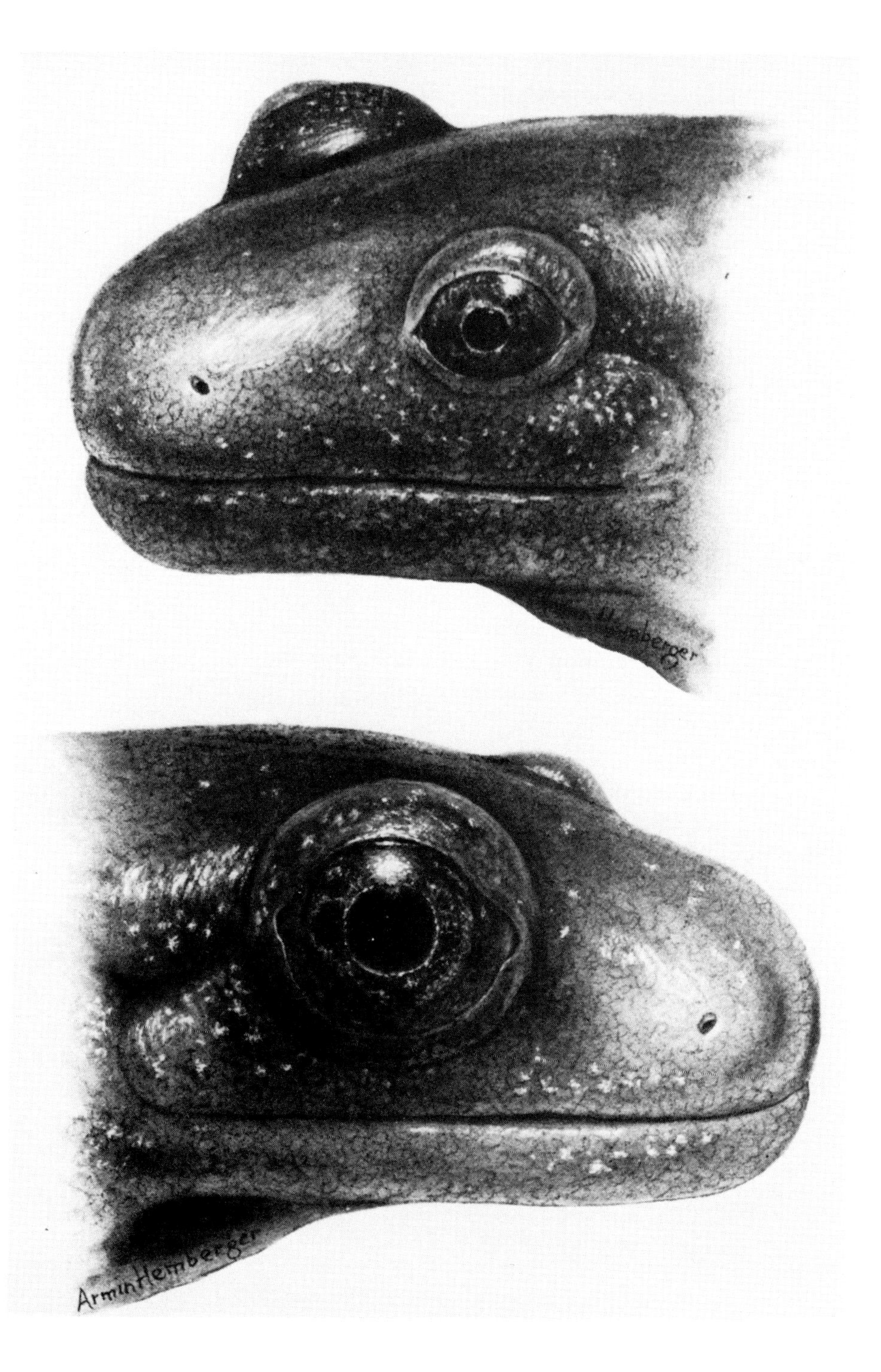
ArminHemberger

seemingly in order to catch up. When caught up, it again proceeds at its species-intrinsic rate. The developmental stage of the host thus seems to be an important factor in regulating eye growth.

Another important factor for retinal development is the lens. When a potential size maladjustment is made by combining the lens from one species with the optic vesicle of another, the two tissues exert a mutual influence on each other and a harmonious mean is reached (Harrison, 1929; Ballard, 1939).[1]

The size and rate of growth of the eye is thus directed by specific intrinsic and extrinsic factors. The eye in turn influences the surrounding tissues. The extrinsic ocular muscles and skeletal structures adjust their size to that of the eye (Twitty, 1932). When a *tigrinum* eye is grafted to a *punctatum*, the host eye muscles grow beyond their normal size, by an increase in number of fibers, in order to keep a reasonable proportion with the giant eye to which they are now attached (Twitty, 1932, 1955). Furthermore, heteroplastic giant (or minature) eyes with greater (or fewer) than the normal host number of retinal ganglion cells lead to cellular hyperplasia (or hypoplasia) in the contralateral midbrain. It is not known whether the midbrain hyperplasia represents induced mitoses or absence of cell death, nor whether there are any trophic substances working in this system.

Ear and Nose

Ears can be heteroplastically and xenoplastically induced (Yntema, 1955). Once induced, the ear, like the eye, seems to try to maintain its original growth rate and species-specific morphology when grafted heteroplastically, but it is not quite so successful (Richardson, 1932; Andres, 1945, 1949). The cartilagenous ear capsule of the host seems to exhibit a restraining influence on the size of the heteroplastic organ. A *tigrinum* ear in a small *punctatum* head is smaller than its normal control, but much larger than the contralateral host ear. Conversely, the *punctatum* ear in the large *tigrinum* head with its extra space becomes somewhat larger than normal, but not nearly as large as the host's

[1] Like the eye, the lens, when first induced from the epidermis by the neural ectoderm of the optic vesicle, is size-regulated by the inducing agent (Rotmann 1939, 1940). Thus, epidermis from a species that would normally form a small lens, when induced by the optic vesicle of a species with a large eye, forms, at first, a large lens. Rotmann (1940) also showed that haploid epidermis, which has small cells, induced by diploid retina forms a lens of normal size for the host, but with 70% more cells.

own ear (Richardson, 1932). Heteroplastic noses also tend to develop according to donor size (Burr, 1930). In the cases of both ear and nose, again like the eye, heteroplastic giants cause hypertrophy of the associated brain centers (Burr, 1930; Richardson, 1932), while heteroplastic miniatures lead to hypoplasia.

Spinal Cord

From cases like the preceding ones, a rule was formulated that when an organ from one species is induced or supported in its development by tissue from another, this organ will tend to grow with the specific rate and morphology determined by its genotype (Holtfreter, 1935*a*,*b*; Rotmann, 1934, 1939; Spemann, 1938). This rule has many exceptions (Balinsky, 1958). One is the spinal cord.

Heteroplastic grafting of the spinal cord was first carried out by Wiemann (1925, 1926). He showed that such grafts had no effect on the rate and manner of growth of the limbs innervated by them.[2] Detwiler (1931) was interested in studying the size and growth of spinal cord segments. He substituted the brachial region (segments 3, 4, 5) of the *tigrinum* cord for the corresponding portion of the *punctatum* cord. Shortly after the transplantation, the *tigrinum* graft in the *punctatum* embryo grew in length at the regular rate of the donor control. But it grew in thickness and cell number more rapidly than the control cord of either species (Detwiler, 1932*a*,*b*). For 20 or more days the graft was too long, too thick, and had too many cells. In the following 2 weeks, however, regulatory processes began. By 30 days the graft was identical to that of the corresponding region of the normal *punctatum* cord. In the reciprocal experiment, *punctatum* grafts on *tigrinum* embryos were also almost exactly regulated by the host.

This final size regulation might be accounted for by the mechanical constraints on the cord and the amount of sensory and descending innervation of the grafted region. Isolated portions of the *tigrinum* cord grafted lateral to the myotomes in a *punctatum* embryo develop into much larger structures than do similar chunks of *punctatum* cord grafted to the same position (Detwiler, 1936). Yet, electrical activity is probably unimportant in the original growth of the cord, because chunks of *Ambystoma* cord orthotopically transplanted into the tetro-

---

[2] This was also the conclusion of heteroplastic limb transplantation cited above (Harrison, 1924).

dotoxin-containing California newt appear to grow to normal proportions, adjusting perfectly with the newt cord (Twitty, 1937).

What accounts for the early burst in growth seen in Detwiler's experiments? The injury incurred at the time of transplantation might induce compensatory cell proliferation. Homoplastic transplantations, however, do not show any burst in growth. These results seem to indicate the presence of differential growth factors in the early stages of spinal cord development (Detwiler, 1936). These were the first indications of nerve growth factors. Can these results, in fact, be explained by NGF? Or do they imply other kinds of growth factors? Today these questions might be ripe for answering.

Xenoplastic transplantation of the spinal cord between *Bombinator* and *Triton* was performed in the neural plate stage by Roth (1950). In contrast to the above findings, Roth found that the xenoplastic cord segments seemed to maintain their intrinsic speed of development, volume, shape, distribution of ganglion cells, and fissures. He had to study these transplants only at early times, possibly before regulation might set in, because the grafts were usually resorbed by the host larvae.

Heart

Another example of size regulation is afforded by results of exchanging heart rudiments (Copenhaver, 1925, 1927, 1930, 1933, 1939). For a short time after transplantation the heteroplastic larval hearts follow their genetically specified growth rate and thus become either "too large" or "too small" for their hosts. The tiny *punctatum* heart is usually insufficient for the large *tigrinum* host. The animal becomes edematous and dies. In the reverse case there is less of a problem with survival. Those animals of both categories that do survive larval life and go through metamorphosis are found to have hearts of normal size for the host.

Copenhaver (1930) was able to transplant the anterior or posterior heart rudiment orthotopically. Thus chimeric animals were created with composite hearts. Again, at first there was a size discrepancy between the two parts of the heart that later regulated to make a well-proportioned organ suited for the host.

Copenhaver (1930, 1939) studied the rate of beating of such chimeric hearts. *Tigrinum* control hearts beat at around 40 per minute, while *punctatum* hearts beat at around 60 per minute. The composite hearts, no matter what size they were, and no matter what chest cavity they were in, always beat with the frequency inherent in the genotype of the ventricle. Thus, for example, a heart in a *punctatum* host with a *punctatum* atrium and a *tigrinum* ventricle beats at about 40 per minute, as do control *tigrinum* hearts.

These transplantations were done before peripheral innervation had occurred. Thus the basal heart rate must be set up and maintained by the genetically controlled pacemaker properties of cells in the ventricle. The physiological mechanisms underlying these pacemaker activities are only now beginning to be understood (Nobel, 1975).

Summary

These studies on growth rate and size regulation in the limb, eye, ear, nose, spinal cord, and heart tell us that it is impossible to predict a priori what factors are likely to be most important in the development of any particular organ. What is true for one part of the central or peripheral nervous system might not be true for another. These studies also show that chimeras afford a powerful methodology for localizing the intrinsic and extrinsic factors that play a role in the development of any particular structure of interest.

Clonal Analysis

Origin of Nervous Tissue

Amphibian species may differ from each other not only in morphology and rate of growth, but also in pigmentation. Such a difference can be used as a reliable and long-lasting marker between transplant and host tissue, and this makes it possible to decide the species origin of a given cell in a chimeric individual. Even more reliable markers might come from mutants (discussed later) or ploidy differences.[3] Born (1897) was the first to note the autonomy of pigmentation in his original chimeras.

The primary origin of nervous tissue was uncovered by the chimeric analyses of Spemann and Mangold (1924). Before their experiments

---

[3] Jacobson and Hirose (1978), using haploid-triploid mosaics induced at the first mitotic division of the *Xenopus* embryo, recently demonstrated that the entire right (left) half of the adult brain, except for a sector of ventral retina and ventral diencephalon, originates from the right (left) blastomere of the two-cell staged embryo. These exceptional sectors originate from the opposite blastomere.

it was known that the dorsal lip of the blastopore was critical in determining the major axes and development of the embryo. A dorsal lip transplanted ectopically on the embryo, unlike other ectodermal transplants, which join in the development of the new surroundings, results in the formation of a small secondary embryo with a neural tube, notochord, and somites (Lewis, 1907; Spemann, 1918). At first the significance of these experiments was unknown, since it was assumed that the secondary embryo was formed from the material of the implant (Spemann, 1938). That this was not the case was first shown by Spemann and Mangold (1924), who transplanted the dorsal lip heteroplastically between the gastrulae of differently pigmented *Triton* species. The result of this experiment was, of course, the birth of the primary organizer of neural induction. The dorsal lip invaginates and becomes mesodermal or chordomesodermal, and from its position inside the gastrula induces the overlying ectoderm of the host animal to form the spinal cord and brain (see Figure 4).

Clearly, part of the power of chimeras in these studies lies in their ability to show the various potencies of embryonic tissues. Following the experiment of Spemann and Mangold (1924), Marx (1925) made systematic heteroplastic transplantations of portions of the presumptive medullary plate into epidermal regions of differently pigmented hosts. These studies showed that primordial neural tissue becomes irreversibly committed only after contact with the archenteron roof has been established.

Spemann's (1921) study, in which the term "chimera" was first introduced into embryology, showed that presumptive epidermis could develop into brain (Figure 5) and presumptive brain into epidermis, if the transplants were done before neural induction. Mangold (1923) further showed by heteroplastically interchanging different germ layers in the gastrula that presumptive neural ectoderm was capable of taking part in a variety of mesodermal and endodermal organs.

Migrations

Chimeras with pigmentation differences can be useful to study the migration of tissues later in development. Harrison (1898) exchanged the growing tail region between the differently pigmented *Rana palustris* and *Rana virescens*, species that do not differ much in size or morphology. He was able to tell unambiguously the origin of individual cells of the epidermis, endoderm, and axial musculature. With this system he demonstrated the shifting of epidermis over the subjacent muscle plates

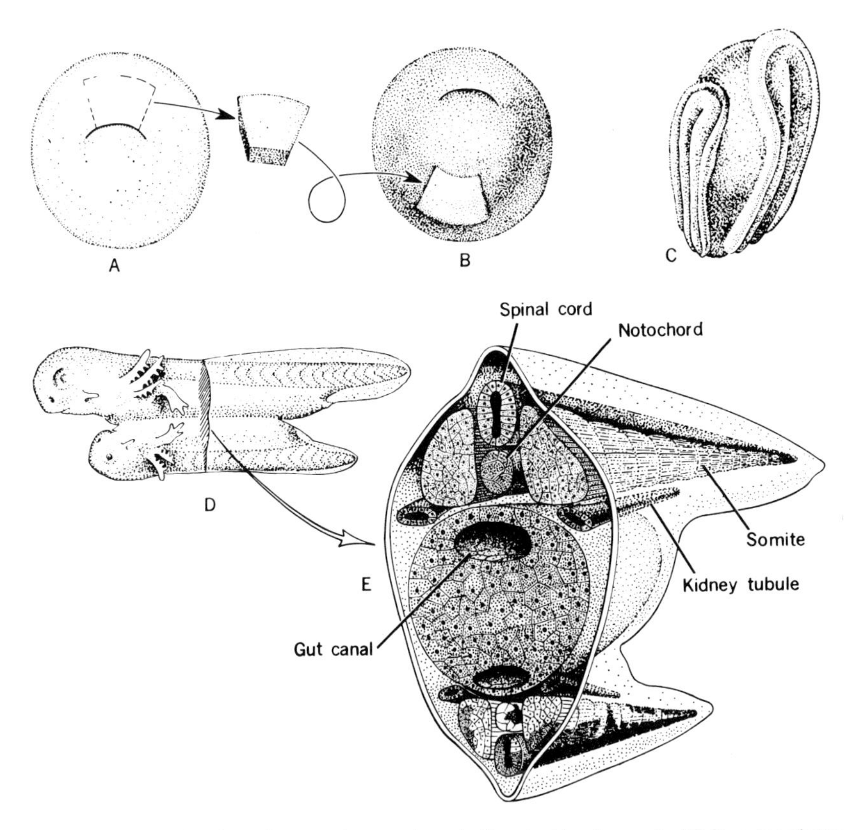

FIGURE 4. Induction of secondary embryo in the darkly pigmented *Triton taeniatus* by transplantation of dorsal lip of the blastopore from the light-colored *Triton cristatus*. This experiment, first performed by Spemann and Mangold (1924), shows that the graft organizes the tissues of the host to participate in the development of the secondary embryo (note the dark tissue in the spinal cord). (From Twitty, V. C. [1966]. *Of Scientists and Salamanders*. With permission of W.H. Freeman and Company.)

during development (Figure 6). These chimeric studies explained the previously known displacement of the sensory area and motor belt of the same segmental nerve in that the epidermal migration corresponds in direction and amount with this displacement (Harrison, 1898). Another early and classic example of the use of chimeras to trace the origin and migration of nervous elements is Harrison's (1903) study of the lateral line. By grafting an anterior part of the early tailbud from *palustris* to *sylvatica*, Harrison proved that the lateral line sensory organs were of placodal origin and was able to follow the path of

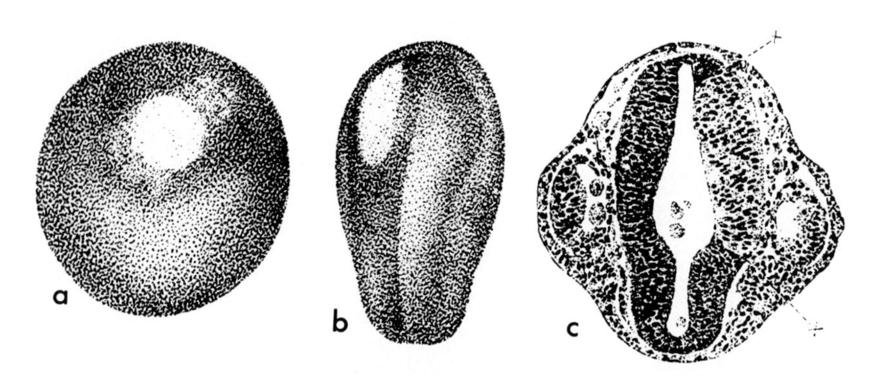

FIGURE 5. Heteroplastic exchange between *Triton taeniatus* and *cristatus*. *a*: The dark *taeniatus* blastula with the light *cristatus* graft (presumptive epidermis). *b*: Same embryo at the neural plate stage of development, with the *cristatus* graft in the region of the left side of the presumptive brain. *c*: Cross section through the same *taeniatus* embryo at tailbud stage, showing part of the wall of the forebrain (between x and x) formed from the grafted light-colored presumptive epidermis of the *cristatus* embryo. (From Spemann, H. [1921]. Die Erzeugung tierischer Chimaeren durch heteroplastische embryonale Transplantation zwischen *Triton cristatus* und *taeniatus, Wilhelm Roux' Arch. Entwicklungsmech. Org.* **48:**533–570. With permission of Springer-Verlag.)

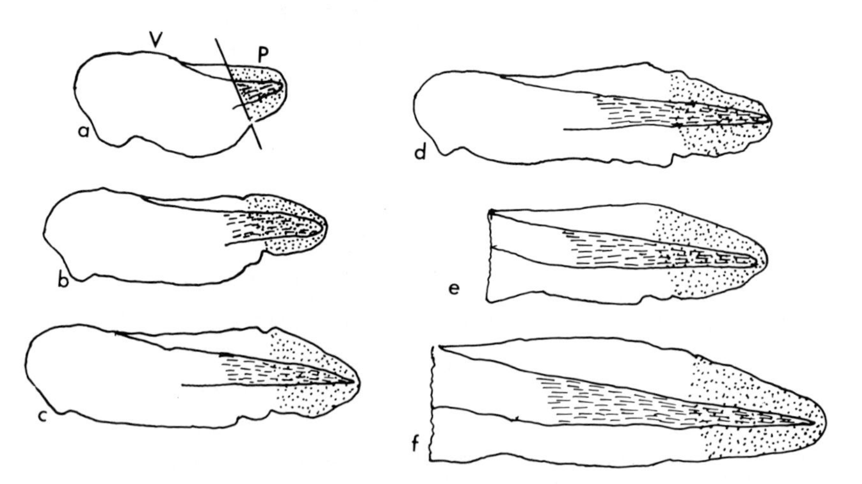

FIGURE 6. Demonstration of how, in the growing tail, the epidermis slides posterior relative to the underlying muscle. *a*: Cut made in embryos when *Rana palustris* tail is to replace *Rana virescens* tail; *b*: 30 h after operation; *c*: 53 h after operation; *d*: 77 h after operation; *e*: 5 days after operation; *f*: 7 days after operation. *Virescens* tissue simply outlined; *palustris* epidermis dotted; *palustris* musculature dashed. (After Harrison, 1898.)

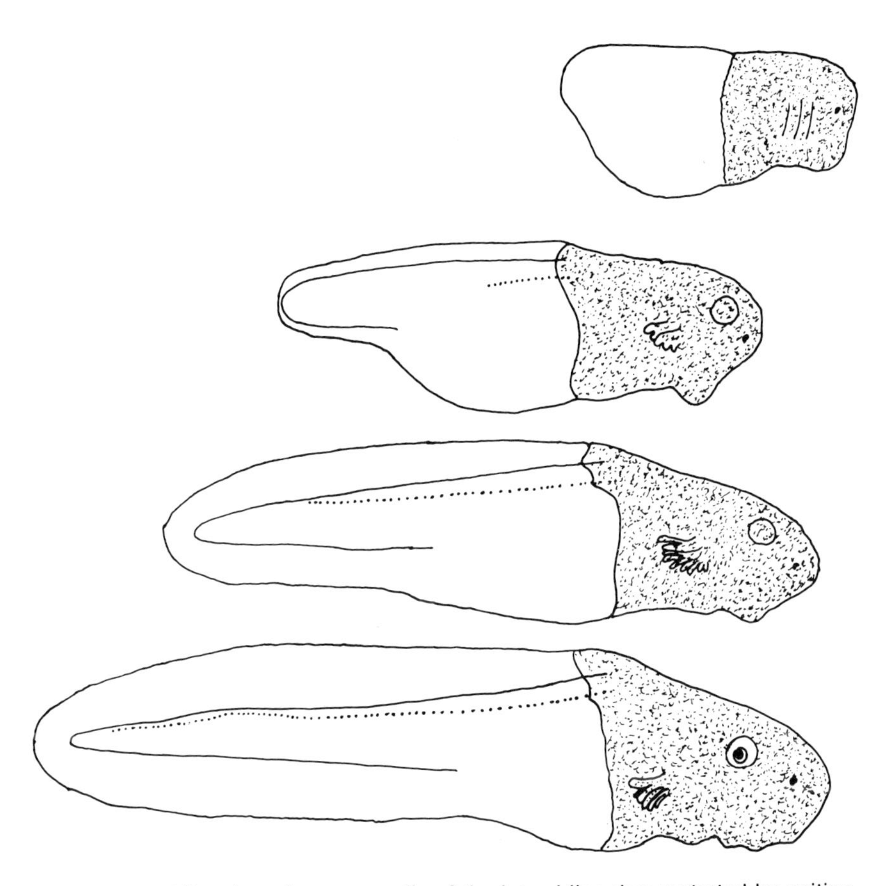

FIGURE 7. Migration of sensory cells of the lateral line demonstrated by uniting the anterior half of the darkly pigmented *Rana sylvatica* embryo with the posterior half of the light-colored *Rana palustris* embryo. Lateral line organs represented by dotted line. (After Harrison, 1903.)

migration of the developing organs (Figure 7). By combining animals in various orientations with respect to each other, he studied the polarizing influences on this patterned migration. Some of these results are summarized in Figure 8. Of this study Spemann (1938) says: "I have learned more than from almost any other investigation, not only for technique, but also for methodically advancing analysis." An extension of Harrison's study was conducted 30 years later by Stone (1935), who transplanted the primary midbody lateral line placodes between *Ambystoma* species. In *tigrinum* the primary organs bud to form groups

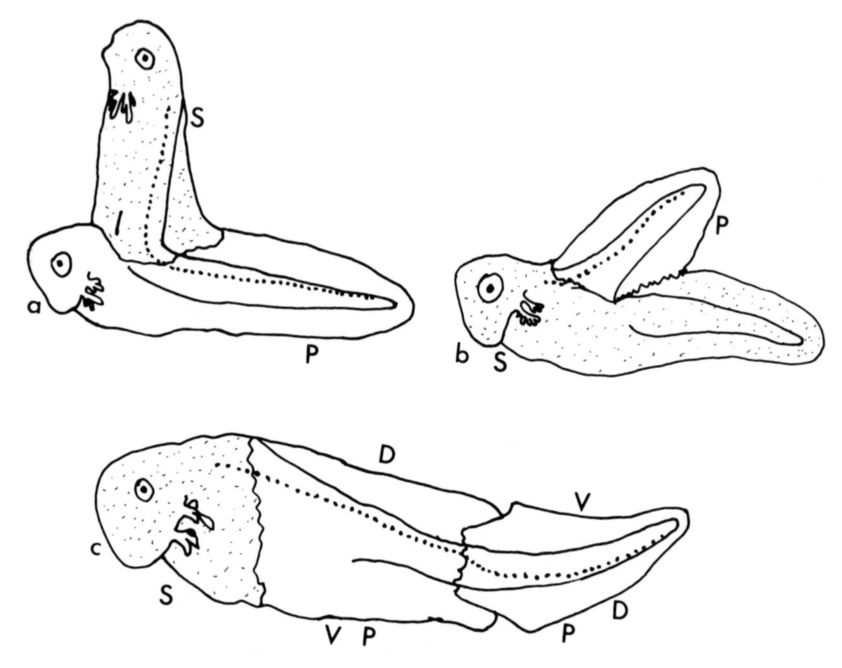

FIGURE 8. *a,b,c*: Experiments showing that the direction and position of lateral line migration is influenced by the host tissue through which the migration occurs. In all these cases the lateral lines are of dark *sylvatica* origin so that they can be visualized on the light *palustris* host. S, *sylvatica*; P, *palustris*; D, dorsal; V, ventral. (After Harrison, 1903.)

of four to eight organs per cluster. In *punctatum* there are two or three per cluster. Although the path of migration of lateral line organs is dictated by the host environment (Harrison, 1903), the final budding pattern of the organ is determined by the donor species (Stone, 1935).

Summary

Pigmentation differences in chimeras have given us important information, then, on the origin, inductive powers, development capacities, and migration patterns of different parts of the nervous system at different moments in development. Much of this information might have been otherwise inaccessible.

Mutants and Chimeras

Mutant-nonmutant chimeras can be used to localize a genetic defect to a specific faulty tissue (a "focus" in *Drosophila* terminology). Mutants can also be used in chimeras as markers in the study of clonal analysis. In mutants with a particular, well-understood defect, chimeras can be used to make localized deficits in various parts of an animal.

Compared to *Drosophila* or the mouse, amphibia are not a rich source of neurological mutants. Nevertheless, many of the mutants that do exist in amphibia have been analyzed in detail. The accessability of the amphibian embryo to experimental manipulation has advanced our understanding of some of these mutants beyond what is possible in some more genetically advanced organisms.

Eyeless

The eyeless (*e*) mutant was discovered as a simple recessive (Humphrey, 1969). The most outstanding feature of the mutant syndrome is the failure of optic vesicles to form (Humphrey, 1969; Van Deusen, 1973). These mutants are also darkly pigmented and sterile. Van Deusen (1973) undertook a series of experiments involving transplantation of embryonic tissue between eyeless mutants and normal axolotls in order to identify which component of the inductive system for optic vesicle formation is directly affected by the mutation. His most instructive experiment is outlined in Figure 9. Presumptive ectoderm or presumptive chordomesoderm was grafted reciprocally between white eyeless and dark normal early gastrulae.[4] The results of these experiments placed the fault with the ectoderm and not the inducing mesoderm, since normal mesoderm could not induce eye formation in eyeless ectoderm, and eyeless mesoderm was capable of inducing normal ectoderm. There remained one final distinction to be made: was the neural ectoderm failing to be induced or was the overlying epidermis (transplanted simultaneously with the neural ectoderm) somehow suppressing optic vesicle formation? Brun (1978) showed the latter alternative to be the case by transplanting presumptive normal or mutant lens tissue over presumptive eye neural ectoderm at the neural fold stage of development. The result is that the fault seems to lie in the epidermis. Mutant epidermis will suppress eye formation when grafted over normal diencephalon, and normal epidermis will allow eye development of mutant ectoderm (Brun, 1978). This result might make some sense in light of Ulshafer and Hibbard's (1976) electron microscopic finding that the space between the diencephalon wall and head ectoderm is

[4] Here the white-colored pigmentation mutant serves as a marker for genetically eyeless ectodermal tissue. Mutants used as markers are described in more detail later.

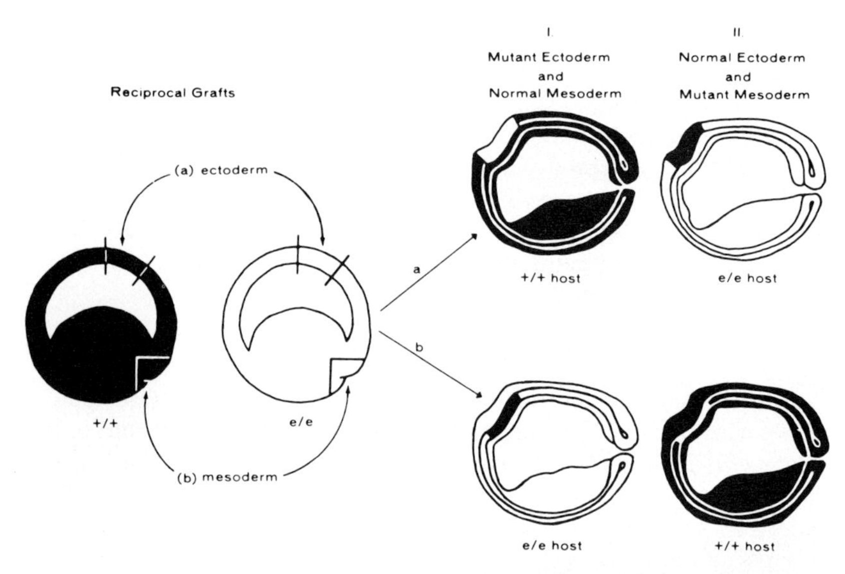

FIGURE 9. Experimental scheme for determining which component(s) of the eye-forming complex is primarily affected by the eyeless gene. Reciprocal orthotopic exchanges of either the presumptive anterior neural plate ectoderm (a) or presumptive mesoderm (b) between white eyeless (*d/d*; *e/e*) and dark normal (+/+; +/+) early gastrula produce chimeras in which normal mesoderm is tested for its ability to induce mutant ectoderm (Ia, Ib) and mutant mesoderm is tested for its ability to induce normal ectoderm (IIa, IIb) to differentiate eyes. (From Van Deusen, E. [1973]. Experimental studies on a mutant gene (*e*) preventing the differentiation of eye and normal hypothalamus primordia in the axolotl, *Dev. Biol.* **34:**135–158. With permission of Academic Press.)

filled with mesenchyme in normal embryos. For example, it is possible that the epidermal cells are responsible for clearing away the mesenchymal cells over the incipient optic vesicle, and that the absence of overlying mesenchyme is necessary for the evagination of the vesicle. The final molecular explanation of the mutant defect is, of course, unknown at present. Nevertheless the localization of the defect to the epithelium by chimeric experiments is a very useful beginning.

Other grafting experiments on the eyeless axolotl show that the sterility is localizable to a hypothalamic dysfunction (Van Deusen, 1973). The pigmentation defect in eyeless was shown to be secondary to loss of vision, because grafting an eye into eyeless restored normal pigmentation, and a phenocopy of eyeless pigmentation was achieved by removing the eyes of normal axolotls (Epp, 1972; Van Deusen, 1973).

Spastic

The spastic (*sp*) mutation in the axolotl leads to bizarre coiling and thrashing locomotor patterns, inability to swim with sinusoidal oscillations, and equilibrium dysfunction (Humphrey, 1975; Ide and Tompkins, 1975; Ide, Miszkowski, Kimmel, Schabtach, Tompkins, Elbert, and Duda, 1976). A battery of different approaches have been used to localize the focus of the defect to the cerebellum. First, the onset in development of defective swimming behavior in the spastic mutant is at the stage classically corresponding to the onset of function in the midbrain and cerebellum (Coghill, 1924, 1926; Ide and Tompkins, 1975; Ide et al., 1977). Second, phenocopies of the mutant can be produced by lesioning the vestibular root or the vestibulocerebellum in normal animals (Ide et al., 1977). Third, the location of Purkinje cells in the mutant vestibulocerebellum is shifted ventrally (Ide, Tompkins, and Miszkowski, 1977). Fourth, electrophysiological studies have shown that a certain class of vestibulocerebellar units responding to tilt are present but displaced from their normal positions (Ide, 1977).

Chimeric studies with spastics have not been published, but preliminary work indicates that the focus is in the cerebellum. Unilateral grafting of normal presumptive cerebellum into the neural plate of the spastic neurula seems to correct the equilibrium dysfunction on the ipsilateral side (Ide, personal communication). Bilateral cerebellar transplantation corrects both the equilibrium and swimming disabilities (Hunt and Ide, personal communication). Unfortunately these transplantations do not increase the already extremely low viability of the spastic mutants, and none of the behaviorally rescued spastics have reached adulthood.

Cardiac

Humphrey (1968, 1972) discovered and performed preliminary studies on the cardiac lethal (*c*) mutation of axolotl. Homozygotes lack functional hearts. Affected embryos develop without apparent abnormalities to the heartbeat stage. At this time the heart fails to beat and looks atrophic by morphological, biochemical, and immunohistochemical criteria (Lemanski, 1978). Cardiac action potentials are absent and direct electrical stimulation fails to elicit contraction (Justus and Hollander, 1971).

Using chimeras, Humphrey (1972) showed that mesodermal heart rudiments from cardiac embryos develop into normal functional hearts when transplanted into nonmutant hosts. Reciprocal transplants don't. He suggested that cardiac embryos had either a defect in the anterior endoderm, which induces heart, a hypothesis that received some support from morphological studies on anterior endoderm (Lemanski, Marx, and Hill, 1977), or that mutant embryos produce a substance that inhibits function. The latter alternative appears closer to the truth, since hearts from mutant embryos immediately begin to beat normally when isolated in organ culture or washed thoroughly in amphibian saline (Kilikowski and Manasek, 1978). Thus heart induction must be essentially normal in the mutant, but cardiac function is inhibited, possibly by something as trivial as an altered ionic environment.

Other Possible Neurological Mutants

Other possible neurological mutants discovered by Humphrey (1975) have yet to be studied in detail. Two of these are quivering lethal (*q*), in which the young larvae shake vigorously until dead, and microphthalmia (*micro*), which develops with tiny eyes. Another mutant that originally seemed neurological is microphthalmia lethal (*mi*). As with *micro*, *mi* embryos develop small eyes, but unlike *micro*, they soon die. This mutation was shown to be lethal to all cells containing the homozygous mutant genes, because even small grafts of *mi* tissue onto normal embryos die (Humphrey and Chung, 1977).

Marker Mutants

Mutants of no neurological interest can be used as markers in homoplastic grafts. Albino mutants,[5] for instance, are available in a number of amphibian species. This mutation in axolotl was introduced into the genome in three stages (Humphrey, 1967; Hennen, 1977). A spontaneous occurrence of albinism in the closely related species *Ambystoma tigrinum* was discovered by a high school teacher in Minnesota. By artificial insemination this albino *tigrinum* was mated with an axolotl. The resulting embryos failed to develop properly, but hybrid blastula nuclei were transplanted into enucleated axolotl eggs. These embryos developed further but still did not survive. Humphrey, however, was

---

[5] In all mammalian species so far studied, the albino mutation is of neurological interest, causing an abnormally large number of retinal ganglion cell axons to decussate at the chiasm. One study on the axolotl albino by Guillery and Updyke (1976) suggests that the neurological abnormality is not present in this amphibian mutant.

able to transplant the presumptive gonadal region from these into similar stages of normal axolotl embryos. Some of the F2 generation from these chimeric individuals were albino axolotls.

Albino (*a*) axolotls, like albino mutants in other organisms, fail to synthesize melanin (Benjamin, 1970) but do differentiate melanoblasts (Dalton, unpublished). Although tyrosine dopa oxidase (TDO) is present in these mutants, it is inactive. The fault seems to lie in another factor that normally activates the TDO and that fails to do so in the mutant (Horsa-King, 1978). Chimeric studies using the albino mutant show that it acts autonomously and is therefore an excellent marker for donor or host-derived tissue.

Another useful pigmentation marker in axolotl is the recessive white (*d*). This mutation is semiautonomous and acts to restrict the number and distribution of melanophores and xanthophores in the skin. Other skin color mutants are also available: melanoid (*m*), axanthic (*ax*), and iridophore (*irid*) (Malacinski and Brothers, 1974).

The nucleolar mutants in axolotl and *Xenopus* are also useful as cell markers in chimeras. The mutations either reduce the size of the nucleoli (n^1, n^2, n^3, n^4 in axolotl; Humphrey, 1975) or eliminate them (*nu* in *Xenopus*; Elsdale, Fischberg, and Smith, 1958). The *Xenopus* mutant and one of the axolotl mutants (n^4) act by deleting rDNA in the nucleolar organizer region of the genome (Wallace and Birnstiel, 1966; Sinclair, Carroll, and Humphrey, 1974). These mutants are healthy in the heterozygous state, in which only one of the two nucleoli per cell is affected. They act in a strictly cellular autonomous way. Using these mutants in chimeras, one should theoretically be able to identify the clonal origin of every cell in the adult organism.

Marker mutants have been used by Hunt and Ide (1977) to study retinotectal specificity in *Xenopus*. A chunk of presumptive anterior retina and pigment epithelium, marked with normal pigmentation and a single nucleolus per cell, from a stage 31 *Xenopus* was inverted and placed in the dorsal retina of a similar-stage host marked with albinism and the normal two nucleoli per cell. As the animal grew, the patch of retina derived from the transplant grew radially. There is a rough correspondence between the shape of the neural retina patch and that of the clonally related pigment epithelium. In the adult, the transplanted sector of retina projects to the same appropriate quadrant of the tectum as the anterior sector of the host retina, but in a reversed retinotopic way consistent with the transposition of the marked clone. This experiment shows that retinal growth is radial, and positional information in the specified retina is passed on clonally or radially, yet the positional information itself is Cartesian.

Neurotoxic Chimeras

In an amazing set of chimeric experiments by Twitty (1937), tetrodotoxin[6] was shown to act specifically on nervous tissue. Parabiosis of the California newt *Taricha torosa* with several species of salamanders resulted in the paralysis of the parabiotic salamander twin throughout embryogenesis and well into larval life. The toxin did not seem to affect the *Taricha* embryo or larva in any way. By transplanting various tissues from *Taricha* into other embryos, Twitty showed that the toxin was present in all embryonic newt tissues, but concentrated in the yolk cells.

To locate the components of the neuromuscular complex affected by the toxin, Twitty (1937) carried out a series of experiments, one of which is outlined in Figure 10. By transplantation, he forced a piece of ectopic *Taricha* spinal cord to innervate an *Ambystoma* muscle in a *Taricha* host. "Although," says Twitty (1937), "these questions are generally attacked by the special methods of neurophysiology, in the present case the problem may be approached, preliminarily at least, by simple experiments in embryonic transplantation." The results of his experiments show that embryonic *Ambystoma* nerves, but not muscles, are paralyzed by the toxin.

By transplanting either dorsal or ventral spinal cord from *Ambystoma* to *Taricha*, Twitty (1937) next showed that all motor components of *Ambystoma* are paralyzed by the toxin, but that at least some sensory components are not. Recently, it has been shown that Rohon-Beard cells, large sensory neurons in the dorsal spinal cord of young amphibia, and some embryonic dorsal root ganglion cells first develop Ca^{2+} action potentials that are tetrodotoxin-resistant (Baccaglini and Spitzer, 1977; Baccaglini, 1978). It would be interesting to know if these were the cells responsible for the sensory activity in Twitty's chimeras. Another related question is whether the normal time course of a cell's switching

---

[6] It was more than 25 years later that the physiological action of tetrodotoxin was understood to be a blocker of the electrogenic sodium channel (Narahashi, Moore, and Scott, 1964). In the same year, it was shown that tetrodotoxin, originally isolated from puffer fish, was identical to the toxin contained in the newts studied by Twitty (Mosher, Fuhrman, Buchwald, and Fischer, 1964).

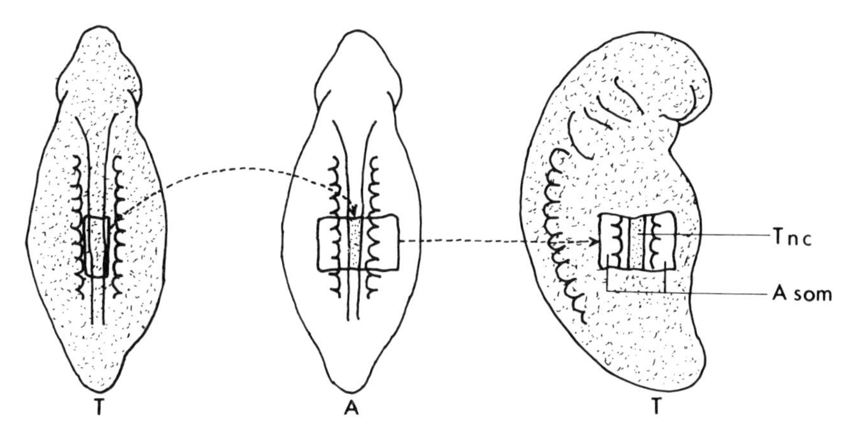

FIGURE 10. The procedure by which muscles and nerves are combined heteroplastically. T, *Taricha torosa*; A, *Ambystoma*; nc, nerve cord; som, somites. (After Twitty, 1937.)

from a Ca^{2+} to a Na^{+} action potential is affected by the constant exposure to tetrodotoxin in these chimeras.

One can use *Taricha* embryos as hosts for grafts of nervous tissues from other salamander embryos to study the role of electrical excitability in the development of the nervous system. For instance, one can ask if the ability to make action potentials is important when two neurons are competing for the same postsynaptic target. Some preliminary experiments I have performed on this question are outlined below.

If an extra optic vesicle is implanted at an embryonic stage into a frog or an axolotl, an extra eye can develop. And if this extra eye is put next to the host's own eye and care is taken in the grafting to attach the optic stalk to the brain, this extra eye will develop connections with the host's optic tectum (Constantine-Paton and Law, 1978). This is a competitive situation. The transplanted eye is trying to innervate the same target as the host eye. A question of particular interest here is how well the transplanted eye does, and what factors are of primary importance in successful competition. One possibility is that electrical activity may play a role. If the extra, transplanted eye were electrically silenced, how well would it compete with the normal, electrically active host eye?

To answer this question I have been transplanting extra optic vesicles from stage 22–23 axolotls to *Taricha torosa* hosts, and vice versa. It is assumed, but not yet tested, that the developing retinal ganglion

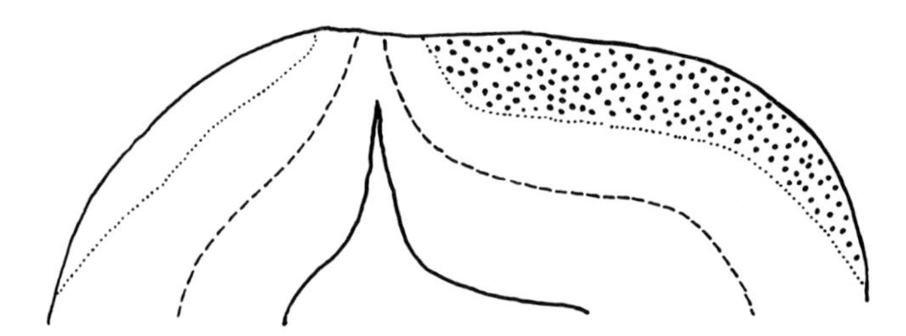

FIGURE 11. Termination zone of retinal ganglion cells from an extra *Ambystoma mexicanum* (axolotl) eye embryonically grafted onto a *Taricha torosa* and injected with tritium-labeled proline. This is from a camera lucida drawing of an autoradiograph of a transverse section through the dorsal midbrain or tectum. The dotted line separates the (upper) stratum opticum from the rest of the tectal neuropil, while the dashed line separates the cellular layer from the neuropil. The heavy dots filling the right stratum opticum represent the area of termination of the tritium-labeled ganglion cells of the axolotl eye.

cells are electrically silenced by the action of the host's tetrodotoxin. If action potentials during development are crucial in successful competition, one might imagine that the transplanted axolotl eye competing with the host's electrically normal eye would be at a disadvantage. This does *not* seem to be the case. Figure 11 shows that neurons from an axolotl eye labeled with tritiated proline by injection into the bulb can extensively invade a *Taricha* tectum even in a competitive situation.

To test this further, I labeled the retinal ganglion cells of the transplanted eye with tritiated proline and both of the host eyes with horseradish peroxidase. The results (Figure 12) show that an axolotl eye can invade a *Taricha* tectum so extensively that it even seems to displace the host's own retinal terminals. Thus it is probably the case that electrical activity throughout early development is not the most critical factor in successful competitive innervation in the central nervous system. Perhaps cell size is more important (axolotl ganglion cells are bigger than *Taricha* ones). Perhaps timing is more important (axolotl ganglion cells may arrive at the *Taricha* tectum earlier than the host's own). These possibilities can also be attacked with chimeras.

DISCUSSION

If one theme emerges from these various studies of the nervous system using chimeras, it is the following: in the differences and similarities between species and mutants lie clues to the development of the nervous system. To put it simply, not much would be learned by making recip-

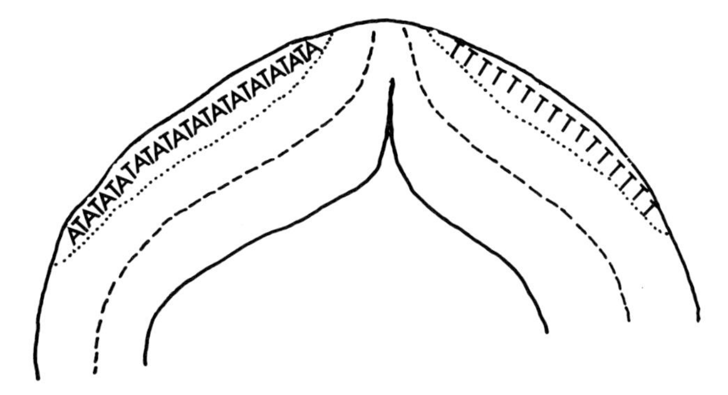

FIGURE 12. Termination zones of the extra axolotl eye and both *Taricha* eyes in another *Taricha* tectum. Similar to Figure 11 except that the tectal projections of both *Taricha* eyes have been labeled with horseradish peroxidase (HRP), while the axolotl eye projection is labeled with tritiated proline. A, axolotl label (i.e., silver grains); T, *Taricha* label (i.e., HRP reaction product). The alternating pattern of T's and A's in this figure is not meant to indicate alternating columns of termination. Rather, there was extensive overlapping with only a small degree of clumping in the original micrograph. There appears to be less dense labeling of *Taricha* terminals in the tectum innervated with the competing axolotl eye than in the contralateral tectum.

rocal orthotopic grafts between genetically identical individuals, yet such grafts between genetically different animals can point to factors that might not be revealed by conventional neurobiological techniques. For example, the presence of intrinsic and extrinsic growth regulators for different parts of the nervous system were hypothesized, and eventually some growth factors were found; among these was NGF (Levi-Montalcini, 1952).

Differences in morphology, for instance in the spinal cord between species and in the presence or absence of structures between mutant and normal animals, can be the basis of chimeric experiments designed to decide which tissues are responsible for these differences. Once the tissue of interest has been localized, biochemistry, anatomy, and physiology can advance the study further. Unexpected similarities between organisms of different species are also sometimes found in chimeras. For instance, Wagner (1949) xenoplastically transplanted anterior neural crest and studied the resulting head structures. He found that bones of toad origin normally absent in toads and present in newts can be induced by newt mesoderm, and other structures present in toads and absent in newts can be induced by newt mesoderm. Various heteroplastic and xenoplastic inductions are successful, and even the axes of retinal polarity can be specified xenoplastically (Hunt and Piatt, 1978). Such studies point to the existence of molecular mechanisms of general biological interest.

Finally, known differences between tissues, such as pigmentation or the presence of a neurotoxin, can be used as tools to answer specific questions about the development of the nervous system.

ACKNOWLEDGMENTS

I would like to thank Eric Frank, Martha Constantine-Paton, and Peggy Hollyday for advice on the manuscript, Karen Larsen and Amy Halio for technical assistance, Marc Peloquin for photography, and the Indiana axolotl colony for animals. This work was largely supported by the Harvard Society of Fellows and the Milton Fund.

REFERENCES

Andres, G. (1945). Über die Entwicklung des Anurenlabyrinths in Urodelen (Xenoplastischer Austausch zwischen *Bombinator* und *Triton alpestris*), *Rev. Suisse Zool.* **52**:400–406.

Andres, G. (1949). Untersuchungen an Chimaeren von Triton und Bombinator. Teil I. Entwicklung Xenoplastischer Labyrinthe und Kopfganglien, *Genetica* **24**:387–534.

Baccaglini, P. I. (1978). Action potentials of embryonic dorsal root ganglion neurones in *Xenopus* tadpoles, *J. Physiol. (London)* **283**:585–604.

Baccaglini, P. I. and N. C. Spitzer (1977). Developmental changes in the inward current of the action potential of Rohon-Beard neurones, *J. Physiol. (London)* **271**:93–117.

Balinsky, B. I. (1958). On the factors controlling the size of the brain and eyes in Anuran embryos, *J. Exp. Zool.* **139**:403–442.

Ballard, W. W. (1939). Mutual size regulation between eye-ball and lens in Amblystoma studied by means of heteroplastic transplantation, *J. Exp. Zool.* **81**:261–285.

Benjamin, C. P. (1970). The biochemical effects of the *d, m,* and *a* genes on pigment cell differentiation in the axolotl, *Dev. Biol.* **23**:62–85.

Born, G. (1894). Die Künstliche Vereinigung lebender Theilstücke von Amphibien-Larven, *Jahresber. Schles. Ges. Vaterländische Cultur. Medicinische Section,* p. 11.

Born, G. (1897). Über Verwachsungsversuche mit Amphibienlarven, *Wilhelm Roux' Arch. Entwicklungsmech. Org.* **4**:349–465, 517–623.

Briggs, R. (1973). Development genetics of the axolotl, pp. 169–199 in *Genetic Mechanisms of Development,* Ruddle, F. H., ed. Academic Press, New York.

Briggs, R. and T. J. King (1952). Transplantation of living nuclei from blastula cells into enucleated frog's eggs, *Proc. Natl. Acad. Sci. USA* **38**:455–463.

Browder, L. W. (1975). Frogs of the genus *Rana*, pp. 19–33 in *Handbook of Genetics,* Vol. 4, King, R. C., ed. Plenum Press, New York.
Brun, R. B. (1978). Experimental analysis of the eyeless mutant in the Mexican axolotl (*Amblystoma mexicanum*), *Am. Zool.* **18:**273–279.
Burr, H. S. (1930). Hyperplasia in the brain of Amblystoma, *J. Exp. Zool.* **55:**171–191.
Coghill, G. E. (1924). IV. Rates of proliferation and differentiation in the central nervous system of Amblystoma, *J. Comp. Neurol.* **37:**71–121.
Coghill, G. E. (1926). V. The growth of the motor mechanism of *Amblystoma punctatum*, *J. Comp. Neurol.* **40:**47–93.
Constantine-Paton, M. and M. I. Law (1978). Eye-specific termination bands in tecta of three-eyed frogs, *Science* **202:**639–641.
Copenhaver, W. M. (1925). Heteroplastic transplantation of the heart of Amblystoma embryos, *Proc. Am. Soc. Zool. Anat. Rec.* **31:**299.
Copenhaver, W. M. (1927). Results of heteroplastic transplantations of the heart rudiment in Amblystoma embryos, *Proc. Natl. Acad. Sci. USA* **13:**484–488.
Copenhaver, W. M. (1930). Results of heteroplastic transplantations of anterior and posterior parts of the heart rudiment in Amblystoma embryos, *J. Exp. Zool.* **55:**293–318.
Copenhaver, W. M. (1933). Transplantation of heart and limb rudiments between Amblystoma and Triton embryos, *J. Exp. Zool.* **65:**131–157.
Copenhaver, W. M. (1939). Some observations on the growth and function of heteroplastic heart grafts, *J. Exp. Zool.* **82:**239–271.
Delanney, L. E. (1978). Immunogenetic profile of the axolotl: 1977, *Am. Zool.* **18:**289–299.
Detwiler, S. R. (1931). Heteroplastic transplantations of embryonic spinal cord segments in Amblystoma, *J. Exp. Zool.* **60:**141–171.
Detwiler, S. R. (1932*a*). Growth acceleration and regulation in heteroplastic spinal cord grafts, *J. Exp. Zool.* **61:**245–277.
Detwiler, S. R. (1932*b*). Further experiments upon accelerated growth in heteroplastic spinal cord grafts, *J. Comp. Neurol.* **56:**465–502.
Detwiler, S. R. (1936). *Neuroembryology.* Macmillan Publishing Co., New York.
Eakin, R. M. and M. Harris (1945). Incompatibility between amphibian hosts and xenoplastic grafts related to host age, *J. Exp. Zool.* **98:**35–64.
Elsdale, T. R., M. Fischberg, and S. Smith (1958). A mutation that reduces nucleolar number in *Xenopus laevis, Exp. Cell Res.* **14:**642–643.
Epp, L. G. (1972). Development of pigmentation in the eyeless mutant of the Mexican axolotl, *Ambystoma mexicanum* (Shaw), *J. Exp. Zool.* **181:**169–180.
Guillery, R. W. and B. V. Updyke (1976). Retinofugal pathways in normal and albino axolotls, *Brain Res.* **109:**235–244.
Gurdon, J. B. (1977). Egg cytoplasm and gene control in development, *Proc. R. Soc. London B Biol. Sci.* **198:**211–247.
Gurdon, J. B. and H. R. Woodland (1975). *Xenopus,* pp. 35–50 in *Handbook of Genetics,* Vol. 4, King, R. C., ed. Plenum Press, New York.
Harrison, R. G. (1898). The growth and regeneration of the tail of the frog larva:

studied with the aid of Born's method of grafting, *Wilhelm Roux' Arch. Entwicklungsmech. Org.* **7**:430–485.
Harrison, R. G. (1903). Experimentelle Untersuchungen über die Entwicklung der Sinnesorgane der Seitenlinie bei den Amphibien, *Arch. Mikrosk. Anat.* **63**:35–149.
Harrison, R. G. (1924). Some unexpected results of heteroplastic transplantation of limbs, *Proc. Natl. Acad. Sci. USA* **10**:69–74.
Harrison, R. G. (1929). Correlation in the development and growth of the eye studied by means of heteroplastic transplantation, *Wilhelm Roux' Arch. Entwicklungsmech. Org.* **120**:1–55.
Harrison, R. G. (1931). Experiments on the development and growth of limbs in the Amphibia, *Science* **74**:575–576.
Harrison, R. G. (1934). Heteroplastic grafting in embryology, *Harvey Lect.* **29**:116–157.
Hennen, S. (1967). Nuclear transplantation studies of nucleocytoplasmic interactions in amphibian hybrids, pp. 353–375 in *The Control of Nuclear Activity*, Goldstein, L., ed. Prentice-Hall, Englewood Cliffs, N.J.
Hennen, S. (1977). Everything you wanted to know about creating a strain of axolotls carrying a tiger salamander gene but were afraid to ask, *Axolotl News.* **3**:1–2.
Holtfreter, J. (1935*a*). Morphologische Beeinflussung von Urodelenektoderm bei Xenoplastischer Transplantation, *Wilhelm Roux' Arch. Entwicklungsmech. Org.* **133**:367–426.
Holtfreter, J. (1935*b*). Über das Verhalten von Anurenektoderm in Urodelenkeimen, *Wilhelm Roux' Arch. Entwicklungsmech. Org.* **134**:427–494.
Horsa-King, M. L. (1978). Experimental studies on a mutant gene, *a*, causing albinism in the axolotl, *Ambystoma mexicanum*, *Dev. Biol.* **62**:370–388.
Humphrey, R. R. (1967). Albino axolotls from an albino tiger salamander through hybridization, *J. Hered.* **58**:95–101.
Humphrey, R. R. (1968). A genetically determined absence of heart function in embryos of the Mexican axolotl (*Ambystoma mexicanum*), *Anat. Rec.* **160**:475.
Humphrey, R. R. (1969). A recently discovered mutant "eyeless" in the Mexican axolotl, *Ambystoma mexicanum*, *Anat. Rec.* **163**:206.
Humphrey, R. R. (1972). Genetic and experimental studies on a mutant gene (*c*) determining absence of heart action in embryos of the Mexican axolotl *Ambystoma mexicanum*, *Dev. Biol.* **27**:365–375.
Humphrey, R. R. (1975). The axolotl, *Ambystoma mexicanum*, pp. 3–17 in *Handbook of Genetics*, Vol. 4, King, R. C., ed. Plenum Press, New York.
Humphrey, R. R. and H.-M. Chung (1977). Genetic and experimental studies on three associated mutant genes in the Mexican axolotl: *st* (for stasis), *mi* (for microphthalmic) and *h* (for hand lethal), *J. Exp. Zool.* **202**:195–202.
Hunt, R. K. and C. F. Ide (1977). Radial propagation for positional signals for retinotectal patterns in *Xenopus*, *Biol. Bull. (Woods Hole)* **153**:430–431.
Hunt, R. K. and J. Piatt (1978). Cross-species axial signalling with realignment of retinal axes, in embryonic *Xenopus* eyes, *Dev. Biol.* **62**:44–51.
Ide, C. F. (1977). Neurophysiology of *spastic*, a behavior mutant of the Mexican axolotl: altered vestibular projection to cerebellar auricle and area acousticolateralis, *J. Comp. Neurol.* **176**:359–372.

Ide, C. F., N. Miszkowski, C. B. Kimmel, E. Schabtach, R. Tompkins, O. Elbert, and M. Duda (1977). Analysis of spastic: a neurological mutant of the Mexican axolotl, pp. 267–287 in *Cellular Neurobiology: Proceedings* (ICN-UCLA Symposium on Neurobiology, Squaw Valley, Calif., 1976), Hall, Z. W., F. C. Fox, and R. Kelley, eds. Alan R. Liss, New York.

Ide, C. F. and R. Tompkins (1975). Development of locomotor behavior in wild type and spastic (sp/sp) axolotls, *Ambystoma mexicanum*, *J. Exp. Zool.* **194:**467–478.

Ide, C. F., R. Tompkins, and N. Miszkowski (1977). Neuroanatomy of *spastic*, a behavior mutant of the Mexican axolotl: Purkinje cell distribution in the adult cerebellum, *J. Comp. Neurol.* **176:**373–386.

Jacobson, M. and G. Hirose (1978). Origin of the retina from both sides of the embryonic brain: a contribution to the problem of crossing at the optic chiasm, *Science* **202:**637–639.

Justus, J. T. and P. B. Hollander (1971). Electrophysiology studies on the cardiac non-function mutation in the Mexican axolotl, *Ambystoma mexicanum*, *Experientia* **27:**1040–1041.

Kilikowski, R. R. and F. J. Manasek (1978). The cardiac lethal mutant of *Ambystoma mexicanum*: a re-examination, *Am. Zool.* **18:**349–358.

Lemanski, L. F. (1978). Morphological, biochemical and immunohistochemical studies on heart development in cardiac mutant axolotls, *Ambystoma mexicanum*, *Am. Zool.* **18:**327–348.

Lemanski, L. F., B. S. Marx, and C. S. Hill (1977). Evidence for abnormal heart induction in cardiac mutant salamanders *Ambystoma mexicanum*, *Science* **196:**894–896.

Levi-Montalcini, R. (1952). Effects of mouse tumor transplantation on the nervous system, *Ann. N.Y. Acad. Sci.* **55:**330–343.

Lewis, W. H. (1907). Transplantation of the lips of the blastopore in *Rana palustris*, *Am. J. Anat.* **7:**137–143.

Malacinski, G. M. and A. J. Brothers (1974). Mutant genes in the Mexican axolotl, *Science* **184:**1142–1147.

Mangold, O. (1923). Transplantationsversuche zur Frage der Spezifität und der Bildung der Keimblätter bei Triton, *Arch. Mikrosk. Anat.* **100:**198–301.

Marx, A. (1925). Experimentelle Untersuchungen zur Frage der Determination der Medullarplatte, *Wilhelm Roux' Arch. Entwicklungsmech. Org.* **105:**19–44.

Mosher, H. S., F. A. Fuhrman, H. D. Buchwald, and H. G. Fischer (1964). Tarichatoxin-tetrodotoxin: a potent neurotoxin, *Science* **144:**1100–1110.

Narahashi, T., J. W. Moore, and W. R. Scott (1964). Tetrodotoxin blockage of sodium conductance increase in lobster giant axons, *J. Gen. Physiol.* **47:**965–974.

Nobel, D. (1975). *The Initiation of the Heartbeat.* Clarendon Press, Oxford.

Piatt, J. (1948). Form and causality in neurogenesis, *Biol. Rev. Camb. Philos. Soc.* **23:**1–45.

Richardson, D. (1932). Some effects of heteroplastic transplantation of the ear vesicle in Amblystoma, *J. Exp. Zool.* **63:**413–445.

Roth, H. (1950). Die Entwicklung Xenoplastischer Neuralchimaeren, *Rev. Suisse Zool.* **57:**621–686.

Rotmann, E. (1929). Über den Anteil Mesoderms und Ektoderms an der Formbildung der Amphibien-Extremität, *Naturwissenschaften* **17**:878.

Rotmann, E. (1931). Die Rolle des Ektoderms und Mesoderms bei der Formbildung der Kiemen und Extremitäten von Triton, *Wilhelm Roux' Arch. Entwicklungsmech. Org.* **124**:747–794.

Rotmann, E. (1933). Die Rolle des Ektoderms und Mesoderms bei der Formbildung der Extremitäten von Triton. II. Operationen im Gastrula und Schwanzknospenstadium, *Wilhelm Roux' Arch. Entwicklungsmech. Org.* **129**:85–119.

Rotmann, E. (1934). Heteroplastischer Austausch einiger inductiv entstehender Organanlagen (Kurze Mitteilung), *Wilhelm Roux' Arch. Entwicklungsmech. Org.* **31**:702–704.

Rotmann, E. (1939). Der Anteil von Inductor und reagierendem Gewebe an der Entwicklung der Amphibienlinse, *Wilhelm Roux' Arch. Entwicklungsmech. Org.* **139**:1–49.

Rotmann, E. (1940). Die Bedeutung der Zellgrösse für die Entwicklung der Amphibienlinse, *Wilhelm Roux' Arch. Entwicklungsmech. Org.* **140**:124–156.

Sinclair, J. H., C. R. Carroll, and R. R. Humphrey (1974). Variation in rDNA redundancy level and nucleolar organizer length in normal and variant lines of the Mexican axolotl, *J. Cell Sci.* **15**:239–257.

Spemann, H. (1918). Über die Determination der ersten Organanlagen des Amphibienembryo I–VI, *Wilhelm Roux' Arch. Entwicklungsmech. Org.* **43**:448–555.

Spemann, H. (1921). Die Erzeugung tierischer Chimaeren durch heteroplastische embryonale Transplantation zwischen *Triton cristatus* und *taeniatus, Wilhelm Roux' Arch. Entwicklungsmech. Org.* **48**:533–570.

Spemann, H. (1938). *Embryonic Development and Induction.* Hafner Publishing Co., New York.

Spemann, H. and H. Mangold (1924). Über Induktion von Embryonalanlagen durch Implantation artfremder Organisatoren, *Arch. Mikrosk. Anat.* **100**: 599–638.

Stone, L. S. (1930). Heteroplastic transplantation of eyes between larvae of two species of Amblystoma, *J. Exp. Zool.* **55**:193–261.

Stone, L. S. (1935). Experimental formation of accessory organs in mid-body lateral-line of amphibians, *Proc. Soc. Exp. Biol. Med.* **33**:80–82.

Twitty, V. C. (1930). Regulation in the growth of transplanted eyes, *J. Exp. Zool.* **55**:43–52.

Twitty, V. C. (1932). Influence of the eye on the growth of its associated structures, studied by means of heteroplastic transplantation, *J. Exp. Zool.* **61**: 333–374.

Twitty, V. C. (1937). Experiments on the phenomenon of paralysis produced by a toxin occurring in Triturus embryos, *J. Exp. Zool.* **76**:67–104.

Twitty, V. C. (1940). Size-controlling factors, *Growth (Suppl.)* **4**:109–120.

Twitty, V. C. (1955). Eye, pp. 402–414 in *Analysis of Development,* Willier, B. H., P. A. Weiss, and V. Hamburger, eds. W. B. Saunders, Philadelphia.

Twitty, V. C. (1966). *Of Scientists and Salamanders.* W. H. Freeman, San Francisco.

Twitty, V. C. and H. A. Elliott (1934). The relative growth of the amphibian eye studied by means of transplantation, *J. Exp. Zool.* **68**:247–291.

Twitty, V. C. and J. L. Schwind (1931). The growth of eyes and limbs transplanted heteroplastically between two species of Amblystoma, *J. Exp. Zool.* **59**:61–86.

Ulshafer, R. J. and E. Hibbard (1976). Morphology of the optic rudiment in eyed and eyeless axolotls, *Anat. Rec.* **184**:552.

Van Deusen, E. (1973). Experimental studies on a mutant gene (*e*) preventing the differentiation of eye and normal hypothalamus primordia in the axolotl, *Dev. Biol.* **34**:135–158.

Wagner, G. (1949). Die Bedeutung der Neuraliste für die Kopfgestaltung der Amphibienlarve. Untersuchungen an Chimaeren von Triton und Bombinator, *Rev. Suisse Zool.* **56**:520–620.

Wallace, H. R. and M. L. Birnstiel (1966). Ribosomal cistrons and the nucleolar organizer, *Biochim. Biophys. Acta* **114**:296–310.

Wiemann, H. L. (1925). Heteroplastic grafts of the anterior limb-level of the cord in Ambystoma embryos, *Science* **61**:422–423.

Wiemann, H. L. (1926). The effect of heteroplastic grafts of the spinal cord on the development of the limb of Ambystoma, *J. Exp. Zool.* **45**:335–348.

Yntema, C. L. (1955). Ear and nose, pp. 415–428 in *Analysis of Development,* Willier, B. H., P. A. Weiss, and V. Hamburger, eds. W. B. Saunders, Philadelphia.

DEVELOPMENT AND DISEASE IN THE NEUROMUSCULAR SYSTEM OF MUSCULAR DYSTROPHIC ↔ NORMAL MOUSE CHIMERAS

Alan C. Peterson, Patricia M. Frair, Helen R. Rayburn, and David P. Cross

McMaster University, Hamilton, Ontario, Canada

Primary mouse chimeras are composite organisms in which two or more genotypically distinct cell lines coexist throughout development (Ford, 1969). By exploring the interaction between these cooperating "clones," unique insights have been gained into the pathogenesis of numerous inherited diseases and also into several aspects of developmental biology. Comprehensive reviews of these chimera studies have been presented by Mintz (1974) and McLaren (1976), and Mullen (1977) has reviewed the particular application of chimeras to issues in the development and pathogenesis of the central nervous system. The chimera preparation also provides a powerful means of investigating the interaction of cells in the peripheral neuromuscular system, and we are investigating the relationship of disease expression and tissue genotype in muscular dystrophic ↔ normal mouse chimeras.

With the straightforward genetics of an autosomal recessive factor (Michelson, Russell, and Harman, 1955; Russell, Silvers, Loosli, Wolfe, and Southard, 1962), murine muscular dystrophy is a disease characterized by a consistent syndrome of diverse abnormalities. Associated with the complex and progressive disease of skeletal musculature is a remarkable failure in the early development of Schwann cell-axon relationships in peripheral nerves (Bradley and Jenkison, 1973; Bray, Perkins, Peterson, and Aguayo, 1977). In certain spinal roots and cranial nerves, and at particular locations in more peripheral nerves, there is a marked deficiency of normal Schwann cells: the majority of the axons as they course through these areas are completely naked.

Two mutations have occurred at the *dy* locus (*dy* and dy^{2J}) (Meier and Southard, 1970), and although affected mice of both types express the spinal root anomaly (Stirling, 1975), the muscle diseases in the two are dramatically different (Butler and Cosmos, 1977). In 129/ReJ *dy/dy* mice the majority of the glycolytic muscle fibers degenerate prior to extensive loss of oxidative fibers, whereas in C57BL/6J dy^{2J}/dy^{2J} mice the oxidative fibers appear most susceptible to the disease. However, the nature of the genetic defect in mice of either genotype has not yet been established. Indeed, the primary site of expression of the genetic defect is not yet clear, and despite intensive effort, the exact relationship of the muscle and nerve abnormalities, both to each other and to the underlying genetic defect, remains a complete mystery (Peterson, 1979).

If the muscle disease results exclusively from an intrinsic deficiency of the muscle fibers, then both normal and dystrophic muscle would consistently express the phenotype predicted by their respective genotypes and no influence of an alternate host environment could lead to a significant alteration. Thus, muscle reciprocally transplanted between genotypically different hosts would consistently remain either healthy or diseased. This transplantation approach (Cosmos, 1973, 1974) and numerous other experimental strategies have been attempted (e.g., Hamburgh, Bornstein, Peterson, Crain, Masurovsky, and Kirk, 1973; Douglas and Cosmos, 1974; Hamburgh, Peterson, Bornstein, and Kirk, 1975; Montgomery, 1975; Law, Cosmos, Butler, and McComas, 1976), but the combined results of these experiments have been difficult to interpret, and the exact tissue site of mutant gene expression remains totally unknown.

In the above manipulations, new muscle-host relationships can be established only after the neuromuscular system has undergone substantial, usually postnatal, differentiation. However, gross abnormalities in developing muscles (Platzer, 1979) and nerves (Bray et al., 1977) have been described in late fetal and newborn dystrophic mice, revealing that the genetic function affected by the *dy* mutation is essential for normal prenatal development. Thus, altering the genotype of the source of extrinsic influences at the later juvenile or adult stages may have little or no effect, even though such influences may have a significant role in the earlier pathogenesis of muscular dystrophy.

In the mouse chimera preparation there is an opportunity for cells of differing genotype to interact throughout differentiation; therefore the establishment of relevant relationships of tissue genotype does not depend upon either regenerative responses or invasive techniques. The phenotype of a tissue in a chimera can be assessed at precisely the same anatomical site and at the same stage of development as the corresponding tissue in a normal or diseased mouse. On the basis of these and other potential experimental advantages, we have attempted to determine if disease expression and tissue genotype consistently correlate in chimeras of muscular dystrophic ↔ normal composition (Peterson, 1974, 1979).

MATERIALS AND METHODS

Chimera Production

Chimeras were produced by aggregation of eight-cell preimplantation embryos using techniques described by Mullen and Whitten (1971). Eight-cell embryos of dystrophic genotype were obtained by inducing ovulation in 129 *dy/dy* females and artificially inseminating them with sperm obtained from 129 *dy/dy* males (Leckie, Watson, and Sterling, 1973). All such cleavage eggs were homozygous for the *dy* allele, so no ambiguity of the genotype of the 129 cells could exist in resulting chimeras. For the genotypically normal counterpart, preimplantation embryos were obtained from timed natural matings of unaffected C57BL/6J mice.

Analysis of Muscle Genotype

To determine whether individual chimera skeletal muscles have derived from the normal (C57BL/6J), the dystrophic (129/ReJ), or both cell lines, we have exploited the properties of a number of enzyme variants known to differ between these two inbred strains. The electrophoretic variants of glucosephosphate isomerase (GPI) have proven most useful, because we have been able to develop a highly sensitive system for their electrophoretic analysis that also permits accurate relative quantitation (Peterson, Frair, and Wong, 1978). In addition, the multimeric nature of enzymatically active GPI (Carter and Parr, 1967; DeLorenzo and Ruddle, 1969) provides a basis for detecting those muscle fibers in which both types of myonuclei coexist (Figure 1) (Mintz

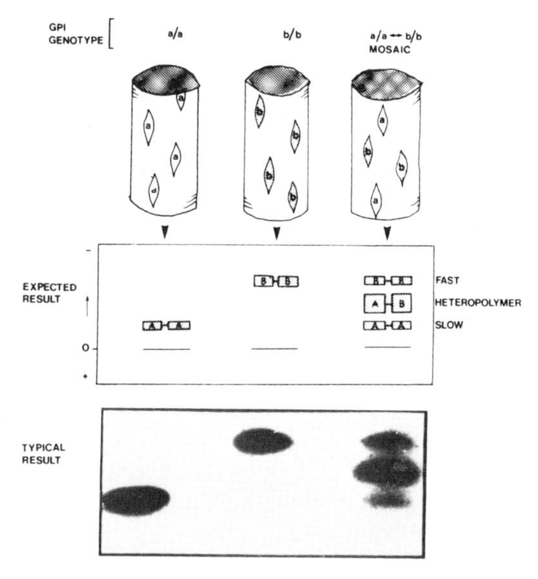

FIGURE 1. Genotype analysis of chimera muscle fibers using electrophoretic variants of glucosephosphate isomerase (GPI). Chimeras are produced by aggregating cells of *Gpi-1*$^{a/a}$ and *Gpi-1*$^{b/b}$ genotype. If myoblasts of both genotypes enter into a common myotube, the heteropolymer, i.e., GPI-1AB, enzyme is formed in the resulting muscle fiber. 129/ReJ *dy/dy* cells are *Gpi-1*$^{a/a}$ and C57BL/6 +/+ cells are *Gpi-1*$^{b/b}$.

and Baker, 1967). However, prior to accepting the validity of this approach, we must consider a number of theoretical issues.

Mammalian skeletal musculature is known to consist of numerous fiber types that have characteristic physiological and biochemical properties. If fibers differed widely in their concentration of GPI, an inaccuracy would be introduced into the quantitative analysis of GPI types in whole muscles. We have demonstrated that GPI is present in all of the individual fibers of a muscle (Frair and Peterson, 1978). Nevertheless, it has not been possible to do relative quantitative analysis on the fibers of a muscle, and no detailed histochemical study of muscle GPI has been reported. Between-fiber variations in GPI concentration must therefore be considered as a potential limitation in the interpretation of the genotype analysis of whole mosaic muscles.

Second, very little is understood about the relationship of the multiple nuclei and the cytoplasm within the skeletal muscle fiber syncitium. For example, each nucleus could preferentially control the local cytoplasm, or all nuclei could direct the synthesis of similar products that are then uniformly distributed throughout the fiber. If the former situation pertains, then the GPI type obtained from a short segment of a muscle fiber would not necessarily reflect the genotype of the nuclei in the remainder of that fiber. To determine whether such nuclear territories exist in skeletal muscle, we have analyzed serial sections of individual chimera muscle fibers for the proportion of GPI types along the length. If the relative proportion of GPI-1AA, GPI-1AB and GPI-1BB varies along the length of an individual fiber, then these "nuclear territories" would appear to exist. If the distribution of GPI variants is homogeneous throughout the fiber, no such territories would be indicated. Figure 2 demonstrates the isolation of a section from an individual fiber, and Figure 3 presents the results of one such experiment: all sections of the same fiber gave identical results, clearly supporting the view that the distribution of GPI is homogeneous in mosaic muscle fibers. Thus, the proportion of GPI isozymes present in a muscle can be predicted from the electrophoretic analysis of a single cross section, and the majority of the muscle remains available for analysis of histochemical phenotype.

A third issue is whether all the myonuclei in a fiber equally direct the synthesis of the same muscle components. Specifically, having determined that all fibers contain GPI and that the GPI types within a mosaic fiber are distributed uniformly throughout that fiber, we must next determine if all active myonuclei uniformly direct the synthesis of muscle GPI. If the nuclei of an individual mosaic fiber were subdivided into specialized groups, GPI could be made by nuclei of one genotype, whereas the synthesis of other products important in preventing the expression of muscle disease could be directed by undetected nuclei of the other genotype. That is, if all active nuclei in a fiber do not equally direct the synthesis of all products, the use of GPI as a genotype marker is invalidated. This issue has not been conclusively resolved. However, we have made observations consistent with the theory that all the myonuclei within a given fiber are functionally equivalent. In chimeras that are mosaic with respect to three electrophoretic markers (variants of glucosephosphate isomerase, malic enzyme, and mitochondrial malic enzyme), we can obtain three independent estimates of the proportions of myonuclei of each genotype. The results of these experiments consistently indicate that the synthesis of all three enzymes is directed by the same population of myonuclei. Although

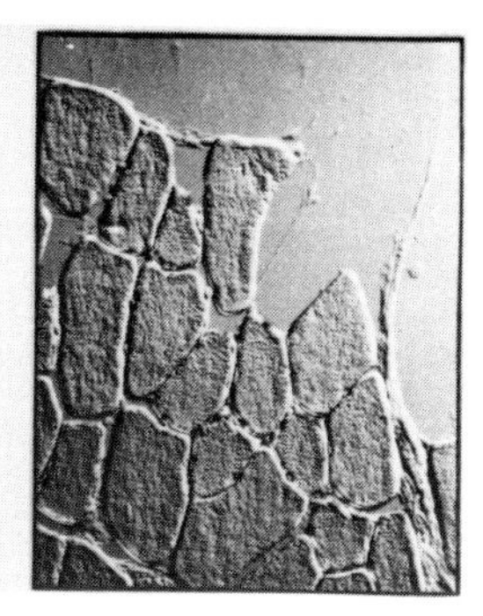

FIGURE 2. Dissection of cryostat section for isolation of individual muscle fiber cross sections. A 20-μm thick section of whole muscle is recovered on a polyvinylpyrrolidone-coated cover slip and immediately covered with silicone oil (Dow Corning 200 Fluid). The section is viewed with differential interference optics and sections of individual fibers are recovered using tungsten needles.

the functional equivalence of myonuclei cannot be definitively established without extending this type of investigation to all gene products, we have accepted our application of GPI variants as a valid and relatively accurate marker of the genotype of chimera muscles.

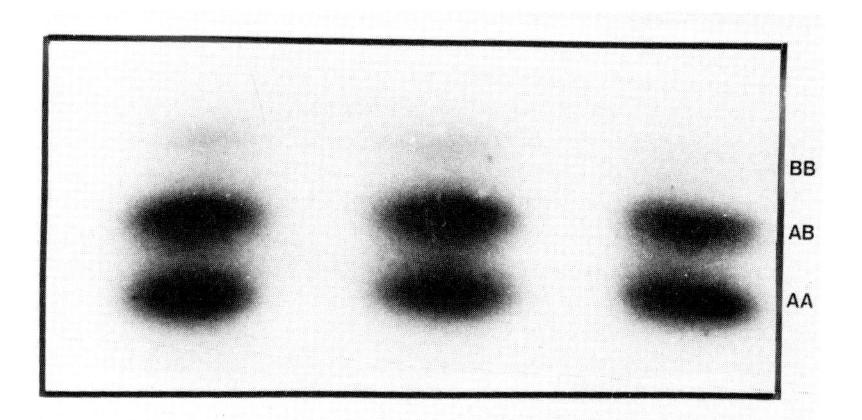

FIGURE 3. Electrophoretic analysis of GPI from an individual chimera muscle fiber (SM/J ↔ C57BL/6J). Sections of the same fiber were isolated from three different cryostat sections and electrophoresed on the same gel. All samples along the length of individual fibers reveal identical proportions of GPI types.

Analysis of Muscle Phenotype

Muscles of interesting genotype were classified as healthy or diseased on the basis of well-documented structural and histochemical characteristics.

RESULTS AND DISCUSSION

Chimera Muscle

During the course of these investigations it became clear that chimeras from a large variety of strain combinations have muscles that develop with the participation of both of the genotypically distinct cell lines. Furthermore, most chimera muscles reveal the heteropolymer band of GPI, indicating that both types of myonuclei have entered into a common cytoplasm (e.g., Figure 3). This circumstance clearly limits the usefulness of the chimera for determining whether the muscle genotype is relevant to the expression of muscle disease. If, for example, the chimera muscles were healthy, this could be due to either an intrinsic effect of the normal myonuclei or some extrinsic influence of genotypically normal cells elsewhere in the chimera. For the ideal experimental situation we require chimera muscles that are not genotypically mosaic.

Beyond aggregating midcleavage eggs of known genotype, the investigator has no further direct control over the eventual composition of the resulting chimeras. Nonetheless, numerous strain combinations are known to result in chimeras in which the two cell lines are not equally represented, i.e., unbalanced chimeras (Mullen and Whitten, 1971). For our purposes we were fortunate to identify chimeras of one strain combination (129/J ↔ C57BL/6J) in which muscle, in particular, preferentially derives from the 129 cell line (Table 1). Although it remained a possibility that early expression of the *dy* gene could interfere with the development of the 129 cell line, we expected that the majority of the muscle in chimeras of 129 *dy*/*dy* ↔ C57BL/6J +/+ genotype would be genotypically dystrophic. Six 129 *dy*/*dy* ↔ C57BL/6J +/+ chimeras have been produced, only one of which revealed significant clinical signs of the disease, and when the genotypes of the muscles in these mice were determined, the expected preponderance of the 129 *dy*/*dy* cells was found. Figure 4 demonstrates the electrophoretic results obtained from an anterior tibialis muscle from one such chimera. Although the electrophoretic system is sensitive enough to

TABLE 1. *Genotype analysis of muscle and other tissues from C57BL/6J ↔ 129/J chimeras*

Chimera[1]	♀A	♀B	♀C	♂D	♂E	♂F
	% GPI-1B (C57BL/6J type)					
Nonmuscle tissues						
Red blood cells	0.0	24.7	0.0	7.7	17.2	4.1
Brain	12.2	45.3	12.2	42.6	54.7	21.0
Liver	6.0	65.8	1.8	69.4	83.8	39.0
Kidneys (2)	17.9	37.2	10.0	40.2[4]	51.7	18.7
Mean result	10.8	42.0	6.8	40.0	54.0	20.3
Muscles						
Forelimb muscles (4)[2]	9.4	19.8[5]	4.9	16.7	31.5	9.4[6]
Hindlimb muscles (6)[3,8]	2.6	6.6	0.2[7]	3.9	21.1	4.6
Mean result	4.8	11.0	2.3	9.0	25.2	5.3

[1] Chimeras were between 945 and 977 days old at time of sacrifice.
[2] Forelimb muscles: triceps and extensor carpi ulnaris of each limb.
[3] Hindlimb muscles: anterior tibialis, gastrocnemius and soleus of each limb.
[4] One sample.
[5] Three samples.
[6] One sample.
[7] Five samples.
[8] In chimeras of this type, the 129 cell line predominates in most tissues, but this bias is extreme in hindlimb muscles.

detect the GPI in a muscle fiber cross section only a few microns thick, no GPI encoded by C57BL/6J nuclei could be detected in the samples of muscle examined here. We estimate that if 0.2% of the nuclei were genotypically normal, heteropolymer and fast electrophoretic bands would have been detected. Therefore we conclude that this chimera muscle is genotypically dystrophic. Nonetheless, when subsequent sections of the same muscle were analyzed for the expression of muscle disease, the normal phenotype was revealed (Figure 5). Thus, the genotype of the muscle does not always correlate with the muscle phenotype. This result clearly defines a significant extramuscular influence, but, by itself, it does not necessarily implicate an extramuscular defect as the only factor in the pathogenesis of murine muscular dystrophy. Muscle that was intrinsically defective could escape its typical pathology if genotypically normal cells elsewhere in the chimera were able to provide a product that is required for the normal development and

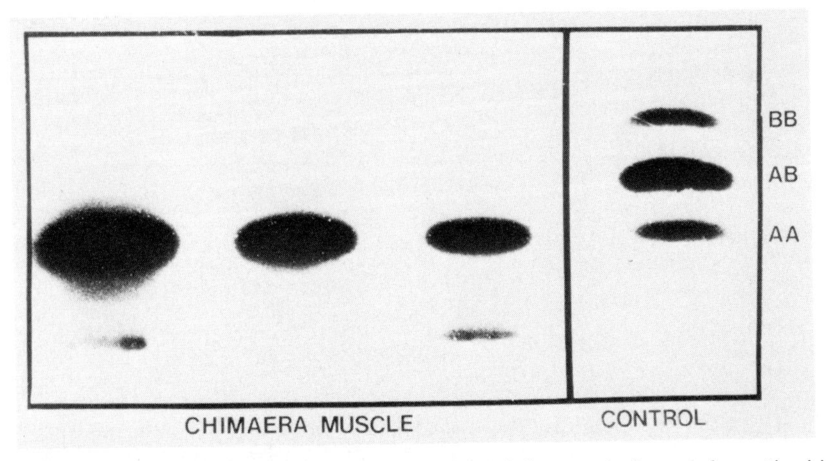

FIGURE 4. Genotype analysis of left anterior tibialis muscle from a 4-month-old healthy chimera of 129/ReJ *dy/dy* ↔ C57BL/6J +/+ composition. Groups of fibers isolated from cryostat cross sections revealed only the GPI type encoded by the 129/ReJ *dy/dy* nuclei, i.e., GPI-1AA. No heteropolymer (GPI-1AB) or fast (GPI-1BB) enzyme was detected; therefore this muscle apparently developed exclusively from the 129/ReJ *dy/dy* cell line.

maintenance of genotypically dystrophic muscle. However, if the disease could be imposed upon genotypically normal muscle, an entirely extramuscular cause of the disease would be indicated.

Chimera muscles of totally normal genotype are of course not likely to develop in chimeras of C57BL/6J +/+ ↔ 129 *dy/dy* composition where 129 cells predominate in myogenesis. Nonetheless, we have recovered muscles with a significant normal component from chimeras of a slightly different combination (i.e., C57BL/6 gus^h/gus^h *le/le* ↔ 129). The C57BL/6 mice used in this combination not only had the normal alleles of the *dy* locus but also were homozygous for two other unrelated recessive mutations: heat-labile glucuronidase (gus^h) and light ears (*le*) (Paigen, Swank, Tomino, and Ganschow, 1975). In chimeras of this type, the 129 myogenic advantage is not expressed to the same extent. Thus, from a further six chimeras of 129 *dy/dy* ↔ C57BL/6 gus^h/gus^h *le/le* composition we expected to recover some muscles with a significant normal component. Figure 6 demonstrates the results of electrophoresis of GPI from an anterior tibialis of one such chimera: 18.5% of the GPI present was encoded by genotypically normal nuclei, and the presence of a heteropolymer band indicates

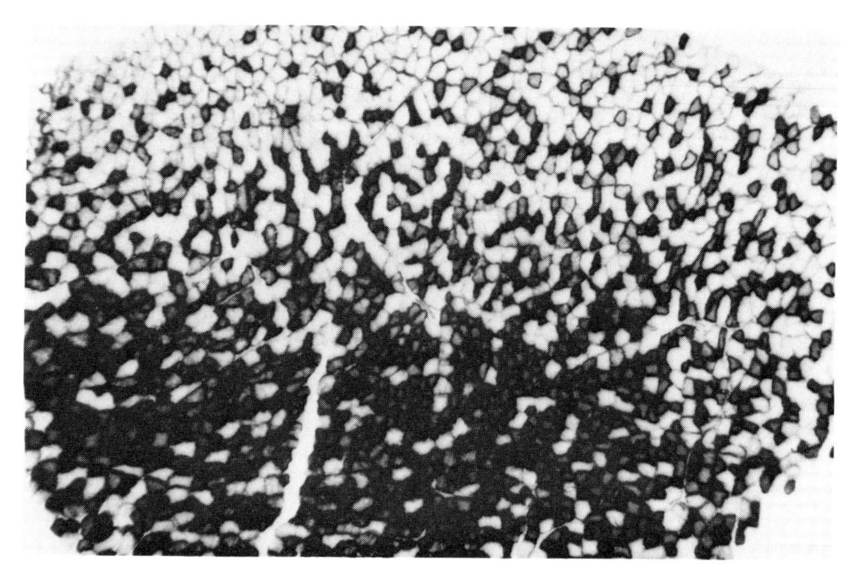

FIGURE 5. Normal phenotype expressed by the left anterior tibialis muscle of dystrophic genotype from a healthy 4-month-old chimera. Cryostat section stained for succinic dehydrogenase activity reveals the normal characteristics and distribution of all fiber types.

that individual fibers contained nuclei of both normal and dystrophic genotypes. However, when groups of fibers were dissected from cryostat sections, the distribution of normal nuclei was not uniform: some groups of muscle fibers were found to be completely devoid of the GPI type encoded by the normal nuclei, whereas no entirely genotypically normal fibers were identified. Thus, this particular muscle appeared to consist of fibers of only mosaic or dystrophic genotype.

The 8-month-old chimera from which the above muscle was taken expressed the full clinical disease, and histochemical examination of the muscle showed it to be grossly abnormal, i.e., the glycolytic fibers normally present in the crown of the anterior tibialis muscle had evidently degenerated, and the remaining oxidative fibers expressed a wide range of abnormalities typical of the muscle disease in 129/ReJ *dy/dy* mice (Figure 7). Therefore we have clearly demonstrated that genotypically normal myonuclei are present in a muscle that expresses the typical disease. This observation rules out the possibility that a small proportion of normal myonuclei are capable of rescuing the entire musculature by some form of metabolic cooperation. However, in this anterior tibialis muscle, the fibers of most interest, the glycolytic type, had presumably degenerated during the course of the disease. Establishing the genotype and phenotype of these fibers is particularly important because the abnormalities observable in the relatively resistant oxidative fibers could, at least in part, result as a normal response to the pathological destruction that occurs in the glycolytic fibers. The analysis of affected muscles obtained at a stage prior to the complete destruction of the glycolytic fibers may afford the opportunity to explore this issue.

The full pathology of dystrophic muscle develops through several cycles of destruction and attempted regeneration. Indeed, even in relatively advanced stages, cells with the characteristics of young regenerating fibers can be observed. In affected chimeras, repeated cycles of attempted regeneration might provide an opportunity to select for those fibers with a significant proportion of normal myonuclei and, if so, during the course of the disease the genotype of the muscle could significantly change in favor of the normal component. However, in the case of the 8-month-old chimera, less than 20% of the myonuclei remaining in the left anterior tibialis are genotypically normal. Therefore either a selection for muscle fiber genotype does not occur, or if

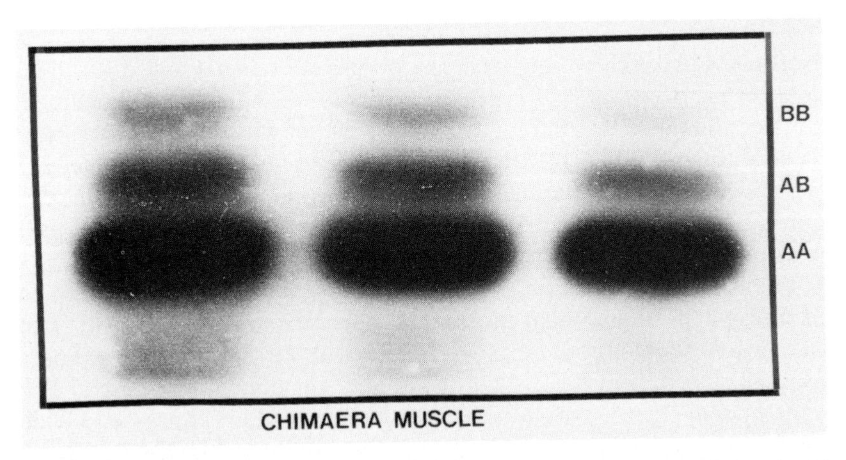

FIGURE 6. Genotype analysis of right anterior tibialis muscle from an 8-month-old affected chimera of 129/ReJ *dy/dy* ↔ C57BL/6 *gus^h/gus^h le/le* composition. Whole cryostat cross sections revealed the presence of heteropolymer (GPI-1AB) and fast (GPI-1BB) enzyme that identifies the presence of C57BL/6 nuclei. Quantitative analysis of the electrophoretogram revealed that 18.5% of the GPI was encoded by the normal C57BL/6 nuclei.

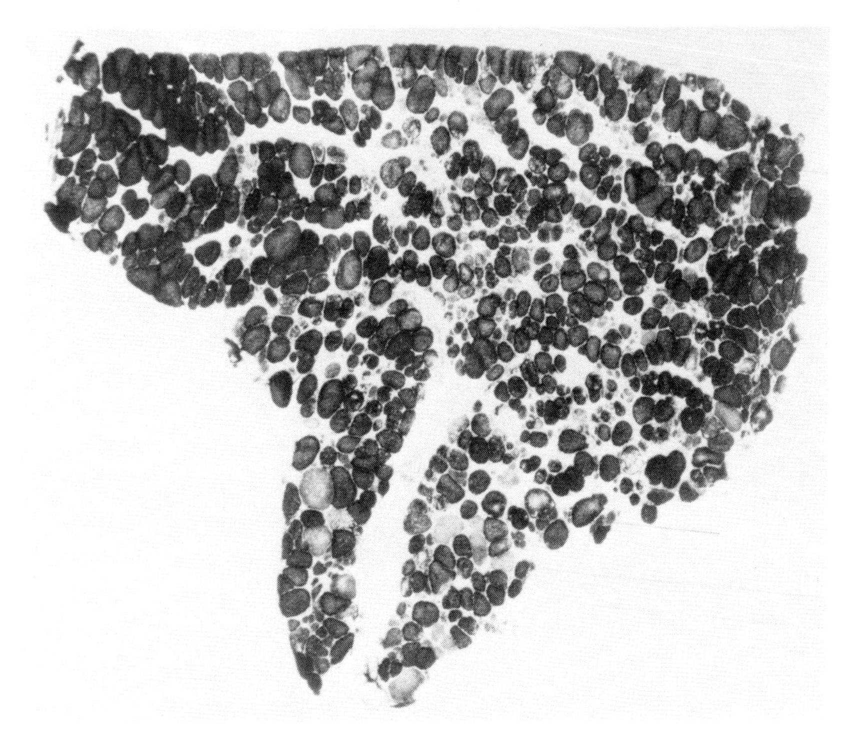

FIGURE 7. Typical dystrophic phenotype expressed by the right anterior tibialis muscle from an affected 8-month-old chimera of 129/ReJ *dy/dy* ↔ C57BL/6 gus^h/gus^h *le/le* composition. Cryostat section stained for succinic dehydrogenase activity reveals that the majority of the surviving fibers are of the resistant oxidative type.

it does, mosaic muscle is highly limited in its ability to respond to that selection pressure.

Chimera Nerves

In addition to investigating the relationship of the muscle genotype to its expressed phenotype, the dystrophic ↔ normal chimera also provides an opportunity to dissect the interrelationship of the neural abnormality and the muscle disease. If, for example, normal chimera nerves were found to be innervating muscle with an abnormal phenotype, then it would appear unlikely that the abnormalities of nerve induce abnormalities of muscle or vice versa. In the small number of chimeras in which the phenotype of both nerve and muscle has been

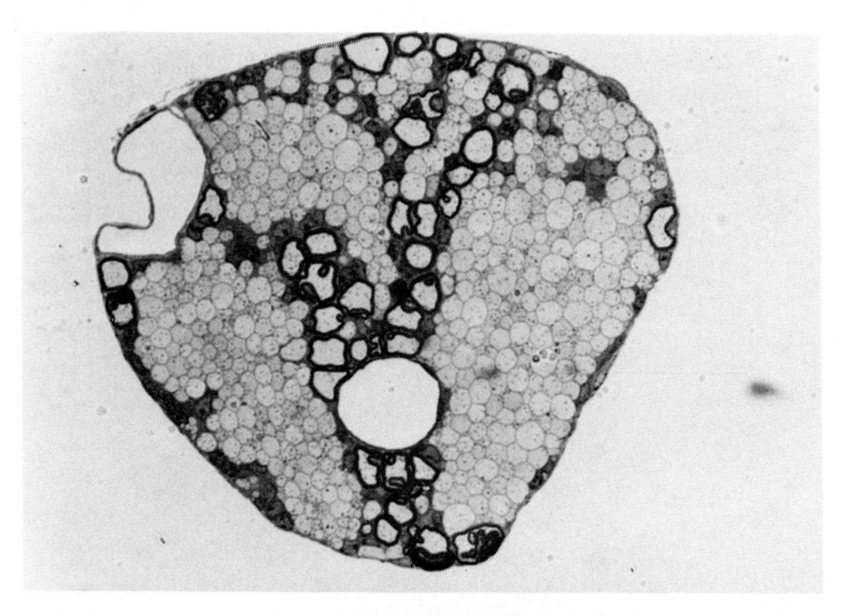

FIGURE 8. Dystrophic phenotype expressed in the middle of the left L4 ventral root of an affected chimera. The majority of the axons are not ensheathed by Schwann cells, and less than 40 have myelin sheaths. One-μm thick cross section of Epon-embedded root stained with toluidine blue.

established, an apparent correlation exists: healthy chimeras had normal roots and muscles, whereas in an affected chimera, naked axons (Figure 8) and severe muscle disease (Figure 7) coexisted. Thus, the two abnormalities could be interdependent or they could equally result from a primary defect expressed at yet a third site.

Although it is of interest that the neural deficit can be expressed in chimeras, it will be necessary to develop an accurate means of determining the genotype of axons and Schwann cells before the analysis of this abnormality will be particularly enlightening. Nonetheless, if the defect were intrinsic to the Schwann cell, it is somewhat surprising that a coexisting population of genotypically normal Schwann cells did not proliferate and correct the deficiency.

SUMMARY

An inherited disease of the neuromuscular system, muscular dystrophy (*dy*), has been investigated in the mouse chimera preparation. The majority of these chimeras are healthy, but some express the disease with variable severity. Analysis of the muscle reveals that genotypically mutant muscle can completely escape the disease and muscle with a significant component of normal genotype can express the typical disease. These results implicate an important and possibly primary extramuscular defect in the pathogenesis of the muscle disease.

In addition, the neural abnormality of dystrophic animals, which consists of naked axons in some spinal roots and elsewhere, is expressed in an affected chimera. Current chimera experiments are designed to assess whether genotypically normal muscle can express the disease, whether abnormal spinal roots and muscle disease are consistently coexpressed, and whether the Schwann cell abnormality is the result of an intrinsic Schwann cell deficit.

ACKNOWLEDGMENTS

This research was supported by the Medical Research Council of Canada and the Muscular Dystrophy Association of Canada. P. Frair, H. Rayburn, and D. Cross are recipients of support from the MDAC. We are particularly grateful for the excellent technical assistance provided by Rita Campbell. E. Cosmos, McMaster University, generously assisted by producing and interpreting some of the muscle histochemical preparations. K. Paigen and V. Chapman, Roswell Park, Buffalo, generously provided the C57BL/6 *gus*^{h}/*gus*^{h} *le*/*le* mice.

REFERENCES

Bradley, W. G. and M. Jenkison (1973). Abnormalities of peripheral nerves in murine muscular dystrophy, *J. Neurol. Sci.* **18**:227–247.

Bray, G. M., S. Perkins, A. C. Peterson, and A. J. Aguayo (1977). Schwann cell multiplication deficit in nerve roots of newborn dystrophic mice. A radioautographic and ultrastructural study, *J. Neurol. Sci.* **32**:203–212.

Butler, J. and E. Cosmos (1977). Histochemical and structural analyses of the phenotypic expression of the dystrophic gene in the 129/ReJ *dy*/*dy* and the C57BL/6J *dy*^{2J}/*dy*^{2J} mice, *Exp. Neurol.* **57**:666–681.

Carter, N. D. and C. W. Parr (1967). Isoenzymes of phosphoglucose isomerase in mice, *Nature (London)* **216**:511.

Cosmos, E. (1973). Muscle-nerve transplants: experimental models to study influences on differentiation, *Physiologist* **16**:167–177.

Cosmos, E. (1974). Muscle transplants: role in the etiology of hereditary muscular dystrophy, pp. 368–373 in *Exploratory Concepts in Muscular Dystrophy,* Vol. 2, Milhorat, A. T., ed. Excerpta Medica, New York.

DeLorenzo, R. L. and F. H. Ruddle (1969). Genetic control of two electrophoretic variants of glucosephosphate isomerase in the mouse (*Mus musculus*), *Biochem. Genet.* **3**:151.

Douglas, W. G. and E. Cosmos (1974). Histochemical responses of dystrophic murine muscles cross-innervated by sciatic nerves of normal mice, pp. 374–380 in *Exploratory Concepts in Muscular Dystrophy,* Vol. 2, Milhorat, A. T., ed. Excerpta Medica, New York.

Ford, C. E. (1969). Mosaics and chimaeras, *Br. Med. Bull.* **25:**104–109.

Frair, P. and A. C. Peterson (1978). Cytoplasmic territories of myonuclei. *Abstracts, 4th International Congress on Neuromuscular Diseases.*

Hamburgh, M., M. B. Bornstein, E. R. Peterson, S. M. Crain, E. B. Masurovsky, and C. Kirk (1973). *In vitro* studies of regeneration and innervation of muscle from dystrophic (dy^{2J}) mutant mice, *Prog. Brain Res.* **40:**497–508.

Hamburgh, M., E. R. Peterson, M. B. Bornstein, and C. Kirk (1975). Capacity of foetal spinal cord obtained from dystrophic mice (dy^{2J}) to promote muscle regeneration, *Nature (London)* **256:**219–220.

Law, P. K., E. Cosmos, J. Butler, and A. McComas (1976). The absence of dystrophic characteristics in normal muscles successfully cross-reinnervated by nerves of dystrophic genotype: physiological and cytochemical study of crossed solei of normal and dystrophic parabiotic mice, *Exp. Neurol.* **51:**1–21.

Leckie, P. A., J. G. Watson, and C. Sterling (1973). An improved method for the artificial insemination of the mouse (*Mus musculus*), *Biol. Reprod.* **9:**420–425.

McLaren, A. (1976). *Development of Cell Biology,* Vol. 4: *A Million Chimaeras,* Abercrombie, M., D. R. Newth, and J. G. Torrey, eds. Cambridge University Press, Cambridge.

Meier, H. and J. L. Southard (1970). Muscular dystrophy in the mouse caused by an allele at the *dy* locus, *Life Sci.* **9:**137–144.

Michelson, A. M., E. S. Russell, and P. J. Harman (1955). Dystrophia muscularis: a hereditary primary myopathy in the house mouse, *Proc. Natl. Acad. Sci. USA* **41:**1079–1084.

Mintz, B. (1974). Gene control of mammalian differentiation, *Annu. Rev. Genet.* **8:**411–470.

Mintz, B. and W. Baker (1967). Normal mammalian muscle differentiation and gene control of isocitrate dehydrogenase synthesis, *Proc. Natl. Acad. Sci. USA* **58:**592–598.

Montgomery, A. (1975). Parabiotic reinnervation in normal and dystrophic mice. I. Muscle weight and physiologic studies, *J. Neurol. Sci.* **26:**401–423.

Mullen, R. J. (1977). Genetic dissection of the CNS with mutant-normal mouse and rat chimeras, *Soc. Neurosci. Symp.* **2:**47–65.

Mullen, R. J. and W. K. Whitten (1971). Relationship of genotype and degree of chimerism in coat color to sex ratios and gametogenesis in chimaeric mice, *J. Exp. Zool.* **178:**165–176.

Paigen, K., R. Swank, S. Tomino, and R. Ganschow (1975). The molecular genetics of mammalian glucuronidase, *J. Cell. Physiol.* **85:**379–392.

Peterson, A. C. (1974). Chimaera mouse study shows absence of disease in genetically dystrophic muscle, *Nature (London)* **248:**561–564.

Peterson, A. C. (1979). Mosaic analysis of dystrophic-normal chimaeras: an approach to mapping the site of gene expression, pp. 630–647 in *Muscular*

Dystrophy and Other Inherited Diseases of Skeletal Muscle in Animals, J. B. Harris, ed. New York Academy of Sciences.

Peterson, A. C., P. M. Frair, and G. G. Wong (1978). A technique for detection and relative quantitative analysis of glucosephosphate isomerase isozymes from nanogram tissue samples, *Biochem. Genet.* **16:**681–690.

Platzer, A. (1979). Embryology of two murine muscle diseases: muscular dystrophy and muscular dysgenesis, pp. 94–113 in *Muscular Dystrophy and Other Inherited Diseases of Skeletal Muscle in Animals,* J. B. Harris, ed. New York Academy of Sciences.

Russell, E. S., W. K. Silvers, R. Loosli, H. G. Wolfe, and J. L. Southard (1962). New genetically homogeneous background for dystrophic mice and their normal counterparts, *Science* **135:**1061–1062.

Stirling, C. A. (1975). Abnormalities in Schwann cell sheaths in spinal nerve roots of dystrophic mice, *J. Anat.* **119:**169–180.

AXON-GLIA INTERACTIONS

NEURONAL REGULATION OF MYELINATING CELL FUNCTION

Peter S. Spencer

Albert Einstein College of Medicine, Bronx, New York

INTRODUCTION

Several laboratories are currently attempting to define the biological mechanisms regulating the function of myelinating cells in the central and peripheral nervous systems. This paper describes our investigations of this problem; these studies were conducted in collaboration with Harold Weinberg and several other colleagues identified below.

Initially, attention was directed to the possibility that the neuron controlled the biological expression of the cells that ensheath its axon. Using previously established techniques of nerve cross-anastomosis and radioactive Schwann cell labeling, we demonstrated that axons regenerating from myelinated nerves caused Schwann cells derived from unmyelinated nerves to initiate myelinogenesis. Concurrently, Aguayo and his colleagues (this volume), using nerve transplantation techniques, demonstrated the same principle, namely, that peripheral axons signal Schwann cells to commence myelinogenesis. Subsequently, both Aguayo's group and ours employed the nerve grafting technique to demonstrate the ability of quiescent oligodendrocytes to myelinate regenerating peripheral axons, an experiment that raised the possibility of a common axonal signal for myelinogenesis in the central and peripheral nervous systems.

In 1971 we postulated that the signaling mechanism was a negative feedback system that utilized bidirectional molecular signals carried by anterograde and retrograde axonal transport between the neuron and the myelinating cell. With focal loss of a myelinating cell (demyelination), the depletion of a steady-state signal to the neuron would precipitate the export of a neuronal signal that would stimulate Schwann cells to occupy the denuded axon segment and to start remyelination (Spencer, 1971). This concept was abandoned when we developed and studied a new model of focal demyelination and remyelination known as the perineurial window. From these experiments, it seemed more likely the axonal signal for myelination resided permanently within the axolemma, and daughter Schwann cells were signaled to commence myelinogenesis at the level of the plasma membrane (H. J. Weinberg, 1978). This led to a speculative theory of the control of myelinogenesis in normal and pathologic states, published in 1978, a hypothesis that drew heavily on the ideas of Mandel, Nussbaum, Neskovic, Sarlieve, and Kurihara (1972), Brady and Quarles (1973), Roseman (1974), Moscona (1974), and Barondes (1975). A major feature of this theory was the concept that neurons whose axons are to become myelinated permanently differentiate their axonal plasma membrane during development to a pre-myelinated state (cf. Quarles, McIntyre, and Sternberger, this volume). The single population of undifferentiated Schwann cells then randomly associate with these and pre-unmyelinated axons. Depending on the type of axon they envelop, the naive Schwann cells are instructed by the axon via their plasma membranes to develop into myelinated or unmyelinated fibers.

This theory assumed the plasma membrane of Schwann cells would change following contact with the surface membrane of the axon. We therefore set about developing a population of naive Schwann cells from which a plasma membrane-enriched fraction recently was obtained. Biochemical properties of the plasma membrane of the naive Schwann cells could then be compared to PNS myelin, the compacted plasma membranes of the axonally regulated Schwann cell. To date, the comparison has focused on properties of membrane-bound protein kinases in these membranes and their relative sensitivity to cyclic adenosine monophosphate (cyclic AMP). This was examined because of the well-known role of cyclic AMP as a mediator of intracellular events—events that conceivably might include myelinogenesis. Subsequently, Raff, Hornby-Smith, and Brockes (1978) demonstrated that the mitotic response of the Schwann cell is probably mediated by cyclic nucleotides.

DEFINING THE PROBLEM

The earliest electron microscope studies of peripheral nerves demonstrated that nerve fibers are constructed on two basic plans: myelinated fibers, each composed of a single axon enwrapped at any level by a single Schwann cell that elaborates a myelin sheath (Figure 1) (Bischoff and Thomas, 1975), and unmyelinated fibers, each with multiple axons situated in lateral furrows of Schwann cells that fail to develop myelin sheaths (Figure 2) (Ochoa, 1975). Although several studies have defined the developmental stages in the construction of peripheral nerve fibers (Webster, 1975), a number of questions remain unanswered. For example, do the Schwann cells of myelinated and unmyelinated fibers represent two different cell types, or merely different phenotypes of the same cell? Is it the axon or the Schwann cell that determines developmental sorting of axons and triggering of myelin formation, and do other factors such as local environmental and hormonal status play a role? How is the Schwann cell activated to produce myelin? Could there be a common signal for myelination in both the peripheral and central nervous systems? How would this signaling system operate, both in normal development and during repair of pathological states? A few of these questions have been illuminated by the studies described below; others are currently under investigation by many groups, including our own.

ESTABLISHING NEURONAL REGULATION OF MYELINOGENESIS

Our initial investigations sought to examine the proposition that the neuron and its axon direct the myelinating behavior of the Schwann cell. To study this question, we employed a technique of surgical cross-anastomosis between myelinated and unmyelinated nerves. This method was first reported at the turn of the century (Langley, 1898; Langley and Anderson, 1903, 1904) and later utilized to investigate certain features of peripheral nerve regeneration (Simpson and Young, 1945; Hillarp and Olivecrona, 1946). The results of these earlier studies had already suggested the instruction of Schwann cell myelinogenesis by the axon. This view was indirectly supported by the well-established fact that after severance of the normal communication between the axon and neuron, Schwann cells associated with the disconnected axon cease to maintain their myelin sheaths (Ramón y Cajal, 1928; Ohmi, 1961; Singer and Steinberg, 1972).

When our studies commenced in 1973, we were mindful that the cross-anastomosis technique would only address the Schwann cell signaling properties of the *regenerating* axon, but we were persuaded that de novo myelination of a regenerating sprout would recapitulate the key steps of developmental myelinogenesis. Aguayo and his colleagues were also studying the problem using similar techniques, and the two

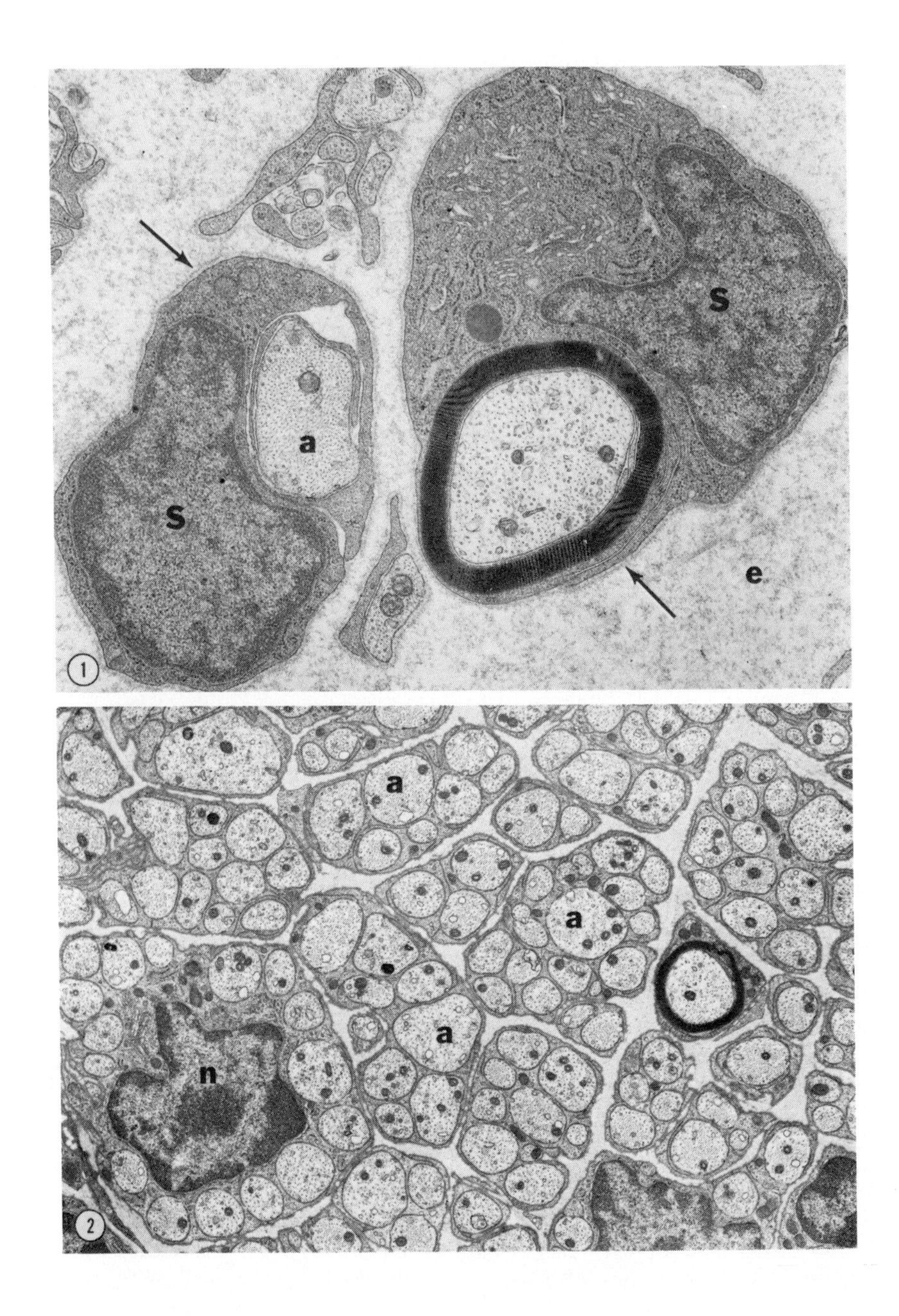
a
S
S
e
1
a
a
a
n
2

laboratories independently published comparable data and similar conclusions on the relationships of axons and myelinating cells. These studies eventually demonstrated that regenerating peripheral axons can stimulate both Schwann cells and oligodendrocytes to initiate myelinogenesis (Aguayo, Charron, and Bray, 1976; H. J. Weinberg and Spencer, 1976; Aguayo, Dickson, Trecarten, Attiwell, Bray, and Richardson, 1978; E. L. Weinberg and Spencer, 1979).

In cross-anastomosis experiments designed to allow axons regenerating from myelinated nerves to confront Schwann cells derived from unmyelinated nerves, two unifascicular nerves of similar size and location but of opposite fiber composition were needed. The predominantly myelinated nerve to the left sternohyoid muscle and the largely unmyelinated cervical sympathetic trunk (CST) in the neck of the rat met these requirements (Figures 3 and 4). To anastomose these two 100-μm diameter nerves, a 10/0 suture was inserted into the epineurium of each nerve prior to severance; the nerves were then cut caudal to the sutures, and the proximal and distal ends of the sternohyoid nerve and CST, respectively, were introduced into a Silastic tube with an internal diameter of 300 or 500 μm. With the aid of the suture material, all the nerves were simultaneously drawn together in the tube to abut the cut surfaces. The two ends of the suture material were then knotted over the tubing to secure the anastomosis (Figure 6). Operated animals were maintained for 2–15 weeks, and then the anastomosed nerves were examined following systemic perfusion with fixatives.

Successful inosculation of the two nerves was found in a total of 15 of 36 operated animals. Examination of the fused nerve stumps demon-

---

FIGURE 1. Myelinated fiber. Cross section of two regenerating fibers 15 days following a focal crush applied to the peroneal nerve of a rat. At left, Schwann cell (S) is enveloping the axon (a) prior to myelination. At right, Schwann cell (S) has elaborated a myelin sheath consisting of about 14 lamellae. Basal laminae (arrow) surround each fiber. e, Endoneurial collagen. ×23,000. This figure and Figure 2 are electron micrographs of epoxy-embedded tissue stained with uranyl acetate and lead citrate. Tissue in all figures was fixed in buffered glutaraldehyde and post-fixed in buffered osmium tetroxide. (Reproduced by permission from Spencer, P. S. [1976]. Experimentally induced nerve injury and its repair, pp. 131–139 in *Symposium on Microsurgery*, Vol. 14, Daniller, A. I. and B. Strauch, eds. C. V. Mosby, St. Louis.)

FIGURE 2. Unmyelinated fiber. Cross section of cervical sympathetic trunk (CST) of the rat showing mature arrangement of unmyelinated fibers and a single myelinated fiber (right). n, Schwann cell nucleus, a, axon. ×12,000. (Reproduced by permission from H. J. Weinberg and Spencer, 1975.)

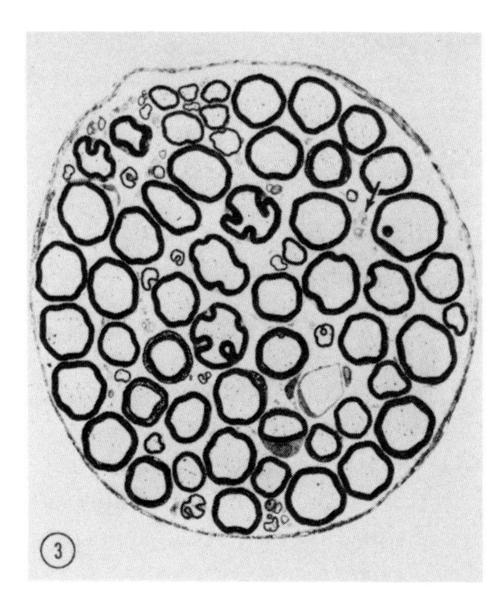

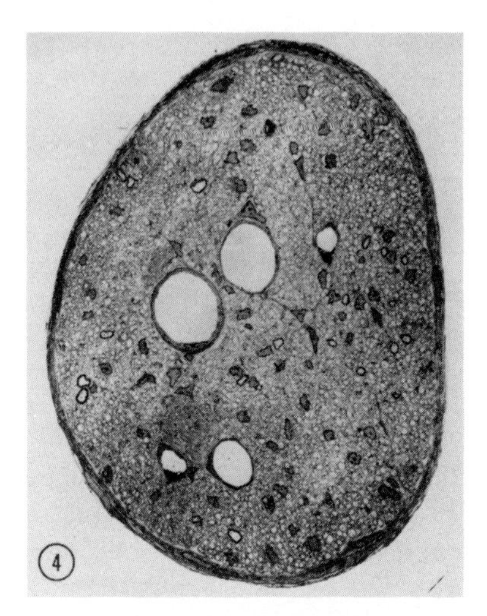

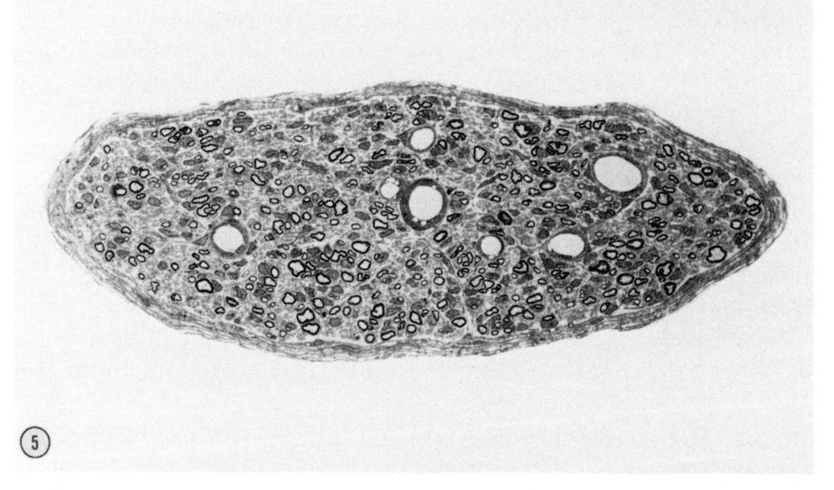

FIGURE 3. Cross-anastomosis experiment. Normal nerve to the caudal extremity of the left sternohyoid muscle ~3 mm caudal to its branch to the rostral end of the muscle. ×1,200. This figure and Figures 4 and 5 are electron micrographs of beam-thinned, unstained Epon ~0.25 μm cross sections. (Reproduced by permission from H. J. Weinberg and Spencer, 1975.)

FIGURE 4. Cross-anastomosis experiment. Normal CST ~10 mm caudal to the superior cervical ganglion. This nerve displays 23 myelinated fibers and 6,648 unmyelinated axons. ×890. (Reproduced by permission from H. J. Weinberg and Spencer, 1975.)

FIGURE 5. Cross-anastomosis experiment. Extratubal portion of the distal stump of a CST after 6 weeks of anastomosis. Note the large number (347) of evenly distributed myelinated fibers. ×855. (Reproduced by permission from H. J. Weinberg and Spencer, 1975.)

strated that fibers had regenerated from the proximal stump of the sternohyoid nerve, and many had penetrated the distal stump of the formerly unmyelinated CST. Counts of myelinated fibers within the preserved fascicle of the reinnervated CST revealed a mean of 164 ± 31, a statistical increase ($P < 0.01$) from that found in the normal CST (Figures 4 and 5). The number of nuclei associated with myelinated fibers was also greater than in the normal CST ($P < 0.001$). These observations indicated that axonal sprouts emerging from a myelinated fiber were able to penetrate a foreign, formerly unmyelinated nerve and establish normal cellular relationships, including myelination (H. J. Weinberg and Spencer, 1975).

Before it could be concluded that the regenerating axons had *induced* myelination by Schwann cells previously associated only with unmyelinated fibers, it was essential to consider the origin of the myelinating cells in the distal stump. If the myelinating cells had been derived from the indigenous Schwann cell population of the CST, it was highly unlikely they were formerly associated with myelinated fibers. After nerve section, each of the few myelinated fibers would have been replaced by longitudinal columns of daughter Schwann cells (Holmes and Young, 1942; P. K. Thomas, 1964). These columns would have formed conduits for a few regenerating axons that subsequently would have become myelinated. However, the majority of regenerating axons would have encountered Schwann cell columns derived from degenerated unmyelinated fibers. Thus, there was a strong possibility the majority of myelinating cells in the distal stump of the CST after anastomosis was derived from Schwann cells formerly associated with unmyelinated fibers. If true, this would have indicated that the regenerating axons had directed the daughter Schwann cells to produce myelin for the first time.

THE PROBLEM OF CELL MIGRATION

Before this issue could be resolved, it was necessary to consider a second possibility: that cells forming myelin in the distal stump of the CST had migrated from the proximal stump of the predominantly myelinated sternohyoid nerve. This was a serious consideration, since the ability of Schwann cells to migrate from a proximal nerve stump to form a neuroma was well known. In fact, this particular phenomenon appeared to explain the presence of miniature fascicles of regenerating fibers around the outside of the fused nerves at the site of anastomosis (Figures 7 and 8). Whether migratory Schwann cells could also penetrate into the distal stump of the CST therefore had to be determined.

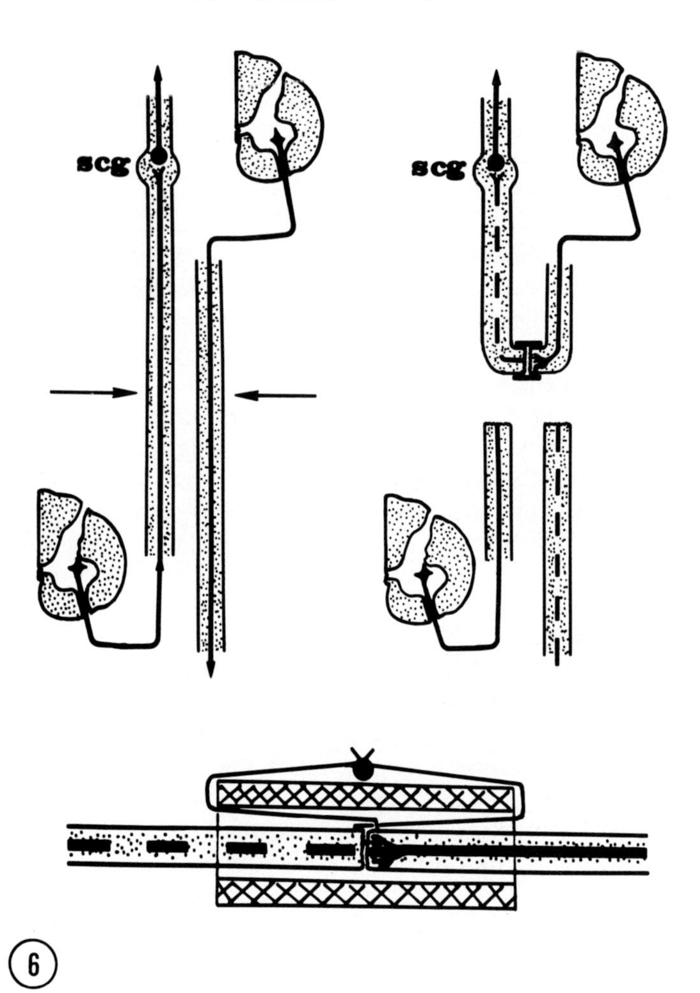

FIGURE 6. Cross-anastomosis experiment. Upper left: diagram illustrating the origin and direction of fiber pathways of the predominantly myelinated sternohyoid nerve on the right and the largely unmyelinated CST on the left. scg, Superior cervical ganglion. Facing arrows indicate the sites of nerve transection prior to cross-anastomosis. Upper right: diagram illustrating the severed proximal stump of the sternohyoid nerve (right) and the severed distal stump of the sympathetic trunk (left) undergoing Wallerian degeneration (broken line) after cross-anastomosis within a Silastic tube (center). scg, Superior cervical ganglion. Lower: diagram showing the anastomotic site. The cut end of the proximal stump of the sternohyoid nerve (right) abuts the cut end of the distal stump of the sympathetic trunk (left) within a piece of tubing (cross-hatched). Epineurial sutures (thin lines), placed at each cut end, are knotted over the tube to secure the anastomosis. (Reproduced by permission from H. J. Weinberg and Spencer, 1975.)

Harold Weinberg and I studied this problem by tracing the fate of Schwann cells in the two nerve stumps. The experimental design was developed from the principle that Schwann cells divide following nerve transection, and their nuclei may be labeled with radioactive nucleic acid precursors (tritiated thymidine) prior to cell division (Friede and Johnstone, 1967; Bradley and Asbury, 1970). Schwann cell division would occur at the end of a proximal stump and throughout the length of a distal nerve stump (G. A. Thomas, 1948; Logan, Rossiter, and Barr, 1953; Abercrombie, Evans, and Murray, 1959). Initially, it was necessary to determine the time course of thymidine uptake in the proximal and distal stumps of the sectioned sternohyoid and CST, respectively. For the sternohyoid nerve, animals were provided with a single dose of label 24 hours prior to sacrifice after maintenance for 1–7 days following transection. After systemic perfusion with saline, the proximal stump was removed and the distal 3 mm counted by liquid scintillation. Thymidine uptake was found to occur 1–4 days after transection, with a peak at 3 days (Figure 9). A similar protocol was followed to determine dividing cells in the distal stump of the CST, except such animals were maintained for 1–14 days and a 6-mm length of the nerve was removed, solubilized, and counted. Thymidine incorporation in these nerves occurred over a much longer period (1–12 days), with a peak at 5 days after transection (Figure 10).

Armed with this information, we proceeded to reexamine the cross-anastomosis experiment. To label one of the two nerves selectively, one nerve was cut and labeled prior to severing the second and performing the anastomosis. Thus, to perform an anastomosis between a labeled, myelinated nerve and an unlabeled, unmyelinated nerve, the following procedure was adopted: the sternohyoid nerve was severed and dividing Schwann cells were labeled by systemic injection of thymidine every 12 hours during the period of maximum labeling (Figure 11, left). After an additional 24 hours, during which no precursor was administered, animals received cold thymidine at a concentration 100 times that of the labeled compound to avoid utilization of any recirculating label. Twelve hours later, the unlabeled CST was cut and its distal stump anastomosed in the usual manner to the previously labeled proximal stump of the sternohyoid nerve. A comparable procedure was followed to achieve selective labeling of the Schwann cells of the CST distal stump (Figure 11, right) and its anastomosis to an unlabeled sternohyoid proximal stump.

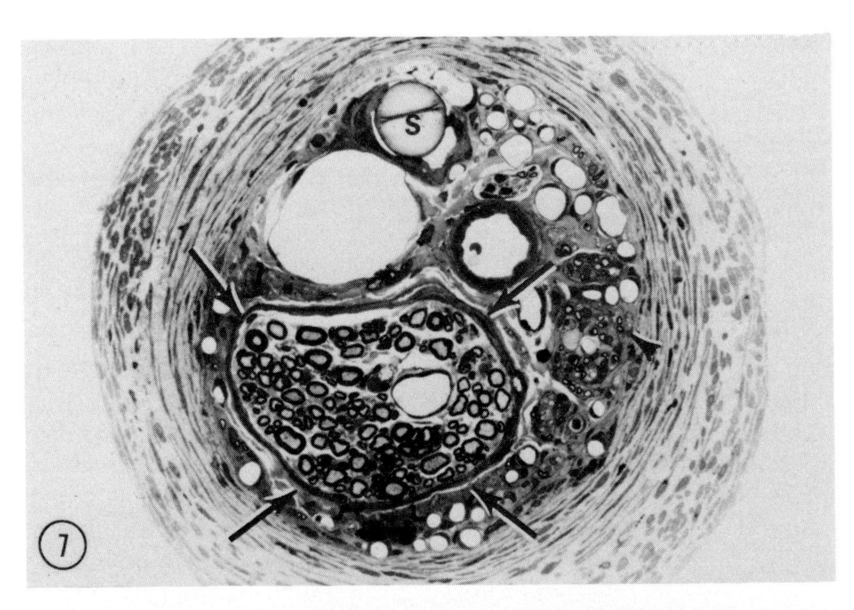

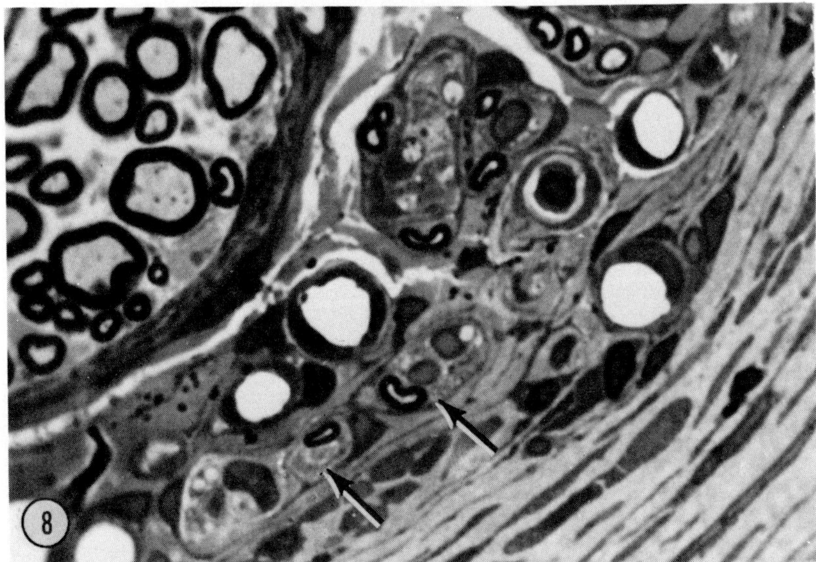

FIGURE 7. Cross-anastomosis experiment. Intratubal proximal stump of a sternohyoid nerve 15 weeks after a successful anastomosis. Note the intact original fascicle (→) containing preserved and regenerating fibers, miniature fascicles (▶), numerous blood vessels, the suture strand (s) in the thickened epineurium, and the cir-

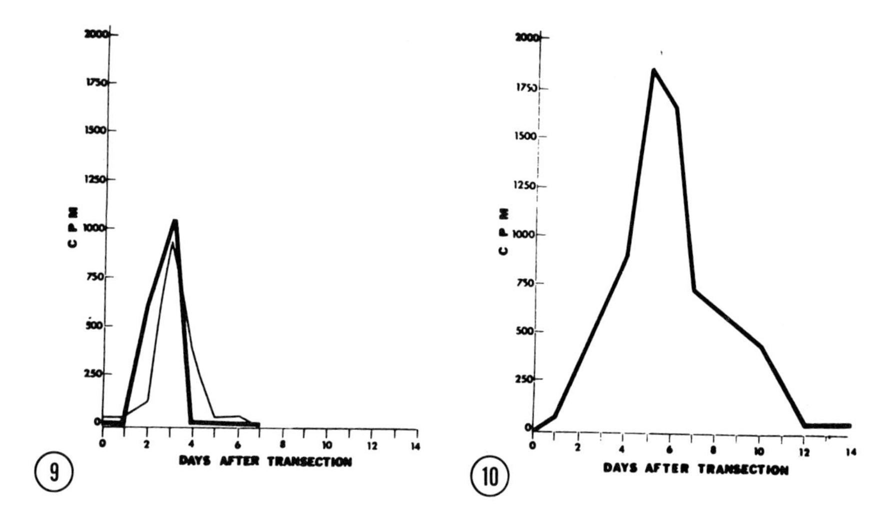

FIGURE 9. Myelinated nerve labeling experiment. Pattern of tritiated thymidine uptake into the terminal 3 mm of the proximal stumps of severed sternohyoid nerves. All animals received a single dose of [^{3}H]thymidine 24 h prior to sacrifice and counting. Broader line represents pattern observed after a dose of 4 mCi/kg. Narrow line represents pattern observed after a dose of 1 mCi/kg. (Reproduced by permission from H. J. Weinberg and Spencer, 1976.)

FIGURE 10. Unmyelinated nerve labeling experiment. Pattern of tritiated thymidine uptake in 6-mm segments of the distal stumps of severed CSTs. All animals received a single dose of 2 mCi/kg 24 h prior to sacrifice and counting. (Reproduced by permission from H. J. Weinberg and Spencer, 1976.)

Animals with selectively labeled anastomosed nerves were allowed to survive for 3 weeks, during which axons from the myelinated nerve penetrated the unmyelinated nerve as before. To determine whether labeled cells had migrated from the proximal stump into the distal stump, a 3-mm portion of anastomosed CST distal to the tubing was counted from animals in which the myelinated sternohyoid nerve had been labeled selectively. This was compared with similar lengths of proximal stump of the sternohyoid and the CST from the unoperated

---

cumscribing strata of connective tissue cells. ×350. (Reproduced by permission from H. J. Weinberg and Spencer, 1975.)

FIGURE 8. Cross-anastomosis experiment. Detail from Figure 7 illustrating part of the main fascicle (upper left), epineurial blood vessels, and the miniature fascicles typical of traumatic neuroma (→). ×1,620. (Reproduced by permission from H. J. Weinberg and Spencer, 1975.)

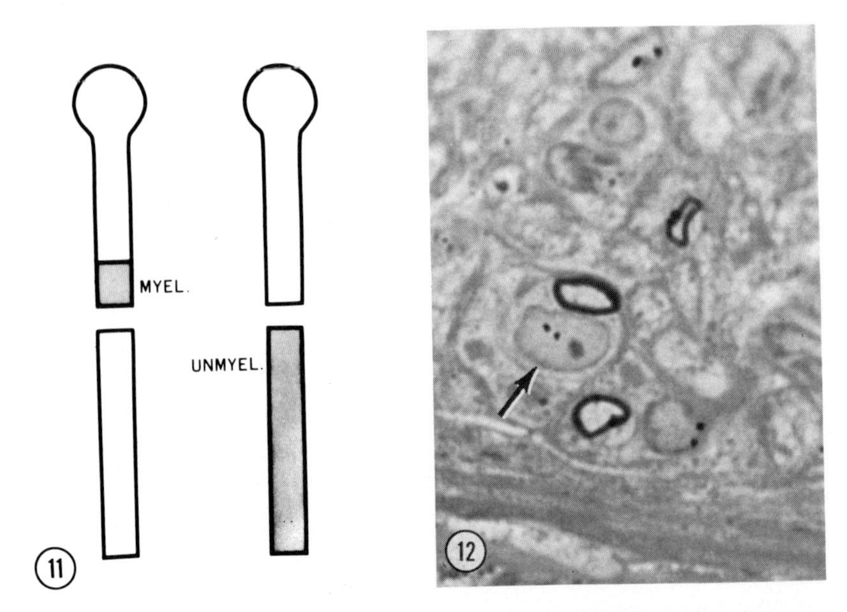

FIGURE 11. Cross-anastomosis labeling experiment. Diagram to show anastomosis of nerves whose Schwann cells were labeled selectively (shaded areas). Left: labeled proximal stump of sternohyoid to unlabeled CST distal stump. Right: unlabeled proximal stump of sternohyoid to labeled CST distal stump.

FIGURE 12. Cross-anastomosis labeling experiment. Cross section of previously labeled CST 3 weeks after anastomosis to an unlabeled sternohyoid proximal stump. Light autoradiogram showing several labeled nuclei. Note the three myelinated fibers, one of which is associated with a labeled Schwann cell nucleus (arrow). ×2,000. (Reproduced by permission from H. J. Weinberg and Spencer, 1976.)

side, the latter to determine background. Most of the label was retained in the proximal stump, but some also appeared in the CST. Autoradiographic analysis of cross sections of the CST revealed label over perineurial cells, endoneurial fibroblasts, and especially cells lining endoneurial blood vessels. No grains were detected over Schwann cells within the original nerve fascicle, although, as expected, nuclear grains occasionally were sited over Schwann cells located in the neuromatous outgrowth of the proximal stump.

Although these results were promising, the reverse experiment provided a definitive answer on the question of Schwann cell migration. When the proximal stump of an unlabeled sternohyoid nerve was anastomosed to the distal stump of a labeled CST, there was consistently a small number of labeled myelinating Schwann cells in the CST 3 weeks after anastomosis (Figure 12). After excluding the possibility that label derived from recirculation, we concluded that the labeled myelinating Schwann cells were part of the indigenous population of the unmyelinated nerve.

Taken together, these studies demonstrated that migration of Schwann cells between the two nerve fascicles was unlikely to occur. We concluded, therefore, that the normal cervical sympathetic trunk of the rat contains a single population of Schwann cells that respond to regenerating axons by forming myelinated or unmyelinated fibers, *depending on the type of axon* with which each cell becomes associated (H. J. Weinberg and Spencer, 1976). An identical conclusion was reached by Aguayo et al. (1976) using nerve grafting techniques in combination with radioactive thymidine labeling. Together, these experiments established that the neuron and axon regulate Schwann cell myelinogenesis.

A COMMON NERVOUS SYSTEM SIGNAL FOR MYELINOGENESIS?

The demonstration of axonal regulation of myelination in the peripheral nervous system led us to inquire if a peripheral axon could also stimulate oligodendrocyte myelination. CNS myelination of PNS axons had been observed previously in the dystrophic mouse, where naturally ectopic oligodendrocytes may elaborate myelin around the PNS portion of spinal roots and trigeminal nerves (Figures 13 and 14) (Bradley and Jenkison, 1973; H. J. Weinberg, Spencer, and Raine, 1975). The converse situation—Schwann cells forming PNS myelin around central axons—was subsequently demonstrated by Blakemore (1977) and Kao, Chang, and Bloodworth (1977). In addition, Schwann cells myelinating CNS axons occasionally had been found within white matter plaques in multiple sclerosis and in spinal cords of animals with chronic experimental allergic encephalomyelitis (Feigin and Popoff, 1966; Feigin and Ogata, 1971; Snyder, Valsamis, Stone, and Raine, 1975).

To address this issue, Ellen Weinberg and I conducted an experiment in which regenerating peripheral axons were allowed to interact with uncommitted oligodendrocytes in degenerated optic nerves. This was achieved by transecting a peripheral nerve and inserting a segment of the autologous optic nerve between the cut ends (Figure 15). Earlier grafting studies had suggested regenerating axons would emerge from the proximal stump of the peripheral nerve and penetrate the graft

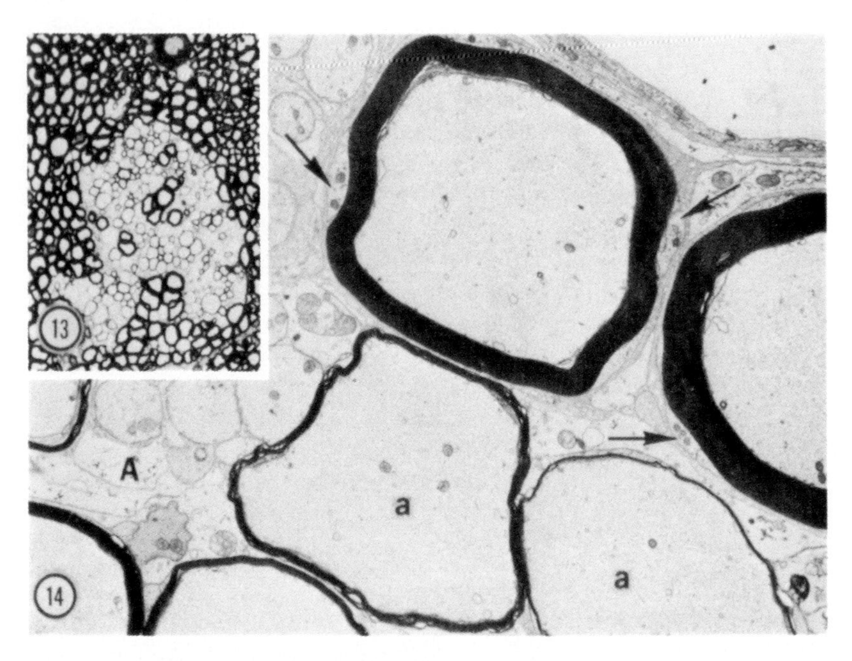

FIGURE 13. CNS myelination of PNS axons. Trigeminal nerve at the level of the Gasserian ganglion of a *dy*^{2J} mouse. An island containing nonmyelinated axons as well as thinly and thickly myelinated fibers is shown. Light micrograph of epoxy section stained with toluidine blue. ×400. (Reproduced by permission from H. J. Weinberg et al., 1975.)

FIGURE 14. CNS myelination of PNS axons. Trigeminal nerve of *dy*^{2J} mouse. Electron micrograph of the region shown in Figure 13. PNS myelinated fibers (Schwann cell cytoplasm at arrows), CNS myelinated axons (a), small naked axons, and astroglial processes (A) are depicted. ×9,600. (Reproduced by permission from H. J. Weinberg et al., 1975.)

before gaining access to the distal stump of the peripheral nerve. Rat optic nerve grafts 1.5 mm in length were transplanted between the cut ends of the peroneal nerve of one limb, or alongside the intact peroneal nerve in the opposite limb as a control, and examined after 4–28 weeks. By 4 weeks, control excised optic nerves displayed a proliferation of astrocytes (Vaughn and Pease, 1970; Cook and Wiśniewski, 1973; Fulcrand and Privat, 1977). Oligodendrocytes remained enmeshed in a complex network of filamentous astrocytic processes whose outer border was lined by a basal lamina, as in a normal optic nerve. Peroneal-optic nerve grafts at all timepoints displayed a stereotyped pattern as

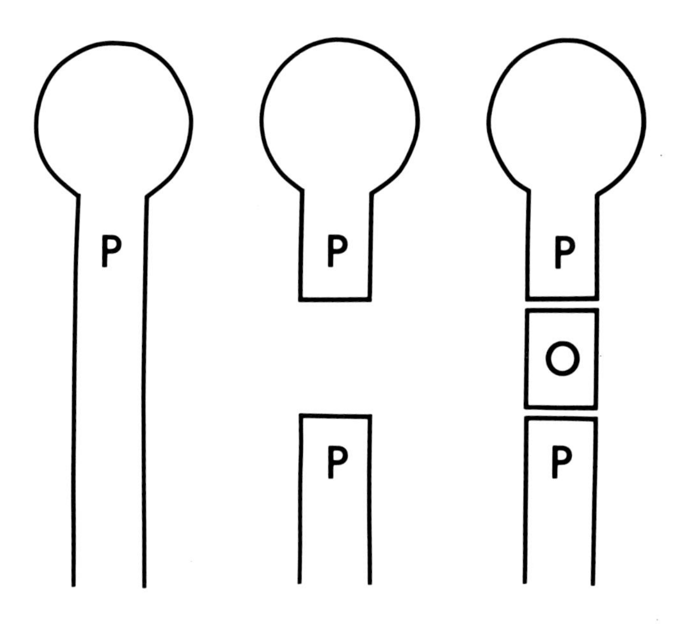

FIGURE 15. CNS myelination of PNS axons. Diagram illustrating technique of grafting optic nerve segment (O) between portions of autologous peroneal nerve (P).

follows. The portion originating from the proximal stump of the peroneal nerve contained preserved and regenerating fibers. The vast majority of the regenerating fibers *failed* to penetrate the optic nerve graft and grew over the surface of the foreign tissue before returning to the distal stump of the parent nerve. These regenerating peripheral axons sometimes developed unusual cellular relationships at the border of the optic nerve: for example, ''unmyelinated'' fibers surrounded by a basal lamina and composed of small axons individually enveloped by filamentous astrocytes.

Our attention focused on two types of myelinated fibers found in the graft at 20 weeks postoperation. The first, located beneath the outer astroglial border, was clearly of peripheral type, being surrounded by a Schwann cell that had elaborated myelin lamellae measuring 16.1 μm across (Figure 16, left). Because these fibers were separated from the encasing astrocytes by two basal laminae—one derived from the myelinating Schwann cell and the second from the astrocytic border—it was clear that these regenerating fibers had either invaginated the astrocytic surface or, alternatively, had been engulfed by an advancing astroglial margin. Such PNS myelinated regenerating axons contrasted with the second type of rare myelinated axon found deep within the astroglial scar (Figure 16, right). These were enveloped by oligodendrocytes that had elaborated CNS myelin with a periodiocity of 12.7 nm. The outermost myelin lamella was focally expanded into a cytoplasmic tongue (typical of an oligodendrocyte) abutting the surface membrane of adjacent astrocytic processes.

These observations suggested that a few regenerating PNS axons had penetrated the optic nerve graft, unaccompanied by Schwann cells, and had become associated with oligodendrocytes that had then commenced myelination. This remarkable result, duplicated in the mouse by Aguayo et al. (1978), indicated that PNS axons could signal oligo-

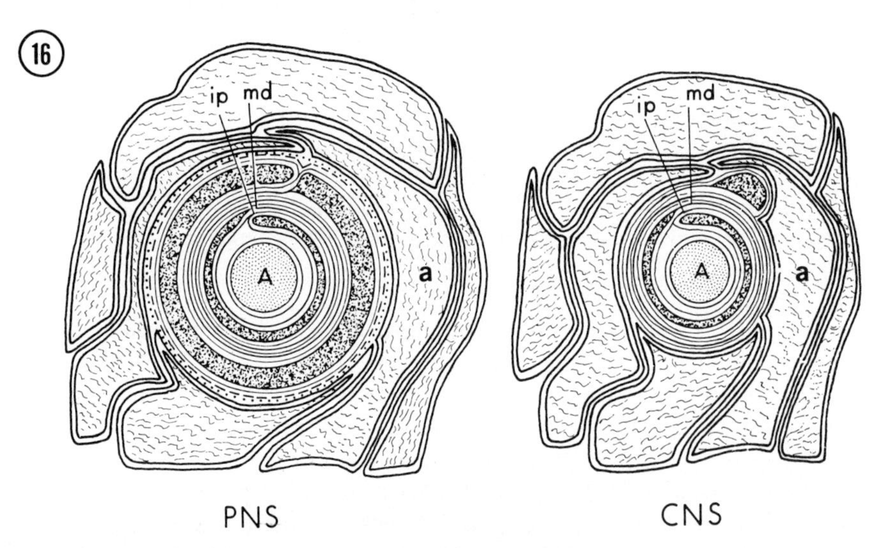

FIGURE 16. CNS myelination of PNS axons. Diagram defining the relationship of astrocytes (a) to PNS- and CNS-myelinated regenerating peripheral axons (A) in the optic nerve graft. The PNS fiber is situated in a narrow channel delimited on one side by the basal lamina of the Schwann cell (broken line) and on the other by a basal lamina bordering the astrocytic processes (dotted line). The peripheral axon with CNS myelin possesses no basal lamina between the outer surface of the fiber and the surrounding astrocytic processes. The major dense lines (md) and intraperiod lines (ip) are also indicated on each fiber. (Reproduced by permission from E. L. Weinberg and Spencer, 1978.)

dendrocytes (as well as Schwann cells) to commence myelination (E. L. Weinberg and Spencer, 1979). Thus, it became clear that (1) peripheral and central axons can both instruct Schwann cells and oligodendrocytes to elaborate myelin, and (2) the type of myelin manufactured by oligodendrocytes or Schwann cells is cell-specific and is not dependent on the location of the myelinating cell within the nervous system or the type of axon enveloped.

One important hypothesis developed from this experiment was the idea that axons might employ a common myelinating signal for both Schwann cells and oligodendrocytes. It was argued that a conservative system of this type might suit the needs of neurons such as the anterior horn cell, which requires proximal axonal myelination by oligodendrocytes and distal myelination by Schwann cells. However, several additional experiments would have to be performed before this hypothesis could be tenable. For example, it would be helpful to know if a neuron whose axon is restricted to the PNS (e.g., a postganglionic sympathetic neuron) could induce oligodendrocytes to execute central myelination. The converse experiment would also be important in resolving this issue, namely, whether a neuron whose axon is restricted to the CNS could stimulate PNS myelinogenesis around a regenerating axon. Attempts to utilize rat retinal neurons for the latter purpose—by grafting the optic nerve attached to the eye into a distal stump of a peripheral nerve—remained inconclusive. Such an experiment may be more successful in a lower vertebrate species where CNS axonal regeneration is more efficacious.

A number of other significant issues were also raised by the CNS-PNS grafting experiments. First, the presence of astrocytes rather than Schwann cells enveloping regenerating unmyelinated PNS axons suggested that the two cells had similar axon surface recognition properties. An analogous relationship is normally found in the CNS-PNS transitional zone of spinal roots where astrocytes focally ensheath the axon, thereby effecting a spatial separation between oligodendrocyte and Schwann cell (Carlstedt, 1977). Second, astrocytes might have prevented the entry of Schwann cells into the nerve graft, thereby allowing oligodendrocytes to form myelin around the otherwise naked regenerating PNS axons. Blakemore (1975, 1977) had previously suggested that astrocytes of the glial limiting membrane serve to prevent the entry of Schwann cells into normal spinal cord. He argued it was the basal lamina lining the outer surface of these astrocytes that limited Schwann cell migration. However, since (1) Schwann cells accompanying the regenerating PNS axons failed to penetrate the cut surface of the optic nerve graft, despite the presumable absence of an astrocyte basal lamina, and (2) previous studies had demonstrated in CNS-PNS transition zones the presence of a continuous basal lamina between Schwann cell and astrocyte (Maxwell, Kruger, and Pineda, 1969; Meier and Sollmann, 1978), it seemed more likely that contact inhibition between astrocyte and Schwann cell kept the two myelinating cells apart. A comparable astrocyte-induced inhibition of oligodendrocyte migration from the CNS to the PNS was also suggested by the observed containment of oligodendrocytes within the astrocytic network of optic nerve grafts. These considerations therefore led us to suggest that astrocytes play a key role in establishing and maintaining the central and peripheral divisions of the vertebrate nervous system.

THEORIES OF NEURONAL REGULATION OF MYELINATION

Several theories have been proposed to relate neuronal influences to the onset and degree of myelination (Figure 17). One hypothesis developed from the concept that a trophic factor, secreted by the neuron and distributed along the axon, functioned in the maintenance of the myelin sheath. The observation that myelin development in vitro was dependent on neuronal maturation, and that Schwann cells become associated with the growing tips of myelinated axons in preference to those of unmyelinated axons in vivo, suggested that the critical stimulus switching on Schwann cells to form myelin might be a chemical messenger derived from the neuronal perikaryon and transported via the axon to instruct the Schwann cell (Peterson and Murray, 1955; Speidel, 1964; Singer, 1968; Spencer, Raine, and Wiśniewski, 1973). Other hypotheses suggested that alterations in axonal size accompanying growth could mediate neuronal regulation of myelinogenesis (e.g., Duncan, 1934). Robertson (1962) proposed that shrinkage of the axon would remove contact inhibition from the Schwann cell, thereby allowing rotational growth to begin. Another idea was that myelin thickness was regulated by the amount the Schwann cell was stretched during longitudinal growth of the axon. A detailed theory that related incremental changes in axonal caliber to the rate and amount of myelin formation was proposed by Friede (1972). Harold Weinberg and I subsequently published a unitary theory of Schwann cell function in normal and pathological states that incorporated contemporary concepts of intercellular communication (Spencer and H. J. Weinberg, 1978). The

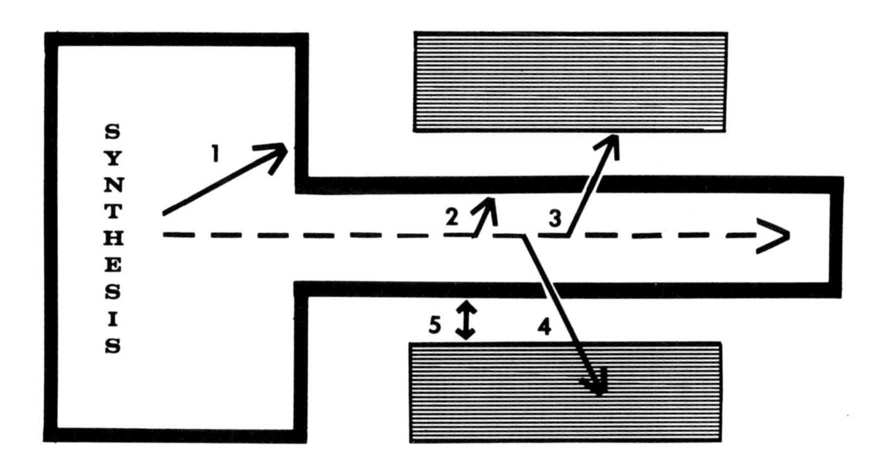

FIGURE 17. Possible signaling mechanisms. Diagram showing possible methods of communication between neuron and myelinating cells (shaded rectangles). 1: Incorporation of signal into plasma membrane of neuron and axon. 2: Incorporation of signal into axolemma after transport from the neuronal perikaryon. 3: Passage of diffusible signal picked up on surface membrane of myelinating cells. 4: Passage of diffusible signal from axon into Schwann cell. 5: Signaling via contact of surface membranes after 1 or 2. (Courtesy of H. J. Weinberg.)

central postulate was that neuronal regulation of myelinogenesis involved a surface membrane interaction between the plasmalemma of the axon and myelinating cell, an idea originally proposed for CNS myelination by Brady and Quarles (1973). This theory replaced our older concept of trophic stimulation of Schwann cells, which was no longer tenable on the basis of theoretical considerations and experimental data.

THE PERINEURIAL WINDOW

One empiric observation that discouraged the idea of trophic signaling was the perineurial window, a study conducted in collaboration with Harold Weinberg, John Prineas, and Cedric Raine. Originally conceived as a method to test the effects of releasing endoneurial pressure, the perineurial window developed into a convenient model to study cellular events involved in demyelination and remyelination. The experiment utilized the peroneal nerve in thc midthigh of adult rats (Figure 18). This nerve was selected because it was composed of a single, rounded fascicle containing myelinated and unmyelinated fibers.

Before producing the perineurial window, we cleared the fascia overlying the nerve and focally excised a portion of the epineurium. Then, by raising the cellular perineurial sheath and cutting horizontally parallel to the nerve, an opening or window was made in the perineurium. This technique produced an oval aperture 1–2 mm long through which underlying nerve fibers and endoneurial tissue immediately herniated (Figure 19). Examination of living and fixed preparations demonstrated that the vast majority of fibers within the window survived the insult but displayed an irregular fiber contour and a focal internodal dilatation of the axon most prominent in the center of the lesion. After a few hours, periodic constrictions appeared along the swollen regions of fibers, and the paranuclear Schwann cell cytoplasm became enlarged. The position of fiber indentations frequently corresponded to Schmidt-Lanterman incisures. These pathologic features became more pronounced with time and, by 5 days, affected regions of fibers appeared swollen, beaded, and twisted. The first sign of myelin loss was seen at 3 days and, by 7 days, the majority of damaged fibers displayed advanced demyelination. The process began with the arrival of phagocytic cells that penetrated affected paranodal regions and removed the damaged myelin segment from opposing ends in a pincerlike maneuver (Figures 20 and 21). As phagocytes removed the myelin, the axon became focally attenuated and new Schwann cells began to invest the demyelinated stretch. By 8 days, these Schwann cells had established their respective axonal territories and commenced myelinogenesis (Figures 22 and 23). The origin of the Schwann cells was not determined, but since they were invested by the basal lamina of the original fiber, their derivation from either the original Schwann cell or from Schwann cells of neighboring internodes seemed likely. Remyelinating Schwann cells occupied short, uniform lengths of axon, 3–5 remyelinating internodes replacing a single original internode. Similar phenomena occur in other remyelinating conditions (P. K. Thomas, 1974). The central portion of the remyelinated fiber retained the serpentine appearance seen prior to demyelination, as if it had been held in this deformed state by surrounding connective tissue. The early appearance of fiber beading, the exacerbation of this phenomenon with time, with indentations being most marked at incisures of Schmidt-Lanterman, coupled with the retention of twisting after remyelination, are consistent with the

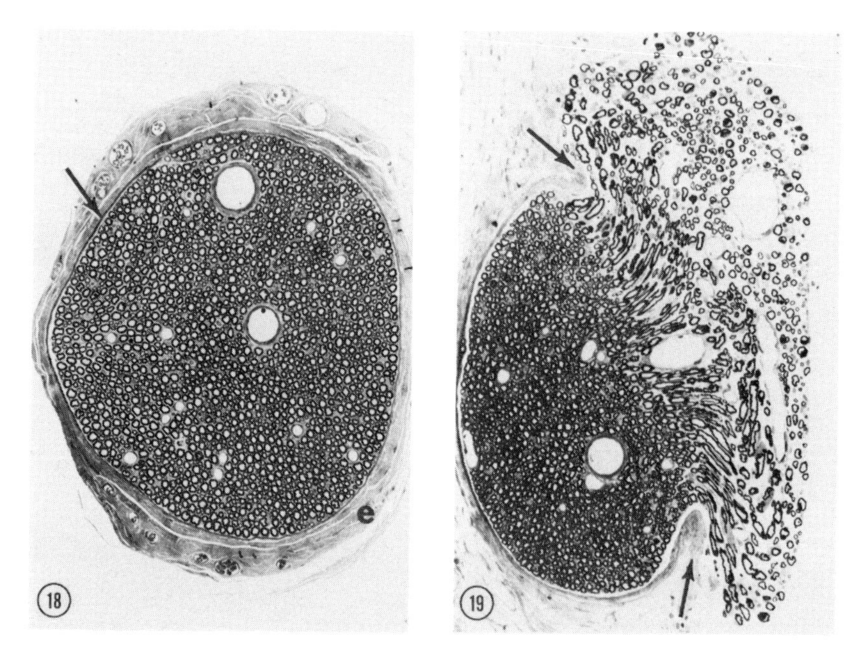

FIGURE 18. Perineurial window experiment. Normal peripheral nerve composed of one fascicle to illustrate epineurium (e), and perineurium (arrow). ×200. This figure and Figure 19 are light micrographs of 1-μm epoxy sections stained with toluidine blue. (Reproduced by permission from Spencer et al., 1975.)

FIGURE 19. Perineurial window experiment. Area of bleb in an operated peroneal nerve 3 days after surgery. Note the widely spaced, abnormally enlarged myelinated fibers in the bleb and the cut edges (arrows) of the perineurium. ×160. (Reproduced by permission from Spencer, P. S. [1976]. Experimentally induced nerve injury and its repair, pp. 131–139 in *Symposium on Microsurgery*, Vol. 14, Daniller, A. I. and B. Strauch, eds. C. V. Mosby, St. Louis.)

idea that the herniated nerve fibers were caught in a collagen network that later retracted as scar formation occurred, the circularly disposed collagen fibers located at incisures promoting the beading phenomenon.

One notable observation was the presence of structures resembling small, regenerating axons adjacent to demyelinated and remyelinating fibers. Initially, these were encased in newly arrived Schwann cells enveloping the demyelinated axon, but subsequently moved radially to form single or multiple nonmyelinated fibers (Figure 24). Because these axons appeared beneath the basal lamina lining the original fiber, it seemed possible they represented collateral sprouts of the demye-

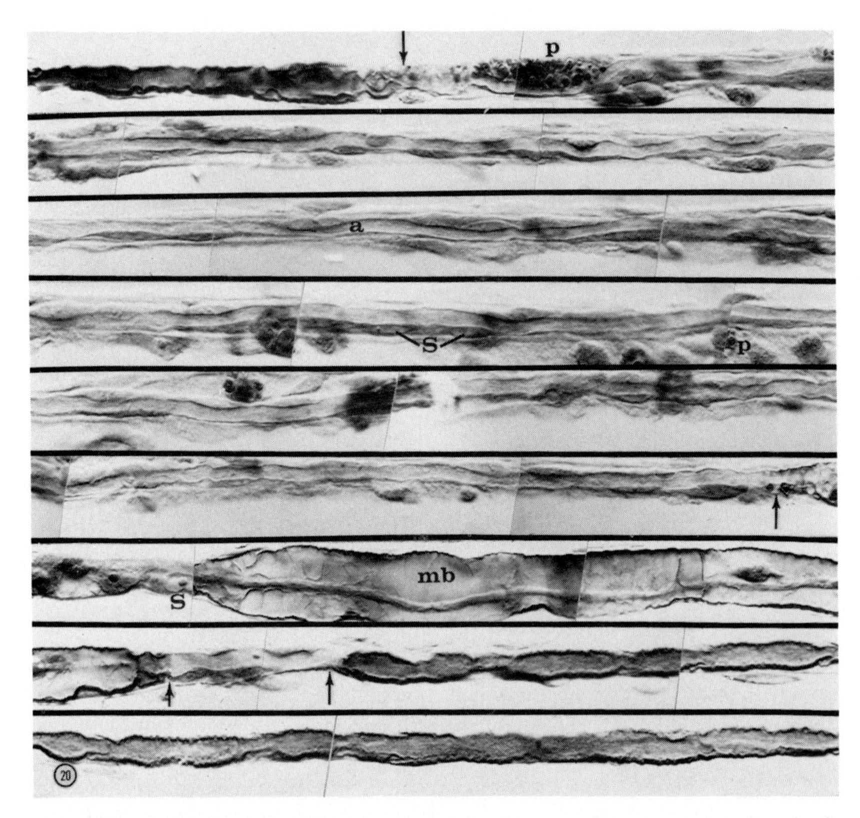

FIGURE 20. Perineurial window experiment. Consecutive segments of a single demyelinated fiber from a 7-day window. Arrows indicate interfaces between demyelinated axon (a) and myelinated axon. S, Schwann cell nuclei; p. phagocytes, mb, myelin bubble associated with a markedly attenuated axon. Teased fiber dissected in epoxy resin; photographed by Nomarski differential interference contrast microscopy. ×1,600.

linated axon, although attempts to demonstrate in serial sections connections between the two met with no success. King, P. K. Thomas, and Pollard (1977) drew attention to comparable axonal sprouts in ventral roots of guinea pigs with experimental allergic neuritis, but suggested they represented axons derived from a distal, degenerated part of the parent fiber that were coursing in a retrograde direction.

Another interesting phenomenon was the deposition of collagen fibers beneath the investing basal lamina before remyelination of the denuded axon had begun (Figure 22). This is consistent with the idea expressed

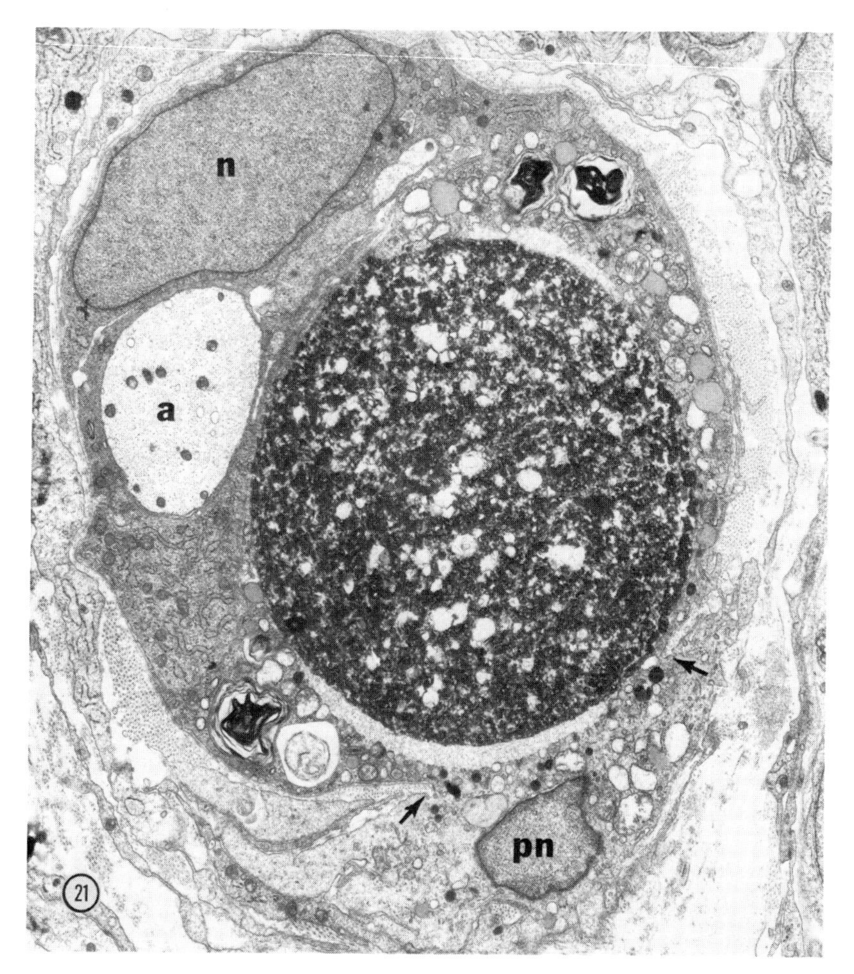

FIGURE 21. Perineurial window experiment. Demyelinated axon (a) associated with an invading phagocyte (pn) penetrating (arrows) the basal lamina of the Schwann cell. n, Schwann cell nucleus. This figure and Figures 22–24 are electron micrographs of cross-sectioned rat peroneal nerves embedded in epoxy resin and stained with uranyl acetate followed by lead citrate. ×21,000.

by R. P. Bunge and M. B. Bunge (1978) that Schwann cells require a suitable substrate to envelop an axon and commence myelination.

Another important aspect of myelinogenesis of the denuded axons was the establishment of longitudinal territories by the newly arrived Schwann cells. This was investigated further, in collaboration with

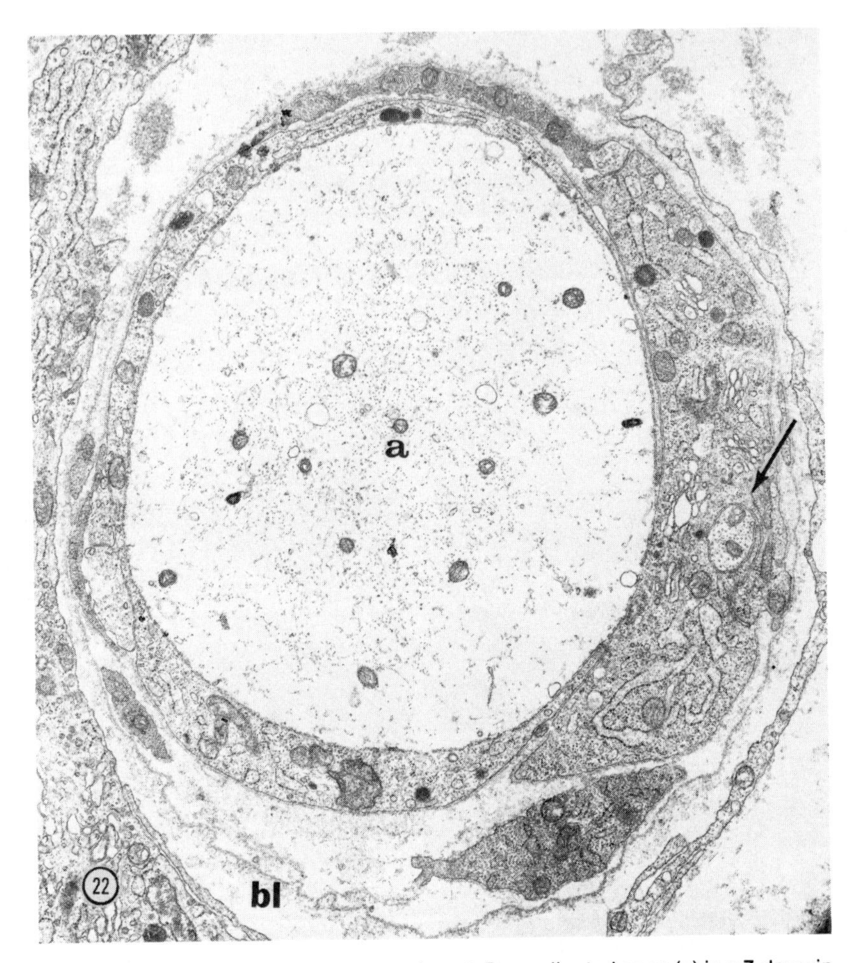

FIGURE 22. Perineurial window experiment. Demyelinated axon (a) in a 7-day window. Note axon sprout (arrow). bl, Basal lamina that invested the original fiber. ×15,250.

Cecile DeBaecque and Cedric Raine, by tracing light microscope changes at nodes of Ranvier during demyelination and remyelination. Nodes of Ranvier were stained histochemically by the metal ion binding technique introduced by Langley and Landon (1969) and subsequently expanded and modified by Quick and Waxman (1977). Before demyelination commenced, severely beaded portions of fibers sited within the window displayed a reduction or loss of nodal precipitate. During the

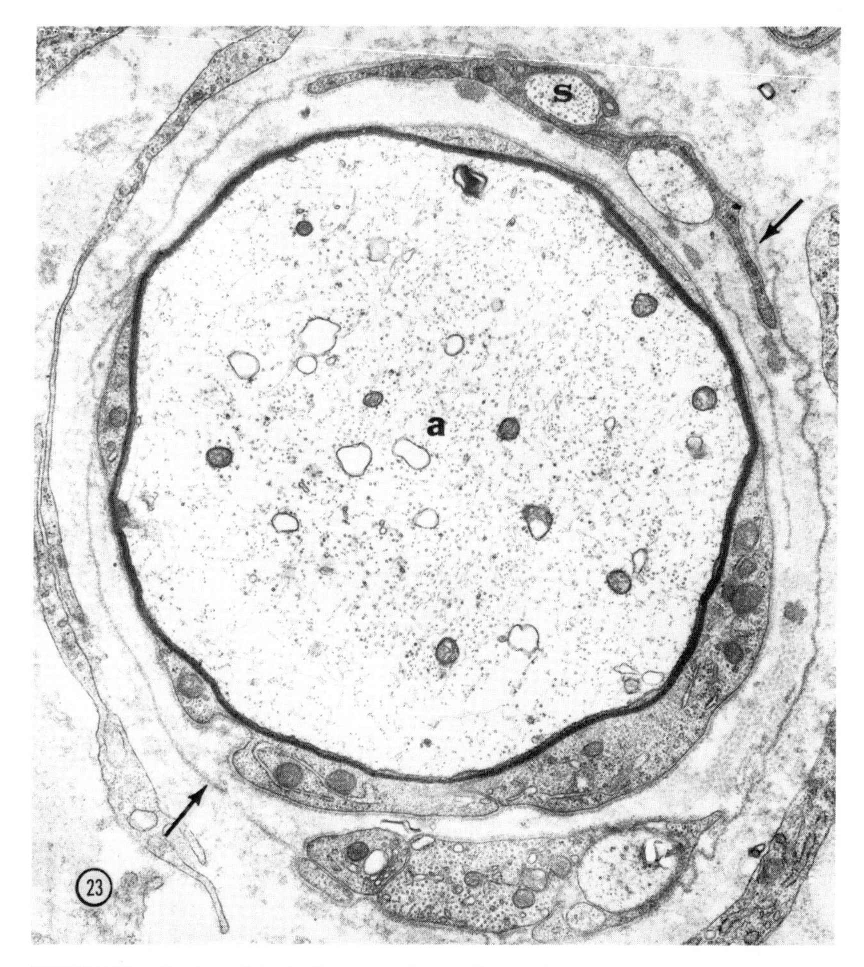

FIGURE 23. Perineurial window experiment. Remyelinated axon (a) in a perineurial window 12 days after surgery. Supernumerary Schwann cell processes envelop axon sprouts (s) within the original basal lamina (arrows). ×16,900.

period of myelin phagocytosis and demyelination, precipitate was absent throughout the denuded axon but present at adjacent, uninvolved nodes. After the commencement of remyelination, a small amount of precipitate appeared at the newly created nodal regions and at interfaces between preserved and remyelinating internodes. The amount of nodal precipitation at new nodes increased in step with the degree

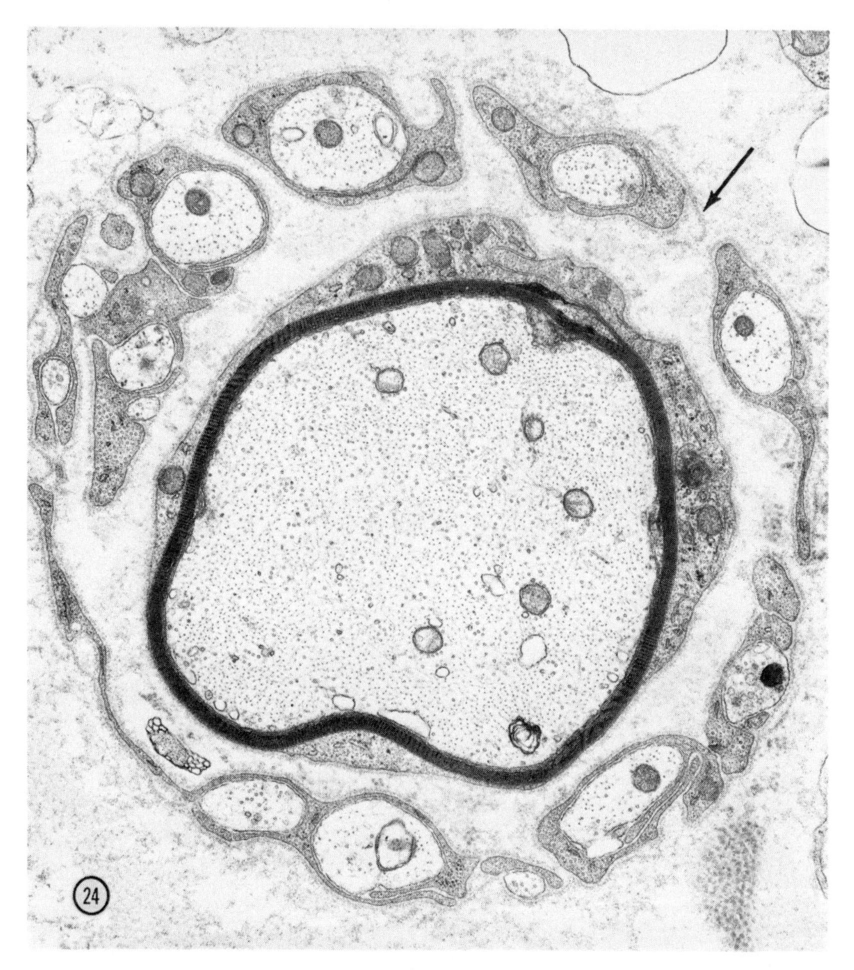

FIGURE 24. Perineurial window experiment. Remyelinated axon in a 14-day window surrounded by numerous nonmyelinated axon sprouts, each enveloped by a Schwann cell process. Remnants of the original investing basal lamina can be seen (arrow). ×22,500.

of remyelination (Figures 25–27). These findings suggested that the functional state of the Schwann cell was responsible for development of the staining property, since cationic binding activity ceased when the original Schwann cell lost its normal contact with the axon and returned only after a new nodal position had been determined. The investigations of Langley and Landon indicated that copper ions predominantly bound to mucopolysaccharides in the nodal gap substance. However, from Quick and Waxman's studies and those of Suzuki and Zagoren (1978), it now seems clear that metal ions bind to the nodal axolemma in preference to the nodal gap substance. The application of these new developments to our light microscope findings in the perineurial window suggests that nodal axon ion binding properties are controlled largely by the interfacing of two Schwann cells. This is reminiscent of the idea advanced by Rosenbluth (1976) that axoglial paranodal specializations in CNS fibers cause a bottleneck for the free diffusion of intramembranous particles, which then accumulate at nodes of Ranvier. The interrelationship of ion binding, particle accumulation, and nodal function is presently under intensive study in several laboratories.

The most striking aspect of the perineurial window study was the restriction of fiber changes exclusively to the area of the window. Axonal swelling, fiber beading, loss of nodal precipitate, demyelination, and remyelination were pathological events that affected the fiber locally as it transversed the blebbed lesion (Figure 28) (Spencer, H. J. Weinberg, Raine, and Prineas, 1975; DeBaecque, Raine, and Spencer, 1976). Harold Weinberg and I confirmed this phenomenon by examining with the electron microscope cross sections of multiple levels of single, teased nerve fibers taken from the window at various stages during damage and repair. That Schwann cells in the window were the only cells on the fiber visibly elaborating myelin suggested to us that the axonal signal for myelinogenesis was restricted to this part of the fiber. Furthermore, the absence of any structural changes in unmyelinated or myelinated fibers adjacent to those undergoing remyelination indicated the signal was not diffusible. These two considerations led us to abandon the concept of a diffusible signal produced by the neuron cell body for export to the axon and thence to the investing Schwann cells. More consistent with experimental observation, and congruent with contemporary concepts of intercellular communication, was the idea of a neuronal signal for myelination permanently resident within the surface membrane of the axon. Daughter Schwann cells encountering the denuded axolemma would be instructed to start myelinogenesis, whereas Schwann cells in proximal and distal regions of the fiber would be unaffected.

REMOVING AXONAL CONTACT FROM SCHWANN CELLS

The idea that Schwann cell function was mediated by axonal contact also had been developed in regard to mitosis. Wood and R. P. Bunge

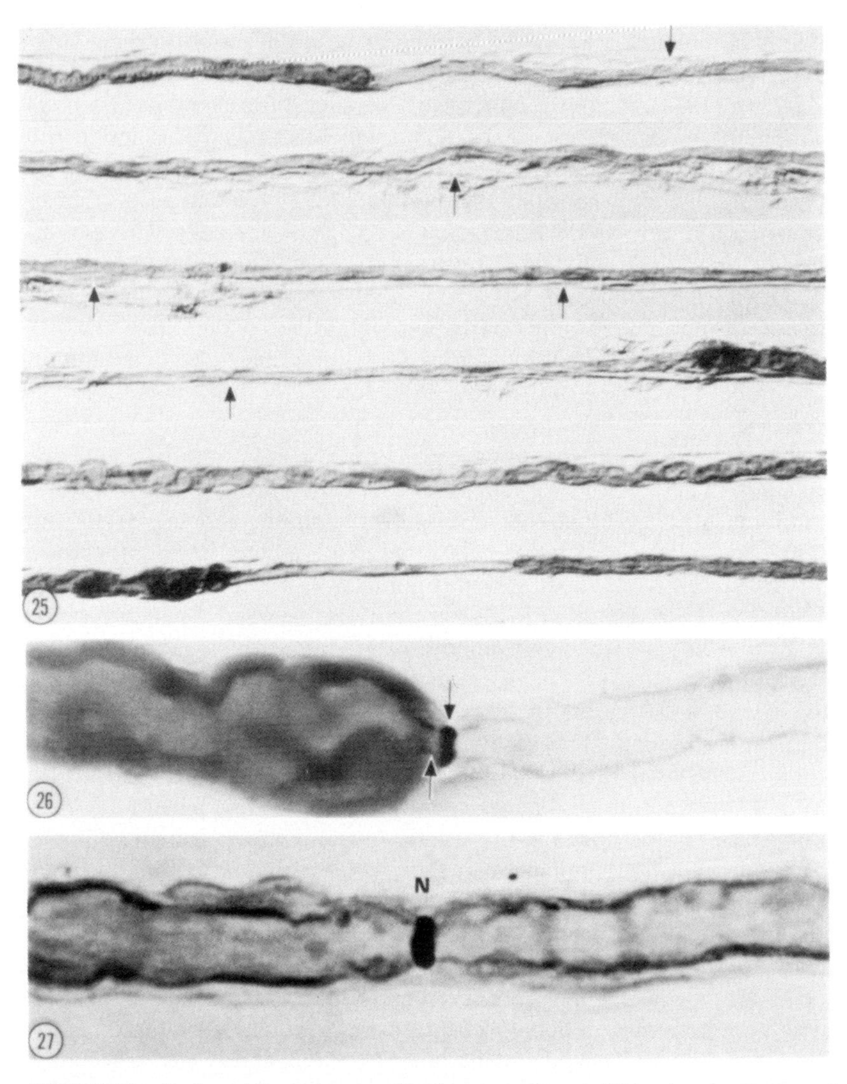

FIGURE 25. Perineurial window ion binding experiment. Fiber showing long and short regions of remyelination between portions of preserved fiber 22 days after surgery. Arrows point to newly created nodes of the long remyelinating portion of fiber. Single teased nerve fiber photographed with Nomarski optics. ×300. (Reproduced by permission from DeBaecque et al., 1976.)

FIGURE 26. Perineurial window ion binding experiment. Enlarged original node from top strip of Figure 25 shows reappearance of reaction product at the original nodal site (right arrow) the denser precipitate on the preserved paranode (left arrow).

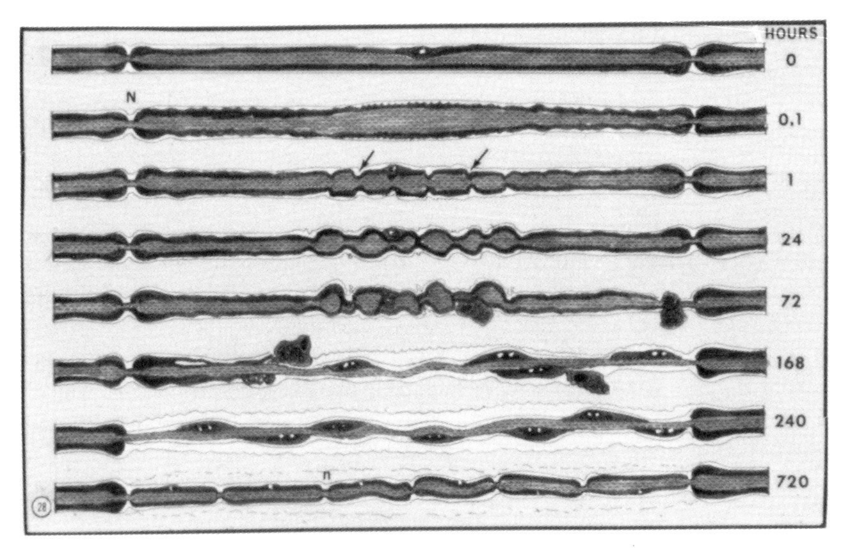

FIGURE 28. Perineurial window experiment. Diagram showing sequence of events with time after creating the perineurial window: 0h, normal nerve fiber; 0.1h, swollen axon and serrations; 1h, corrugation; 24h, early myelin beading; 72h, beading, distortion of entire fiber, and arrival of phagocytes; 168h, phagocytes stripping myelin, leaving attenuated axon, which becomes enveloped by Schwann cells; 240h, complete demyelination and establishment of Schwann cell territory (remyelination begins at this time); 720h, remyelination, with formation of short internodes of myelin. N, original node; n, newly created node.

(1975) and McCarthy and Partlow (1976) demonstrated that contact with sensory and sympathetic axons stimulated Schwann cells to divide in culture. Salzer, Glaser, and R. P. Bunge (1977) subsequently reported that a trypsin-sensitive neurite membrane fraction would stimulate Schwann cell proliferation in vitro. It was also clear from in vivo experiments that *loss* of axonal contact during Wallerian degeneration precipitated a wave of Schwann cell mitosis (Figures 9 and 10). Harold Weinberg and I decided to follow the fate of Schwann cells after pro-

---

This figure and Figure 27 are single teased fibers photographed in bright field. ×1,900. (Reproduced by permission from DeBaecque et al., 1976.)

FIGURE 27. Perineurial window ion binding experiment. Dense precipitate at newly created node of Ranvier (N) in a remyelinated fiber. ×1,900. (Reproduced by permission from DeBaecque et al., 1976.)

longed deprivation of axonal contact in vivo. This was examined in chronically denervated nerve stumps and after transplantation of Schwann cells from these denervated stumps to the isolated environment of the cornea.

Distal tibial nerve stumps were prepared in anesthetized rats and rabbits. The sciatic nerve was ligated with two pairs of monofilament nylon and transected between each pair; the proximal stump of the tibial nerve was anastomosed to the distal stump of the peroneal nerve (Figure 29). These surgical maneuvers were designed to impede sciatic nerve regeneration and direct escaping axon sprouts into the apposed peroneal nerve, away from the tibial nerve stump. Using this technique, regenerating axons were successfully excluded for many weeks from the nerve stump, although a few myelinated and unmyelinated regenerating fibers were encountered at periods after 15 weeks, especially along the periphery of the nerve fascicle.

After nerve transection, all nerve fibers underwent degeneration as expected. This involved the dissolution of the axon and myelin sheath, their degradation and eventual removal from the nerve fascicle. Degenerating fibers were replaced within the basal lamina by columns of daughter Schwann cells whose processes initially contained myelin debris and, subsequently, lipid vacuoles (Figures 30 and 31). In 5-week distal stumps, Schwann cell processes were largely free of lipid vacuoles and formed a serried array. Macrophages and fibroblasts, laden with myelin and lipid and prominent in the early stages of Wallerian degeneration, gradually disappeared from the nerve fascicle. After several weeks of chronic denervation, fibroblasts began to encircle the Schwann cell columns (Figure 32). By 18 weeks, some fibroblasts had adopted the features of perineurial cells; this pattern was markedly developed at 58 weeks after nerve section. During the intervening period, Schwann cell columns had undergone progressive attentuation until few Schwann cells remained. At this stage, the nerve was composed of a series of miniature fascicles containing only collagen and elastin fibers (Figure 33).

Although the sequence of morphological events in Wallerian degeneration had been well established, and the temporal shrinkage of Schwann cell columns previously recorded, the complete loss of columns was a surprising observation. Two possibilities were advanced to explain this phenomenon: either Schwann cells had slowly vacated the tubes of basal lamina, perhaps by migrating proximally to contribute to the formation of the cellular outgrowth at the cut end of the

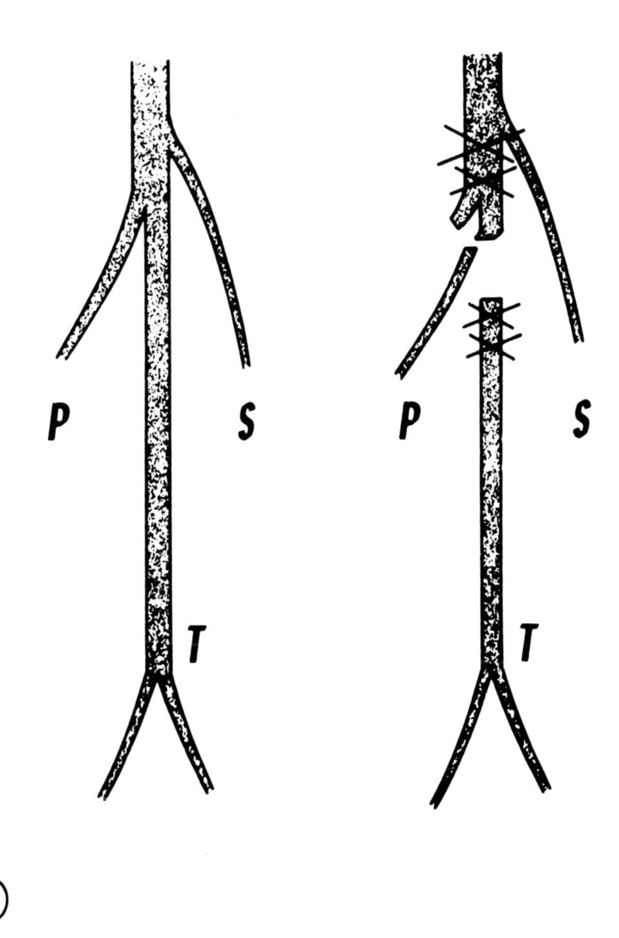

FIGURE 29. Loss of axonal contact. Distal nerve stumps were prepared by severing the sciatic nerve trunk beneath the sural branch (S). To prevent reinnervation of the distal stump of the severed tibial nerve (T), tight ligatures (XX) were placed around the nerves adjacent to the position of transection, and the proximal stumps were ligated to the distal stump of the peroneal nerve (P). (Reproduced by permission from Spencer et al., 1979.)

distal stump, or, alternatively, Schwann cells had undergone complete atrophy after prolonged denervation. The second possibility was strengthened by the presence of Schwann cells in long-term stumps that resembled necrotic Schwann cells seen after transplantation from distal stumps to the cornea, a second site chosen to isolate Schwann

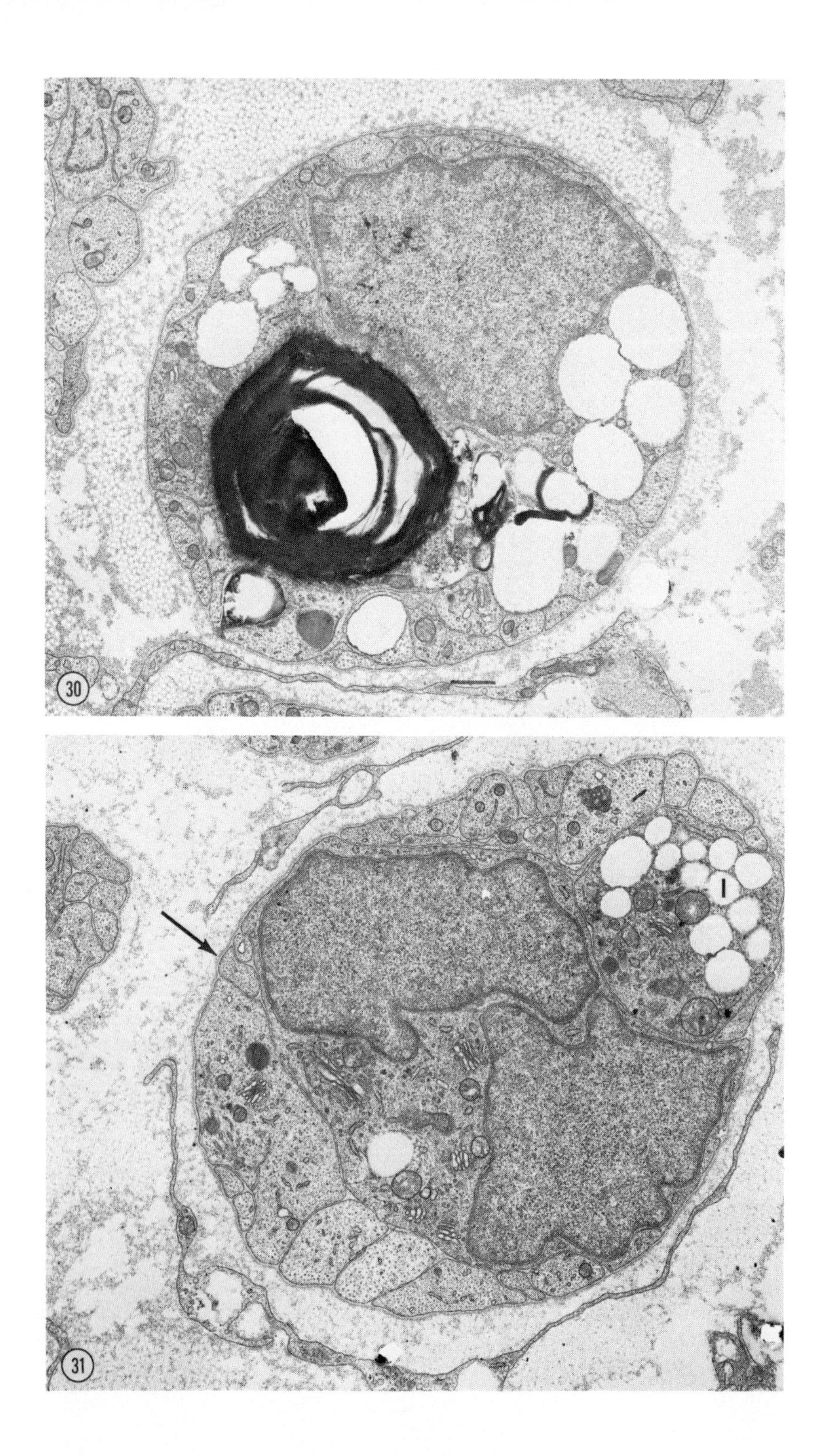
30
31
I

cells from axonal contact. In summary, these studies suggested axonal contact was necessary not only for myelination and mitosis but also for the maintenance of viable Schwann cells in distal nerve stumps (H. J. Weinberg and Spencer, 1978).

HARVESTING THE NAIVE SCHWANN CELL

One approach to the further study of the control of myelinating cell function was to compare the properties of Schwann cells and their plasmalemmae in dedifferentiated (naive) and mature states, i.e., before and after axonal association, and to study the evolution of this change upon contact with the axon. In collaboration with Vivien Krygier-Brévart, Michael Politis, Mohammad Sabri, Ellen Weinberg, and Vivian Zabrenetzky, studies commenced (1) to develop a purified population of naive Schwann cells in vivo, (2) to isolate the plasmalemmae from these cells, (3) to study biochemical parameters of these plasmalemmae and to compare properties with plasmalemmae from mature Schwann cells (i.e., PNS myelin), (4) to induce changes in plasmalemmal function with axolemmal fractions, and (5) to study the relationship of plasmalemmal alteration to changes in Schwann cell function.

Our first task was to develop a purified population of viable, dedifferentiated Schwann cells. We chose to utilize the technique of chronic denervation because it offered important advantages over the elegant tissue culture techniques developed for growing undifferentiated Schwann cells (McCarthy and Partlow, 1976; Wood, 1976; Brockes, Fields, and Raff, 1977). The advantages were that (1) the technique was simple to perform, (2) the yield of tissue was relatively large, (3) cells could be harvested from a number of species or sites for purposes of comparison, (4) the cells behaved normally when challenged

---

FIGURE 30. Loss of axonal contact. Denervated distal stump of a rat tibial nerve 3 weeks after transection, illustrating a large Schwann cell column containing myelin debris as well as lipid droplets. ×18,000. This figure and Figures 30–34 are electron micrographs of cross-sectioned nerves embedded in epoxy resin and stained with uranyl acetate and lead citrate. (Reproduced by permission from H. J. Weinberg, 1978.)

FIGURE 31. Loss of axonal contact. Denervated distal stump of a rat tibial nerve 3 weeks after transection, demonstrating a Schwann cell column composed of many Schwann cell processes of varying density containing nuclear profiles and lipid droplets (l). The column is surrounded by a basal lamina (arrow). ×16,000. (Reproduced in part by permission from H. J. Weinberg and Spencer, 1978.)

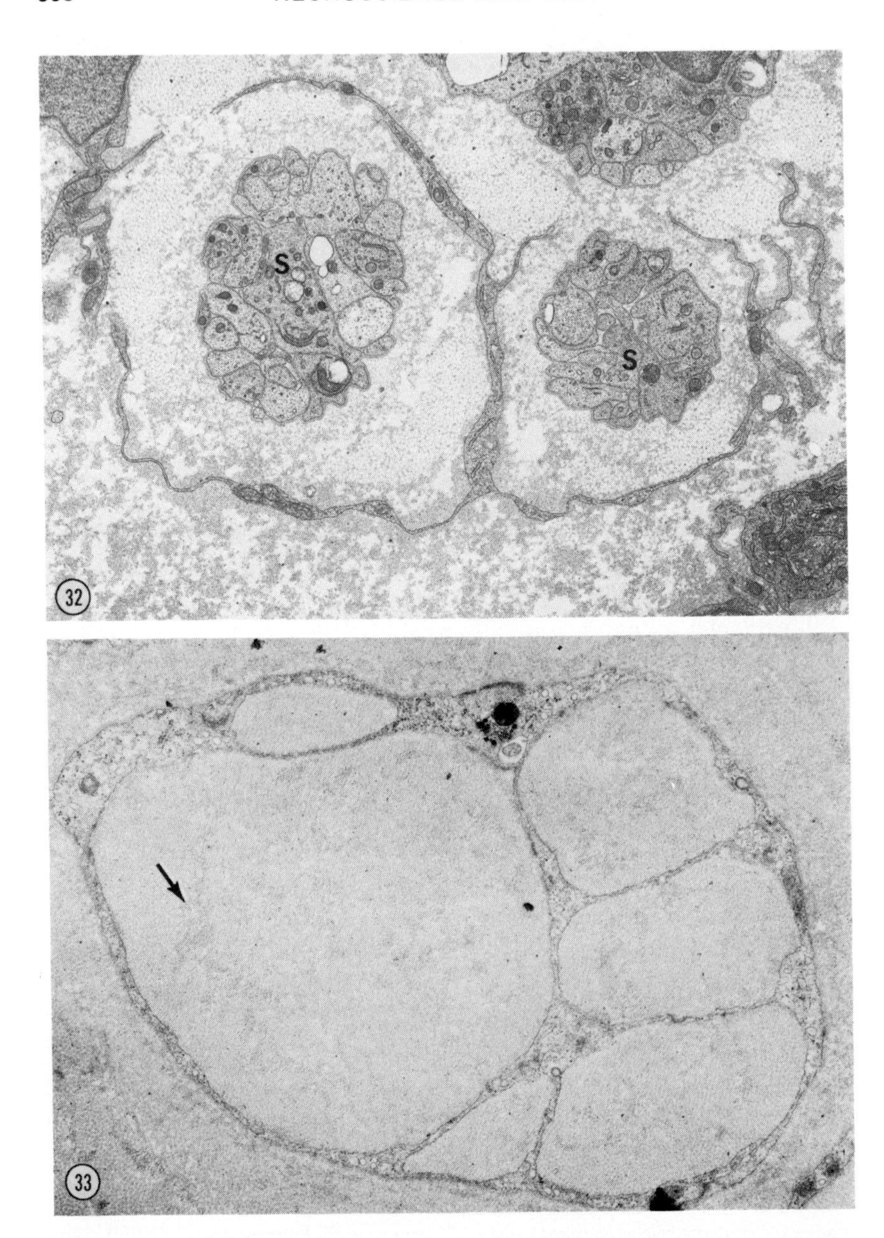

FIGURE 32. Loss of axonal contact. Denervated distal stump of a rat tibial nerve 3 weeks after transection, illustrating two Schwann cell columns (S) partly encircled by fibroblast processes. ×9,600. (Reproduced in part by permission from H. J. Weinberg and Spencer, 1978.)

by axons, and (5) the cells had several months of demonstrated viability in vivo.

The method was developed using the sciatic nerves of rabbits (Spencer, H. J. Weinberg, Krygier-Brévart, and Zabrenetzky, 1979). Animals were operated on using aseptic conditions as described previously (Figure 29). After closure of the wound, rabbits were kept on smooth-floored cages to reduce trauma to anesthetized hindfeet. After 5–12 weeks, the fascicular contents of the distal stump were transformed into a series of Schwann cell columns embedded in a connective tissue matrix (Figure 34). Contamination of the Schwann cell preparation by other cell types was minimized by the following procedures. The most serious contamination was the epineurial and perineurial cellular and connective tissue; this was excluded by removing the hardened terminal Schwannoma and plucking the endoneurial contents containing the Schwann cells from the proximal end of each nerve fascicle (Figure 35) (cf. Brown, Pleasure, and Asbury, 1976). Blood cells were cleared by perfusing the animal with a physiological solution, the temperature and oxygen content of the perfusate being adjusted to obtain viable cells. Endoneurial blood vessels, readily distinguishable from Schwann cell columns, could be removed by microdissection. The remaining acellular (collagen, elastin, amorphous material) and cellular (fibroblasts, phagocytes) components, representing 6–13% of the total cell area, were removed by separation techniques designed to take advantage of the fact that these tissue elements resided within the endoneurial compartment, in contrast to the Schwann cells isolated in columns and surrounded by resilient basal laminae (Figure 36). The remaining Schwann cell columns were free of axons and largely devoid of myelin debris. Cells were spindle-shaped as in vitro (Figure 36, inset), overlapping each other in the Schwann cell columns such that cross sections revealed serried processes, each sectioned at varying distances from its centrally located cell nucleus (Figure 34). The diameter of each Schwann cell column varied in proportion to the diameter of the nerve fiber it had replaced: those derived from myelinated fibers were approximately circular and of various sizes, while those replacing unmyelinated fibers were small and displayed an irregular surface contour (P. K. Thomas, 1974).

FIGURE 33. Loss of axonal contact. Network of cellular processes of perineurial type forming miniature fascicles. These contain collagen and elastin (arrow) fibers. Nerve stump 58 weeks after transection. ×15,000. (Reproduced by permission from H. J. Weinberg and Spencer, 1978.)

The key characteristic of the Schwann cells was that they were "dedifferentiated," being derived by division of the original population of Schwann cells within the nerve stump (cf. Hall, 1978). Our early cross-anastomosis experiments had suggested Schwann cells derived from both myelinated fibers and unmyelinated fibers were of the same genotype. Because such Schwann cells have neither produced myelin before nor received an axonal signal for myelination, we refer to the population as naive.

The possibility that these naive Schwann cells cannot manufacture myelin proteins before establishing contact with an axon is presently under scrutiny. Examination of rabbit Schwann cells by sodium dodecyl sulfate polyacrylamide gel electrophoresis (SDS-PAGE) revealed an almost complete absence of myelin-specific proteins 10 weeks after nerve transection (Figure 37). McDermott and Wiśniewski (1977) similarly demonstrated in rabbit distal nerve stumps a major loss of P_0, P_1 and P_2 proteins 18 days after transection. Their radioimmunoassay indicated a greater loss of P_1 than was apparent from the gels and, by 128 and 420 days, the concentration of P_1 was below the limit of detection. P_0 was hardly detectable on densitometric scans of tissue removed 60 days after transection. Current studies on this problem are testing the synthetic properties of these Schwann cells and tracing how they change following axonal contact.

OBTAINING NAIVE SCHWANN CELL PLASMALEMMAE

Using the purified population of viable, naive Schwann cells, Vivian Zabrenetzky set about obtaining a plasma membrane-enriched fraction. For this purpose, endoneurial tissue removed from denervated nerves, following systemic perfusion with 0.85 M NaCl, was chopped with a razor blade and homogenized in 0.25 M sucrose. Light microscope examination of the tissue after a brief homogenization revealed the contaminating endoneurial cells (fibroblasts, macrophages) to have ruptured, leaving the Schwann cell columns relatively intact within their resilient tubes of basal lamina. After discarding the supernatant and conducting additional homogenizations of increasing intensity, intact nuclei and plasma membrane vesicles were pinched off from the ruptured Schwann cell columns. This sequence of events was determined by examining each homogenate with Nomarski optics. The nuclear fraction was then pelleted by centrifugation of the combined homogenate, and the supernatant layered onto two 17-ml gradients of 0.56 M,

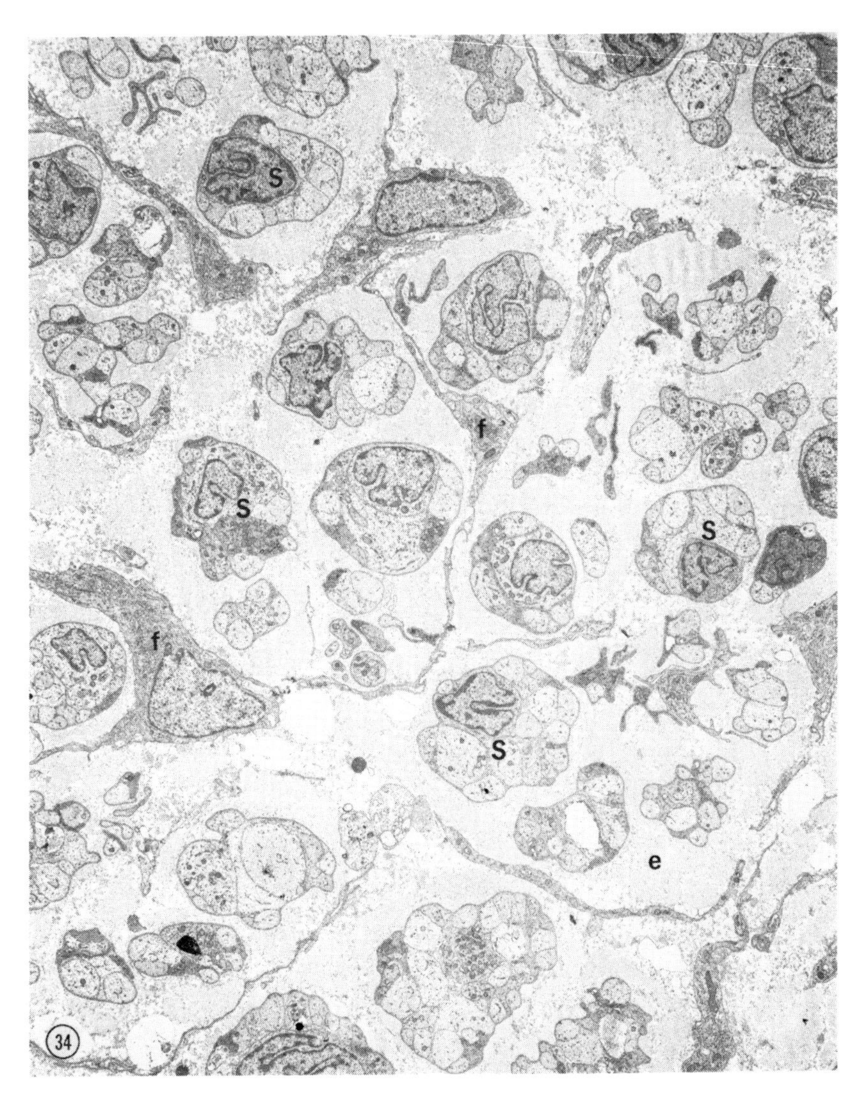

FIGURE 34. Harvesting naive Schwann cells. Denervated distal stump of a rabbit tibial nerve 8–9 weeks after transection. The nerve contains numerous Schwann cell columns (S) embedded in an endoneurial collagen matrix (e). Myelin and axons are absent. f, Fibroblasts. ×2,400. (Reproduced by permission from H. J. Weinberg, 1978.)

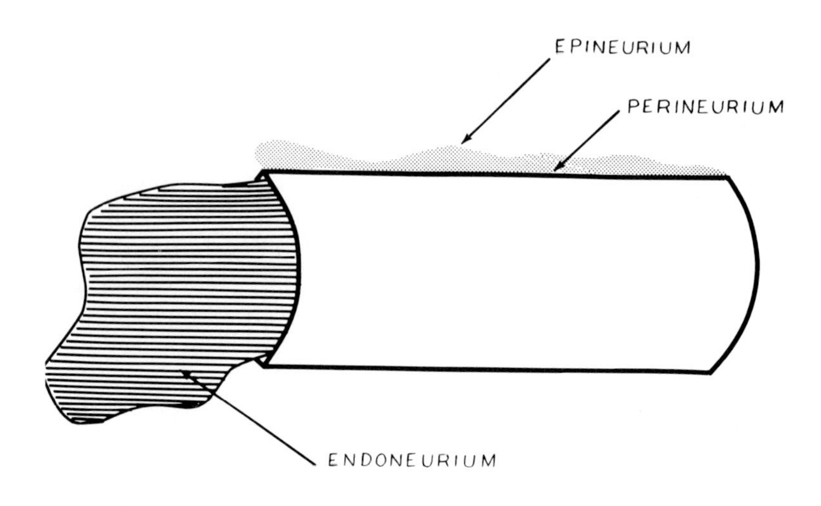

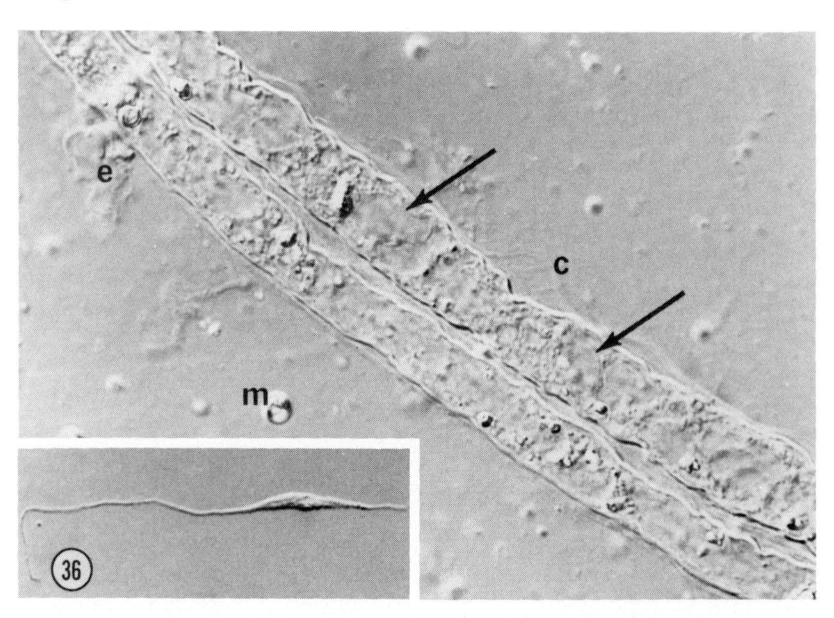

FIGURE 35. Harvesting naive Schwann cells. Diagram illustrating the method by which endoneurial tissue containing naive Schwann cells can be plucked from the enveloping connective tissue sheath.

FIGURE 36. Harvesting naive Schwann cells. Two Schwann cell columns prepared from desheathed endoneurial tissue homogenized in 0.25 M sucrose at 4° C in several steps for 1 min each. Unfixed homogenized tissue photographed with Nomarski

0.85 M, 1.0 M and 1.2 M sucrose. The two gradients were spun in a SW 27 rotor at 0° C for 9 hours at 25 K without the brake. Each fraction and interface was collected and homogenized in double-distilled water as an osmotic shock to release any material trapped within the vesicular structures. Each fraction was then pelleted, taken up and homogenized in 0.2 M sucrose to reform the vesicular structures, and prepared for electron microscope and biochemical study. Electron micrographs of the 0.85–1.0 M and 1.0–1.2 M interfaces showed a homogenous collection of profiles resembling plasma membranes (Figure 38). This morphological identification has been supported by preliminary enzyme profiles in which ouabain-sensitive Na^+K^+-ATPase, a marker of plasma membrane, was found to be enriched in the 0.85–1.0 M and 1.0–1.2 M fractions (Table 1). By contrast, monoamine oxidase was enriched in the mitochondrial pellet. Taken together, these preliminary ultrastructural and biochemical studies show the feasibility of isolating an enriched Schwann cell plasma membrane fraction from the axon/myelin-free Schwann cell preparations.

FUTURE APPROACHES

The control of myelinating cell function is generally assumed to occur at the level of DNA, the cell that forms myelin having differentiated considerably from the nonmyelin-forming cell (Norton, 1977). One view states the biogenesis of myelin is a coordinated phenomenon in which a series of enzymes is transcribed at a particular stage in development. The regulation of this transcription is envisaged either as an induction of myelin enzyme operon, or as a sequential transcription, where the synthesis of one enzyme or its products regulates the transcription of enzymes involved in later stages of biosynthesis (Mandel et al., 1972).

A major hurdle in further understanding the control of myelinating

---

optics to show intact Schwann cell columns composed of nuclei (arrows) and overlapping cytoplasm processes within tubes of basal lamina bordered by endoneurial collagen (c). Contaminating endoneurial cells (e) have ruptured and released myelin debris (m) and intact nuclei into the supernatant. ×1,200.

Inset shows single spindle-shaped cell isolated from Schwann cell columns by incubating desheathed tissue in a medium containing 0.2% collagenase type III, 0.1% balanced salt solution, 0.18 M $CaCl_2$ and 0.1% hyaluronidase in 10 mM phosphate buffer containing 1% Ficoll, 10% fructose, and 10% glucose for 4 h at 37° C. Nomarski optics. ×400.

(Reproduced by permission from Spencer et al., 1979.)

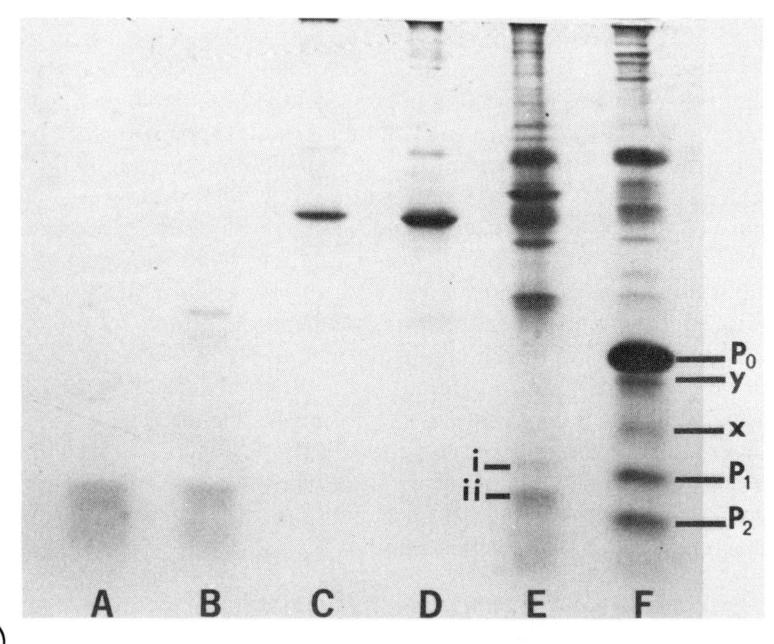

FIGURE 37. SDS-PAGE of naive Schwann cells. Channel F shows the pattern obtained from saline-perfused endoneurial tissue (40 μg protein) of a normal tibial nerve from a rabbit. The myelin-typical proteins P_0, Y, X, P_1, and P_2 are marked. Channel E shows the pattern in a Schwann cell-enriched axon and myelin-free tibial nerve stump (40 μg protein). For comparison, spinal cord filament protein (51,000 daltons) was run in Channel D (14 μg protein) and Channel C (7 μg protein), and two groups of three histone fractions were run in Channels B and A. Part of the latter material comigrates with the material running at i and ii in Channel E. (Courtesy of V. Krygier-Brévart. Spinal cord filament protein was kindly provided by W. Norton.)

cell function is to reveal how information arriving on the cell surface is transmitted to the regulatory centers of the cell. One important and widespread mediator of intracellular metabolism is cyclic AMP. This molecule modulates cell metabolism by stimulating the phosphorylation of proteins by protein kinases, one method known to regulate intracellular function. Conceivably, fluctuating levels of cyclic AMP might regulate the different facets of Schwann cell function such as mitosis (Raff et al., 1978) and myelination. Cyclic AMP-dependent protein kinase stimulation of protein phosphorylation has been found in CNS myelin by Miyamoto and Kakiuchi (1974), and we demonstrated a similar phenomenon in PNS myelin fractions such as that in Figure 39

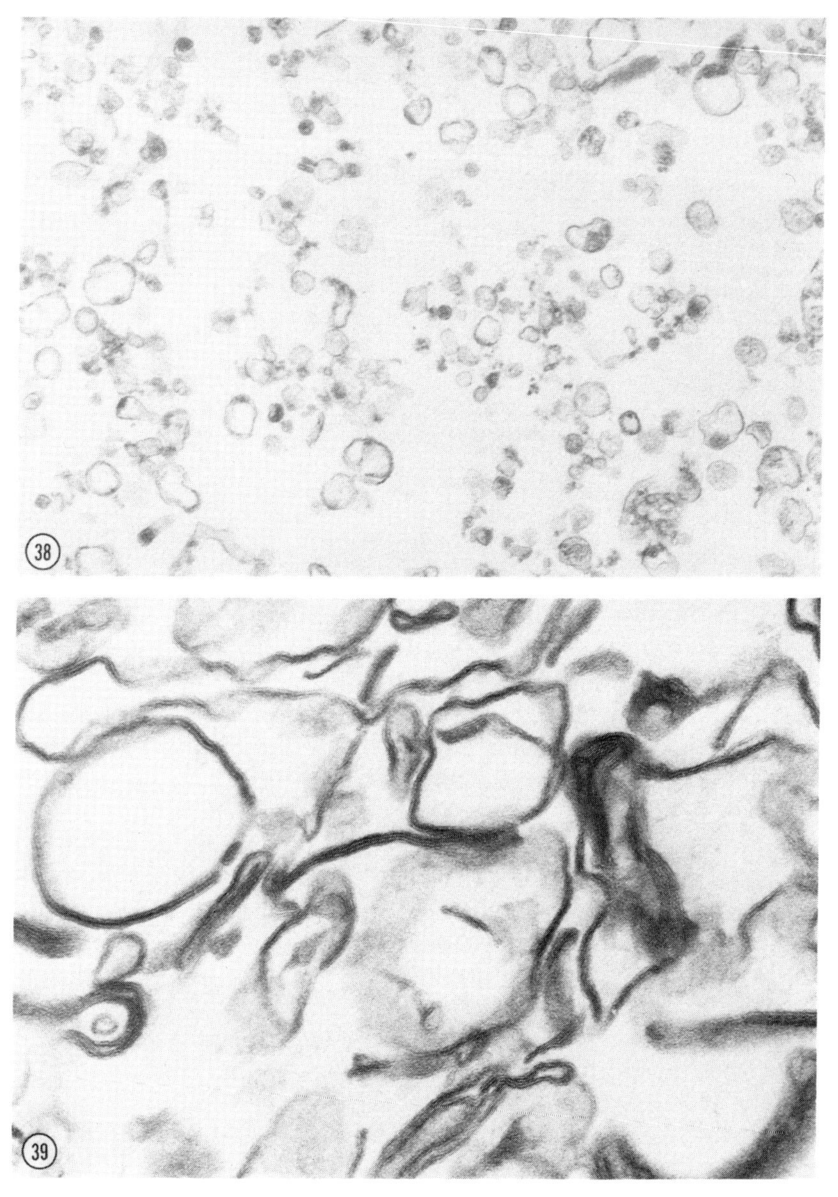

FIGURE 38. Schwann cell-enriched fraction. Electron micrograph of a thin epoxy section of 1.0–1.2 M sucrose fraction showing vesicular profiles enriched in plasma membranes obtained from axon- and myelin-free Schwann cells. ×6,000.

FIGURE 39. PNS myelin fraction. Electron micrograph of a thin epoxy section of fraction containing purified PNS myelin. ×96,600.

TABLE 1. *Marker enzymes in subcellular fractions of Schwann cells*[a]

Subcellular fraction	Ouabain-sensitive Na^+K^+-ATPase		Monamine oxidase	
	Specific activity*	Total activity†	Specific activity*	Total activity†
Homogenate	159	3298	0.29	6.00
Nuclear supernatant	147	2117	0.32	5.00
Nuclear pellet	109	103	0.38	0.40
0.2 M Sucrose	47	181	0.13	0.50
0.25:0.85 M Sucrose	72	80	0.70	0.80
0.85:1.0 M Sucrose	388	105	0.75	0.20
1.0:1.2 M Sucrose	329	37	0.50	0.06
Mitochondrial pellet	139	36	1.02	0.26

[a] Fractions taken from the first discontinuous gradient of Schwann cell homogenate.
* nmol ^{32}P/min/mg protein.
† nmol ^{32}P/min/total mg protein.
Courtesy of V. Zabrenetzky.

(Krygier-Brévart, Zabrenetzky, and Spencer, 1977; Zabrenetzky, 1979). Each myelin-specific protein undergoes enhanced phosphorylation in the presence of cyclic AMP. By contrast, a protein kinase identified by Zabrenetzky in the naive Schwann cell plasma membrane-enriched fraction shows a different response in the presence of cyclic AMP (Zabrenetzky, Krygier-Brévart, and Spencer, 1977). The possibility that these observations have revealed developmental differences in Schwann cell plasma membranes that relate to axon regulation of Schwann cell function remains a subject for future study.

CONCLUSION

This paper describes a number of studies designed to illuminate the control of myelinating cell function in normal and pathologic states. We have demonstrated that myelin formation is under neuronal control and suggested, from studies of demyelination and remyelination in the perineurial window, that the signaling mechanism resides in the axolemma, myelinating cell function changing upon surface contact with the axon. In vivo populations of naive Schwann cells and plasmalemma-enriched fractions of these Schwann cells have been developed to test this thesis.

ACKNOWLEDGMENTS

I acknowledge the incisive and stimulating collaboration of Harold Weinberg, his critical reading of the manuscript, and his permission to reproduce Figures 17 and 30. Vivien Krygier-Brévart and Vivian Zabrenetzky kindly gave permission to reproduce Figure 37 and Table 1. I am also indebted to Elaine Garafola and Monica Bischoff for patiently preparing the manuscript and figures, respectively.

The author gratefully acknowledges grant support from the Joseph P. Kennedy, Jr., Foundation and the National Institutes of Health (NS13106). H. J. Weinberg and E. L. Weinberg were funded by N.I.H. training grants 5T5GM1674 and 5T32GM7288, respectively, and V. Krygier-Brévart by a NIH Interdisciplinary Neurosciences Postdoctoral Training Grant.

REFERENCES

Abercrombie, M., D. H. L. Evans, and J. G. Murray (1959). Nuclear multiplication and cell migration in degenerating unmyelinated nerve fibres, *J. Anat.* **93**:9–14.

Aguayo, A. J., L. Charron, and G. M. Bray (1976). Potential of Schwann cells from unmyelinated nerves to produce myelin: a quantitative ultrastructural and autoradiographic study, *J. Neurocytol.* **5**:565–573.

Aguayo, A. J., R. Dickson, J. Trecarten, J. M. Attiwell, G. M. Bray, and P. Richardson (1978). Ensheathment and myelination of regenerating PNS fibers by transplanted optic nerve glia, *Neurosci. Lett.* **9**:97–104.

Aguayo, A. J., G. M. Bray, C. S. Perkins, and I. D. Duncan (1979). Axon-sheath cell interactions in peripheral and central nervous system transplants, *Soc. Neurosci. Symp.* **4**:361–383.

Barondes, S. H. (1975). Toward a molecular basis of neuronal recognition, pp. 104–130 in *The Nervous System*, Vol. 1: *The Basic Neurosciences*, Tower, D. B., ed. Raven Press, New York.

Bischoff, A. and P. K. Thomas (1975). Microscopic anatomy of myelinated nerve fibres, pp. 104–130 in *Peripheral Neuropathy*, Dyck, P. J., P. K. Thomas, and E. H. Lambert, eds. W. B. Saunders, Philadelphia.

Blakemore, W. F. (1975). Remyelination by Schwann cells of axons demyelinated by intraspinal injection of 6-aminonicotinamide in the rat, *J. Neurocytol.* **4**:745–757.

Blakemore, W. F. (1977). Remyelination of CNS by Schwann cells transplanted from the sciatic nerve, *Nature (London)* **266**:68–70.

Bradley, W. G. and A. K. Asbury (1970). Duration of synthesis phase in neurilemma cells in mouse sciatic nerve during degeneration, *Exp. Neurol.* **26**:275–282.

Bradley, W. G. and M. Jenkison (1973). Abnormalities of peripheral nerves in murine muscular dystrophy, *J. Neurol. Sci.* **18**:227–247.

Brady, R. O. and R. H. Quarles (1973). The enzymology of myelination, *Mol. Cell. Biochem.* **2**:23–29.

Brockes, J. P., K. L. Fields, and M. D. Raff (1977). A surface antigenic marker for rat Schwann cells, *Nature (London)* **266**:364–366.

Brown, M. J., D. E. Pleasure, and A. K. Asbury (1976). Microdissection of peripheral nerve. Collagen and lipid distribution with morphological correlation, *J. Neurol. Sci.* **29**:361–369.

Bunge, R. P. and M. B. Bunge (1978). Evidence that contact with connective tissue matrix is required for normal interaction between Schwann cell and nerve fibers, *J. Cell Biol.* **78**:943–950.

Carlstedt, T. (1977). Observations on the morphology at the transition between the peripheral and the central nervous system in the cat. IV. Unmyelinated fibres in S_1 dorsal rootlets, *Acta Physiol. Scand. Suppl.* **446**:61–71.

Cook, R. D. and H. M. Wiśniewski (1973). The role of oligodendroglia and astroglia in Wallerian degeneration of the optic nerve, *Brain Res.* **61**:191–206.

DeBaecque, C., C. S. Raine, and P. S. Spencer (1976). Copper binding at PNS nodes of Ranvier during demyelination and remyelination in the perineurial window, *Neuropathol. Appl. Neurobiol.* **2**:459–470.

Duncan, D. (1934). A relation between axone diameter and myelination determined by measurements of myelinated spinal root fibers, *J. Comp. Neurol.* **60**:437–471.

Feigin, I. and J. Ogata (1971). Schwann cells and peripheral myelin within human central nervous tissues: the mesenchymal character of Schwann cells, *J. Neuropathol. Exp. Neurol.* **30**:603–612.

Feigin, I. and N. Popoff (1966). Regeneration of myelin in multiple sclerosis, *Neurology* **16**:364–372.

Friede, R. L. (1972). Control of myelin formation by axon caliber (with a model of the control mechanism), *J. Comp. Neurol.* **144**:233–252.

Friede, R. L. and M. A. Johnstone (1967). Responses of thymidine labeling of nuclei in gray matter and nerve following sciatic transection, *Acta Neuropathol.* **7**:218–231.

Fulcrand, J. and A. Privat (1977). Neuroglial reactions secondary to Wallerian degeneration in the optic nerve of the postnatal rat: ultrastructural and quantitative study, *J. Comp. Neurol.* **176**:189–224.

Hall, S. M. (1978). The Schwann cell: a reappraisal of its role in the peripheral nervous system, *Neuropathol. Appl. Neurobiol.* **4**:165–176.

Hillarp, N. Å. and H. Olivecrona (1946). The rôle played by the axon and the Schwann cells in the degree of myelination of the peripheral nerve fibre, *Acta Anat.* **2**:17–32.

Holmes, W. and J. Z. Young (1942). Nerve regeneration after immediate and delayed suture, *J. Anat.* **77**:63–96.

Kao, C. C., L. W. Chang, and J. M. B. Bloodworth, Jr. (1977). Axonal regeneration across transected mammalian spinal cords: an electron microscopic study of delayed microsurgical nerve grafting, *Exp. Neurol.* **54**:591–615.

King, R. H. M., P. K. Thomas, and J. D. Pollard (1977). Axonal and dorsal root ganglion cell changes in experimental allergic neuritis, *Neuropathol. Appl. Neurobiol.* **3**:471–486.

Krygier-Brévart, V., V. S. Zabrenetzky, and P. S. Spencer (1977). Cyclic AMP stimulation of a peripheral myelin protein kinase, *Trans. Am. Soc. Neurochem.* **8**:262.

Langley, J. N. (1898). On the union of cranial autonomic (visceral) fibres with the nerve cells of the superior cervical ganglion, *J. Physiol. (London)* **32:** 240–270.

Langley, J. N. and H. K. Anderson (1903). On the union of the fifth cervical nerve with the superior cervical ganglion, *J. Physiol. (London)* **30:**439–442.

Langley, J. N. and H. K. Anderson (1904). The union of different kinds of nerve fibres, *J. Physiol. (London)* **31:**365–391.

Langley, O. K. and D. N. Landon (1969). Copper binding at nodes of Ranvier: a new electron histochemical technique for the demonstration of polyanions, *J. Histochem. Cytochem.* **17:**66–69.

Logan, J. E., R. J. Rossiter, and M. L. Barr (1953). Cellular proliferation in the proximal segment of a sectioned or crushed peripheral nerve, *J. Anat.* **87:** 419–422.

Mandel, P., J. L. Nussbaum, N. M. Neskovic, L. L. Sarlieve, and T. Kurihara (1972). Regulation of myelinogenesis, *Adv. Enzyme Regul.* **10:**101–117.

Maxwell, D. S., L. Kruger, and A. Pineda (1969). The trigeminal nerve root with special reference to the central-peripheral transition zone: an electron microscopic study in the macaque, *Anat. Rec.* **164:**113–125.

McCarthy, K. D. and L. M. Partlow (1976). Neuronal stimulation of (^{3}H) thymidine incorporation by primary cultures of highly purified non-neuronal cells, *Brain Res.* **114:**415–426.

McDermott, J. R. and H. M. Wiśniewski (1977). Studies on the myelin protein changes and antigenic properties of rabbit sciatic nerves undergoing Wallerian degeneration, *J. Neurol. Sci.* **33:**81–94.

Meier, C. and H. Sollmann (1978). Glial outgrowth and central-type myelination of regenerating axons in spinal nerve roots following transection and suture: light and electron microscopic study in the pig, *Neuropathol. Appl. Neurobiol.* **4:**21–35.

Miyamoto, E. and S. Kakiuchi (1974). In vitro and in vivo phosphorylation of myelin basic protein by exogenous and endogenous adenosine 3′:5′-monophosphate-dependent protein kinases in brain, *J. Biol. Chem.* **249:**2769–2777.

Moscona, A. A. (1974). Surface specification of embryonic cells: lectin receptors, cell recognition, and specific cell ligands, pp. 67–99 in *The Cell Surface in Development*, Moscona, A. A., ed. John Wiley & Sons, New York.

Norton, W. T. (1977). The myelin sheath, pp. 259–298 in *Scientific Approaches to Clinical Neurology*, Vol. 1, Goldensohn, E. S. and S. H. Appel, eds. Lea & Febiger, Philadelphia.

Ochoa, J. (1975). Microscopic anatomy of unmyelinated fibers, pp. 131–150 in *Peripheral Neuropathy*, Dyck, P. J., P. K. Thomas, and E. H. Lambert, eds. W. B. Saunders, Philadelphia.

Ohmi, S. (1961). Electron microscopic study of Wallerian degeneration in the peripheral nerves, *Z. Zellforsch. Mikrosk. Anat.* **54:**39–67.

Peterson, E. R. and M. R. Murray (1955). Myelin sheath formation in cultures of avian spinal ganglia, *Am. J. Anat.* **96:**319–355.

Quarles, R. H., L. J. McIntyre, and N. H. Sternberger (1979). Glycoproteins and cell surface interactions during myelinogenesis, *Soc. Neurosci. Symp.* **4:**322–343.

Raff, M. C., A. Hornby-Smith, and J. Brockes (1978). Cyclic AMP as a mitogenic signal for cultured rat Schwann cells, *Nature (London)* **273:** 672–673.
Ramón y Cajal, S. (1928). *Degeneration and Regeneration of the Nervous System*, Vol. 1, May, R. M., transl. and ed. Hafner Publishing Co., London.
Robertson, J. D. (1962). The unit membrane of cells and mechanisms of myelin formation, pp. 94–158 in *Assoc. Res. Nerv. Ment. Dis.*, Proceedings, Vol. 40: *Ultrastructure and Metabolism of the Nervous System*, Korey, S. R., A. Pope, and E. Robins, eds. Williams & Wilkins, Baltimore.
Roseman, S. (1974). Complex carbohydrates and intercellular adhesion, pp. 255–271 in *The Cell Surface in Development*, Moscona, A. A., ed. John Wiley & Sons, New York.
Rosenbluth, J. (1976). Intramembranous particle distribution at the node of Ranvier and adjacent axolemma in myelinated axons of the frog brain, *J. Neurocytol.* **5:**731–745.
Salzer, J. L., L. Glaser, and R. P. Bunge (1977). Stimulation of Schwann cell proliferation by a neurite membrane fraction, *J. Cell Biol.* **75:**118a.
Simpson, S. A. and J. Z. Young (1945). Regeneration of fibre diameter after cross-unions of visceral and somatic nerves, *J. Anat.* **79:**48–65.
Singer, M. (1968). Penetration of labelled amino acids into the peripheral nerve fibre from surrounding body fluids, pp. 200–215 in *Ciba Foundation Symposium: Growth of the Nervous System*, Wolstenholme, G. E. N. and M. O'Connor, eds. Williams & Wilkins, Baltimore.
Singer, M. and G. M. C. Steinberg (1972). Wallerian degeneration: a reevaluation based on transected and colchicine-poisoned nerves in the amphibian *Triturus*, *Am. J. Anat.* **133:**51–84.
Snyder, D. H., M. P. Valsamis, S. H. Stone, and C. S. Raine (1975). Progressive demyelination and reparative phenomena in chronic experimental allergic encephalomyelitis, *J. Neuropathol. Exp. Neurol.* **34:**209–221.
Speidel, C. C. (1964). *In vivo* studies of myelinated nerve fibres, *Int. Rev. Cytol.* **16:**173–231.
Spencer, P. S. (1971). *Light and Electron Microscopic Observations on Localised Peripheral Nerve Injuries*, Vol. 1. Ph.D. thesis, University of London.
Spencer, P. S. (1976). Experimentally induced nerve injury and its repair, pp. 131–139 in *Symposium on Microsurgery*, Vol. 14, Daniller, A. I. and B. Strauch, eds. C. V. Mosby, St. Louis.
Spencer, P. S., C. S. Raine, and H. Wiśniewski (1973). Axon diameter and myelin thickness—unusual relationships in dorsal root ganglia, *Anat. Rec.* **176:**225–244.
Spencer, P. S. and H. J. Weinberg (1978). Axonal specification of Schwann cell expression and myelination, pp. 389–405 in *The Physiology and Pathobiology of Axons*, Waxman, S., ed. Raven Press, New York.
Spencer, P. S., H. J. Weinberg, V. Krygier-Brévart, and V. Zabrenetzky (1979). An in vivo method to prepare normal Schwann cells free of axons and myelin, *Brain Res.* **165:**119–126.
Spencer, P. S., H. J. Weinberg, C. S. Raine, and J. W. Prineas (1975). The perineurial window—a new model of focal demyelination and remyelination, *Brain Res.* **96:**323–329.

Suzuki, K. and J. C. Zagoren (1978). Studies on the copper binding affinity of fibers in the peripheral nervous system of the quaking mouse, *Neuroscience* **3:**447–455.

Thomas, G. A. (1948). Quantitative histology of Wallerian degeneration. II. Nuclear population in two nerves of different fibre spectrum, *J. Anat.* **82:** 135–145.

Thomas, P. K. (1964). Changes in the endoneurial sheaths of peripheral myelinated nerve fibres during Wallerian degeneration, *J. Anat.* **98:**175–182.

Thomas, P. K. (1974). Nerve injury, pp. 44–70 in *Essays on the Nervous System: A Festschrift for Professor J. Z. Young*, Bellairs, R. and E. G. Gray, eds. Oxford University Press, Oxford.

Vaughn, J. E. and D. C. Pease (1970). Electron microscopic studies of Wallerian degeneration in rat optic nerves. II. Astrocytes, oligodendrocytes, and adventitial cells, *J. Comp. Neurol.* **140:**207–226.

Webster, H. deF. (1975). Development of peripheral myelinated and unmyelinated nerve fibers, pp. 37–61 in *Peripheral Neuropathy*, Dyck, P. J., P. K. Thomas, and E. H. Lambert, eds. W. B. Saunders, Philadelphia.

Weinberg, E. L. and P. S. Spencer (1979). Studies on the control of myelinogenesis. 3. Signalling of oligodendrocyte myelination by regenerating peripheral axons, *Brain Res.* **162:**273–279.

Weinberg, H. J. (1978). *Axonal Regulation of Myelinogenesis.* Ph.D. thesis, Albert Einstein College of Medicine, Yeshiva University, New York.

Weinberg, H. J. and P. S. Spencer (1975). Studies on the control of myelinogenesis. I. Myelination of regenerating axons after entry into a foreign unmyelinated nerve, *J. Neurocytol.* **4:**395–418.

Weinberg, H. J. and P. S. Spencer (1976). Studies on the control of myelinogenesis. II. Evidence for neuronal regulation of myelination, *Brain Res.* **113:**363–378.

Weinberg, H. J. and P. S. Spencer (1978). The fate of Schwann cells isolated from axonal contact, *J. Neurocytol.* **7:**555–569.

Weinberg, H. J., P. S. Spencer, and C. S. Raine (1975). Aberrant PNS development in dystrophic mice, *Brain Res.* **88:**532–537.

Wood, P. M. (1976). Separation of functional Schwann cells and neurons from normal peripheral nervous tissue, *Brain Res.* **115:**361–375.

Wood, P. M. and R. P. Bunge (1975). Evidence that sensory axons are mitogenic for Schwann cells, *Nature (London)* **256:**662–664.

Zabrenetzky, V. (1979). *Characterization of a Protein Kinase in Peripheral Nervous System Myelin.* Ph.D. thesis, New York University.

Zabrenetzky, V. S., V. Krygier-Brévart, and P. S. Spencer (1977). Cyclic AMP stimulated protein phosphorylation in peripheral myelin and Schwann cell plasma membranes, p. 502 in *Proceedings, 6th International Meeting, International Society for Neurochemistry*, Copenhagen.

GLYCOPROTEINS AND CELL SURFACE INTERACTIONS DURING MYELINOGENESIS

Richard H. Quarles, Laurence J. McIntyre, and Nancy H. Sternberger

National Institute of Neurological and Communicative Disorders and Stroke, National Institutes of Health, Bethesda, Maryland

Several lines of experimentation indicate that myelin-forming cells require a signal from the axon in order to produce myelin. The evidence for this is considered in some detail in the papers by Spencer and Aguayo in this volume. Theoretical considerations of possible factors involved have been published by Spencer and Weinberg (1978). A signal on axonal surfaces that causes Schwann cells to proliferate in vitro is sensitive to trypsin (Salzer, Glaser, and Bunge, 1977), but in general very little is known about the chemistry of the molecules that mediate interactions between axons and myelin-forming cells.

There is a rapidly expanding field of research indicating that cell surface glycoproteins are involved in recognition and cell-cell interactions (Hughes, 1976). Therefore, it seems reasonable to suggest that glycoproteins in the surface membranes of Schwann cells or oligodendrocytes, or in the axolemma, could be involved in glial-axonal interactions occurring during the early stages of myelinogenesis (Brady and Quarles, 1973). Since myelin is formed by extension of the plasma membrane of the Schwann cell or oligodendrocyte, it is likely that some of the glycoproteins on the surfaces of these cells will be present in the purified myelin fraction. This paper will briefly review what is known about glycoproteins in myelin and related membranes and consider them in light of research on other cell systems involving the function of membrane glycoproteins in cell-cell interactions.

GLYCOPROTEINS IN PURIFIED MYELIN

Although there were early indications that myelin contained glycoproteins (see Quarles, 1979, for review), the first demonstration of a particular glycoprotein molecule associated with myelin resulted from experiments in which the glycoproteins in various subcellular fractions of rat brain were compared (Quarles, Everly, and Brady, 1972). The glycoproteins in brain can be labeled specifically in vivo by intracranial injection of radioactive fucose. When the CNS myelin fraction is electrophoresed on a discontinuous SDS 10% polyacrylamide gel, at least six fucose-labeled glycoproteins can be discerned (Figure 1). The largest peak is unique for the myelin fraction and corresponds to the major myelin-associated glycoprotein described in previous publications (Quarles, Everly, and Brady, 1973*a*; Quarles, 1979). However, this

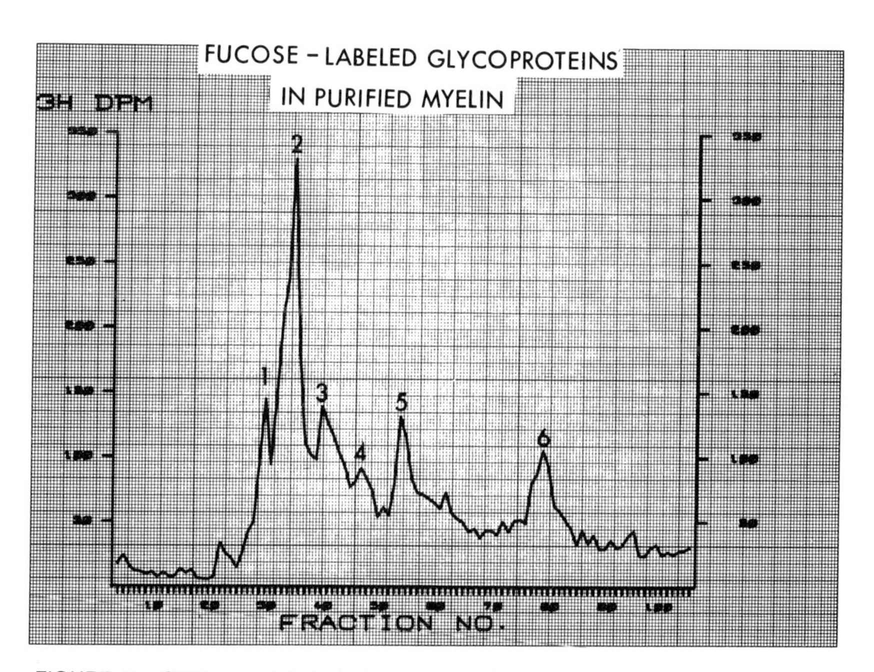

FIGURE 1. [^{3}H]Fucose-labeled glycoproteins in the myelin fraction purified from adult rat brains. The glycoproteins were separated on a discontinuous SDS 10% polyacrylamide gel prepared according to Laemmli and Favre (1973). The gel was cut into 1-mm slices for determination of radioactivity. The six most prominent glycoprotein peaks are numbered in order from large to small molecular weight. Peak 2 is the major myelin-associated glycoprotein described in earlier publications.

gel system gives better resolution of the smaller-molecular-weight glycoproteins than the continuous system used in most earlier work and clearly demonstrates the presence of several other glycoproteins. Even the principal glycoprotein (peak 2) is quantitatively a minor component of the whole myelin fraction. Nevertheless, it is possible to demonstrate it by periodic acid-Schiff staining for carbohydrate on a polyacrylamide gel overloaded with the chloroform-methanol-insoluble proteins of myelin (Figure 2).

Although there is only a small amount of this glycoprotein in the purified myelin fraction, a variety of experiments strongly indicated that it is in CNS myelin or an intimately associated structure and is not in an unrelated contaminant of the isolated fraction. These experiments included white matter-gray matter comparisons and developmental studies (Quarles et al., 1973*a*). However, the most convincing experiments indicating a true association of this glycoprotein with CNS myelin involved the myelin-deficient jimpy and quaking murine mutants (Matthieu, Brady, and Quarles, 1974*a*; Matthieu, Quarles, Webster, Hogan, and Brady, 1974*c*). It should be emphasized that these experiments did not necessarily indicate that this glycoprotein was in compact, multilamellar myelin, but left open the possibility that it could be in other locations in the continuum between the oligodendroglial plasma membrane and compact myelin or even in the axolemma of myelinated axons. Further experiments that will be discussed later suggest that this glycoprotein is not present throughout the myelin sheath, but is selectively localized in particular structures within the myelin-oligodendroglial units. For this reason, we refer to this glycoprotein in isolated CNS myelin as the myelin-associated glycoprotein (MAG). Very recent immunocytochemical studies, described in the final section of this paper, definitively show that MAG is specifically localized in the myelin-oligodendroglial units of the CNS.

It is clear from Figures 1 and 2 that there are also other glycoproteins in purified CNS myelin. The multiplicity of glycoproteins becomes particularly evident when a very sensitive technique of detection such as [^{3}H]concanavalin A binding is used (see next section). The origin of these other glycoproteins is not known. While some of them may be in contaminants of the myelin fraction, there are indications that others may actually be present in myelin or oligodendroglial membranes, as is MAG (Quarles et al., 1973*a*). Some may originate from a small amount of axolemma that is not dissociated from myelin during the purification procedure. Axonally transported glycoproteins are

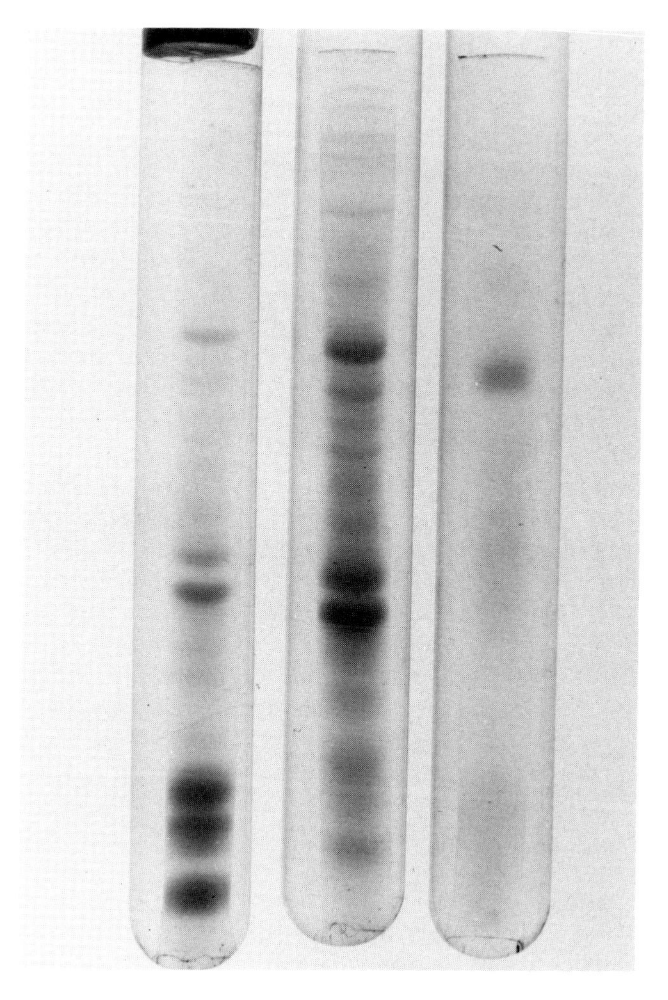

FIGURE 2. Staining of CNS myelin proteins after separation on 5% polyacrylamide gels. The two basic proteins and proteolipid are stained intensely with fast green at the bottom of the left-hand gel, and a number of higher-molecular-weight proteins can also be seen. These high-molecular-weight proteins are enriched in the chloroform-methanol-insoluble fraction of CNS myelin, which is electrophoresed on the center gel and stained with fast green. The right-hand gel shows the same fraction stained with periodic acid-Schiff reagents for carbohydrate. The major myelin-associated glycoprotein is clearly seen about one-third of the way down the gel. (Reproduced with permission from Quarles, R. H. [1975]. Glycoproteins in the nervous system, pp. 493–501 in *The Nervous System*, Vol. 1: *The Basic Neurosciences*, Tower, D. B., ed. Raven Press, New York.)

recovered in purified myelin fractions and are probably present in the axolemma (Autilio-Gambetti, Gambetti, and Shafer, 1975; Monticone and Elam, 1975; Matthieu, Webster, DeVries, Corthay, and Koellreutter, 1978*b*). It is of interest that MAG was not one of the glycoproteins undergoing axoplasmic transport in these studies, indicating that it is in oligodendroglial but not axonal membranes. As mentioned in the introduction, glycoproteins in the axolemma are equally likely to be involved in glia-axon interactions, and radiolabeling by axoplasmic flow appears to be a useful technique for identifying them. Presumably they can be obtained in larger quantity for chemical characterization from axolemma-enriched fractions (DeVries, Matthieu, Beny, Chicheportiche, Lazdunski, and Dolivo, 1978; Matthieu et al., 1978*b*). However, this paper will focus on glycoproteins in myelin or myelin-forming cells.

After identifying MAG in CNS myelin, we examined rat sciatic nerves to see if this glycoprotein was also in PNS myelin. Similar experiments were done in which glycoproteins were labeled in vivo with radioactive fucose or stained with periodic acid-Schiff reagents (Everly, Brady, and Quarles, 1973). These experiments did not demonstrate MAG in sciatic nerve myelin, but surprisingly showed that the major structural protein of PNS myelin is a glycoprotein. A protein called P_0 (molecular weight = 30,000) accounts for about half the protein in peripheral myelin and is not present in central myelin. This PNS-specific protein, which must be the major structural component of peripheral myelin, was extensively labeled by [^{3}H]fucose and stained with periodic acid-Schiff reagents (Figure 3). Wood and Dawson (1973) made a similar observation at about the same time. Recently, this P_0 glycoprotein has been purified and chemically characterized (Kitamura, Suzuki, and Uyemura, 1976; Roomi, Ishaque, Kahn, and Eylar, 1978*a*). The protein contains about 5% carbohydrate, which appears to be present in a single nonasaccharide unit. There are also minor glycoproteins of lower and higher molecular weight in PNS myelin (see Quarles, 1979, for review).

In summary, although there are glycoproteins in both CNS and PNS

FIGURE 3. Staining of PNS myelin proteins after separation on 10% polyacrylamide gels. The proteins of rat sciatic nerve myelin are stained with fast green on the left, and the major P_0 protein is labeled. The right-hand gel shows that the P_0 protein stains with periodic acid-Schiff reagents for carbohydrate.

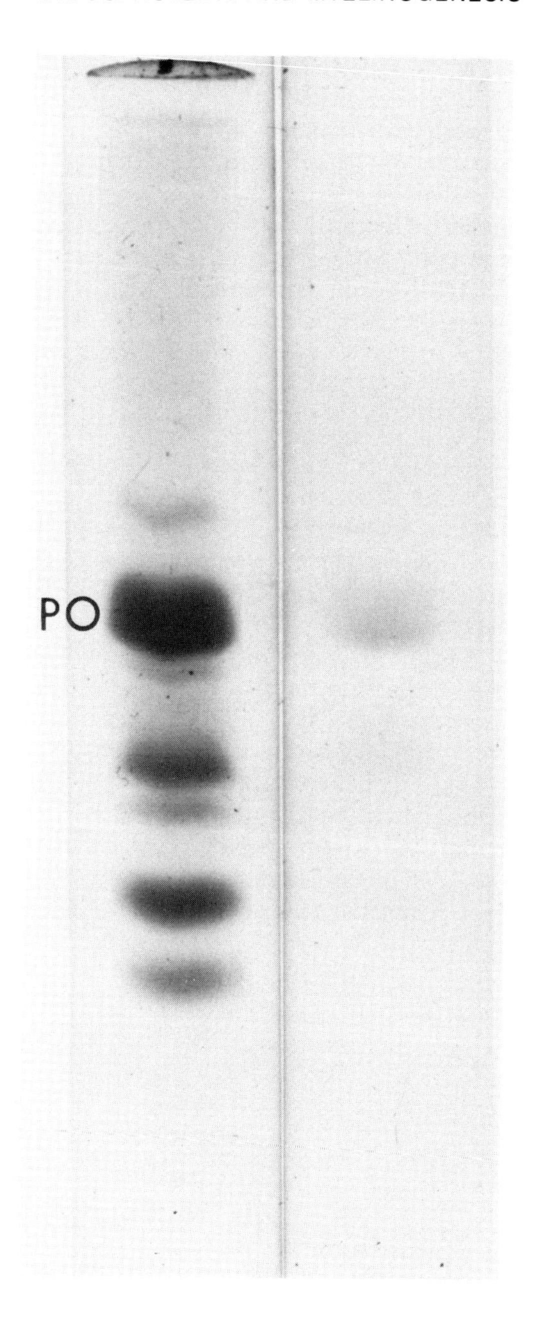
PO

myelin, there are clearly important differences between them. Whereas the major structural protein of PNS myelin is glycosylated, the glycoproteins in CNS myelin are confined to quantitatively minor high-molecular-weight components. Therefore, it was of interest to qualitatively and quantitatively compare the glycoprotein-carbohydrate in central and peripheral myelin. The results of such a comparison are shown in Table 1. Although both types of myelin contain each of the sugars characteristic of glycoproteins, peripheral myelin contains four- to sevenfold more of each individual sugar. This is as expected, since the major protein of peripheral myelin is a glycoprotein.

INVOLVEMENT OF GLYCOPROTEINS IN MYELINOGENESIS

Since myelin is formed as an extension of the plasma membrane of the oligodendrocyte or Schwann cell, some of the glycoproteins recovered in the myelin fractions may be exposed on the surfaces of the myelin-forming cells early in development. In such a location, they could play a critical role in the axon-glia interactions during the initial stages of myelinogenesis. It is not difficult to postulate specific surface interactions in which glycoproteins might be involved. They could function in enabling the glial cell to recognize axons that have reached the stage of development at which they are ready to be myelinated, i.e., the receptive axon could produce a new molecule on its surface that specifically binds particular sugar residues in the glycoprotein. Such a carbohydrate-binding protein would not be unlike the plant lectins that have high affinity for specific sugar residues. Indeed, developmentally regulated lectin-like molecules are present in neural tissue and have been postulated to function in controlling development

TABLE 1. *Glycoprotein-carbohydrate in CNS and PNS myelin*[a]

	CNS	PNS
Fucose	1.3 ± 0.27	9 ± 1.4
Mannose	5.4 ± 0.78	37 ± 3.3
Galactose	2.7 ± 0.37	14 ± 0.9
N-Acetylglucosamine	5.8 ± 0.88	28 ± 4.6
N-Acetylneuraminic acid	1.7 ± 0.14	11 ± 1.2

[a] The glycoproteins in purified rat brain and sciatic nerve myelin were converted to glycopeptides by exhaustive pronase digestion. The glycopeptides were purified by gel filtration and analyzed for sugars by gas-liquid chromatography. Values are given as nmol of each sugar per mg total protein in the myelin fraction. Data are from Quarles and Everly (1977), with permission.

(Simpson, Thorne, and Loh, 1977). Furthermore, the glycoproteins in purified myelin do interact with lectins, much in the manner of those on the surfaces of various cell types (see below). The analogy to the interactions of plant lectins with cell surfaces can be extended further by consideration of the fact that some lectins have a mitogenic effect on the cells to which they bind. It is of interest that contact between the axonal surface and Schwann cells in culture results in the transmission of a signal causing the Schwann cells to proliferate (Wood and Bunge, 1975).

After positive recognition between the axon and glial cell has determined that myelin formation will proceed, a more adhesive or permanent interaction between the cells may develop as the axon is engulfed by the glial processes. The known involvement of high-molecular-weight glycoproteins such as the fibronectins in cellular adhesion (Yamada and Olden, 1978) suggests that glycoproteins on the axolemma or glial plasma membrane could mediate this tighter interaction. Finally, after the axon has been completely surrounded by the myelin-forming cell, a different type of interaction must occur as the surfaces of the glial cell come into apposition and begin to form the intraperiod line. Since the carbohydrates of glycoproteins are present on the outside surfaces of cells, the oligosaccharide units of any glycoproteins retained in compact myelin would end up in the intraperiod line. The P_0 glycoprotein is a major constituent of peripheral myelin, and there is histochemical evidence indicating that its carbohydrate is in the intraperiod line (Peterson and Pease, 1972; Wood and McLaughlin, 1975), suggesting that it could be involved in the compaction process. The presence of hydrophilic oligosaccharide groups at this location could account for the greater separation of the intraperiod surfaces of peripheral myelin in comparison with central myelin, where the more hydrophobic proteolipid molecules would allow closer interaction with lipids and greater fusion of the surfaces (Peterson and Pease, 1972; Wood and McLaughlin, 1975). It is well known that the periodicity of peripheral myelin is greater than that of central myelin, due in part to a greater separation of the intraperiod membrane surfaces. As will be discussed later, it appears that MAG of central myelin is not present in compacted, multilamellar myelin to any significant extent and, therefore, would not affect the periodicity.

Concanavalin A (con A) and a variety of other plant lectins have been used extensively to probe the molecular topography on the surfaces of many mammalian cell types. The capacity of cells to be agglutinated by lectins often relates to their state of differentiation and whether or not they are contact-inhibited (see Hughes, 1976, for review). A similarity between the properties of glycoproteins in myelin and those on cell surfaces was demonstrated by a simple experiment in which con A was added to an aqueous suspension of rat brain myelin (Matthieu, Daniel, Quarles, and Brady, 1974*b*). This resulted in the rapid agglutination of the myelin fragments, and the agglutination was inhibited by the specific inhibitor of con A binding, α-methylmannoside. The con A receptors in CNS myelin are capable of diffusing laterally in the plane of the membrane (Matus, De Petris, and Raff, 1973). Recent experiments in which [^{3}H]con A was allowed to bind to myelin proteins after their separation on polyacrylamide gels revealed that MAG is the major con A binding protein in CNS myelin, but a number of other glycoprotein receptors for the lectin are also present (Figure 4) (McIntyre, Quarles, and Brady, unpublished). Similarly,

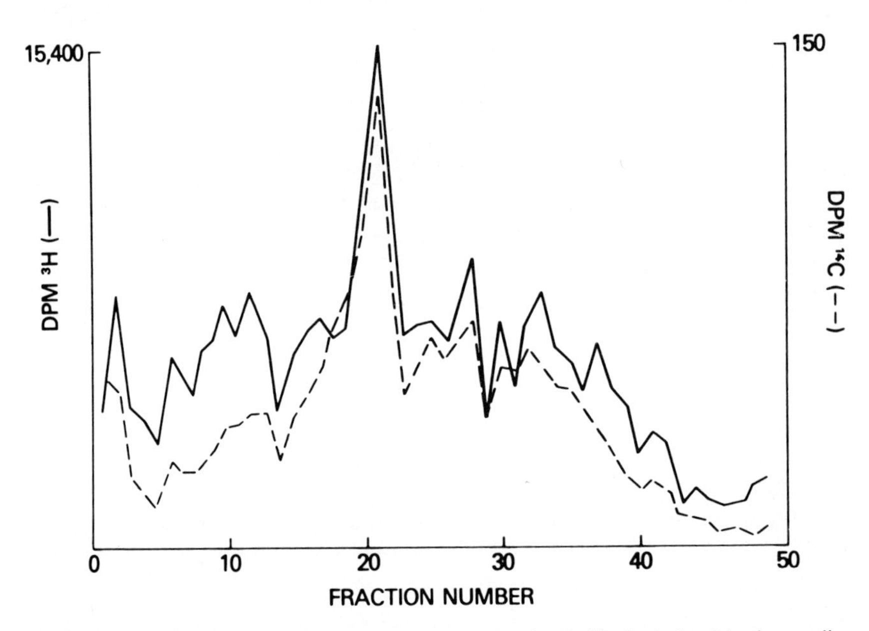

FIGURE 4. Binding of [^{3}H]con A to glycoproteins in purified adult rat brain myelin after separation on a 5% polyacrylamide SDS gel. The glycoproteins had been labeled in vivo with [^{14}C]fucose. After electrophoresis, [^{3}H]con A of high specific activity was allowed to bind to the proteins. The major peak of con A binding coincides with the ^{14}C-labeled MAG, but a large number of other con A binding glycoproteins are also present. All con A binding to proteins was prevented by α-methylmannoside.

MAG was a principal con A binding protein in immature 14-day rat myelin, but at this stage the relative amount of con A bound by other glycoproteins was increased. This developmental change in the lectin-binding proteins might indicate a difference in the carbohydrates that are exposed on the surfaces of myelin and related membranes, which is important for the types of interactions described in the preceding paragraph. Several other lectins with sugar specificities different from con A also interact with the glycoproteins in CNS myelin (Quarles, Foreman, Poduslo, and McIntyre, 1977; Zanetta, Sarlieve, Mandel, Vincendon, and Gombos, 1977), and some agglutinate isolated myelin fragments (Quarles, McIntyre, Pasnak, and Brady, unpublished). Con A also binds to the P_0 glycoprotein of PNS myelin (Wood and McLaughlin, 1975). These experiments on lectin interactions with myelin emphasize the relevance of the broad field of research on cell surface glycoproteins in general to the investigation of myelinogenesis.

An interesting result with regard to the possible involvement of MAG in the early stages of CNS myelin formation was the observation that the glycoprotein in immature myelin has a larger apparent molecular weight on SDS gels than that in mature myelin (Quarles, Everly, and Brady, 1973*b*). This finding was originally made by double label isotope techniques in which the MAG in immature rat myelin was labeled with [^{14}C]fucose and that in mature myelin was labeled with [^{3}H]fucose (Figure 5). However, the difference in molecular weight can be seen more directly by protein staining when partially purified MAG preparations are run side by side on an SDS slab gel (Figure 6). The larger, immature glycoprotein is present during the very early stages of myelinogenesis (12 to 20 days in rat brain), but by the time of most rapid myelin formation (20 to 30 days) the glycoprotein is largely of the mature form (Matthieu, Brady, and Quarles, 1975). This suggests that the larger form of the glycoprotein could function specifically in the early aspects of glia-axon interactions, rather than in the later thickening of myelin sheaths. Furthermore, the transformation from the larger to the smaller mature MAG appears to be a sensitive marker for myelin maturation. Thus, in a number of experimental situations in which myelin formation is reduced or delayed, the MAG of affected animals is shifted toward higher molecular weight in comparison with that of age-matched controls (see Quarles, 1979, for more detailed review). One of the most interesting examples is the myelin-deficient quaking mouse, which forms a small amount of immature, uncompacted myelin. The MAG in this mutant is of higher apparent molecular weight than that in controls (Matthieu et al., 1974*a*; Matthieu, Koellreutter, and Joyet, 1978*a*).

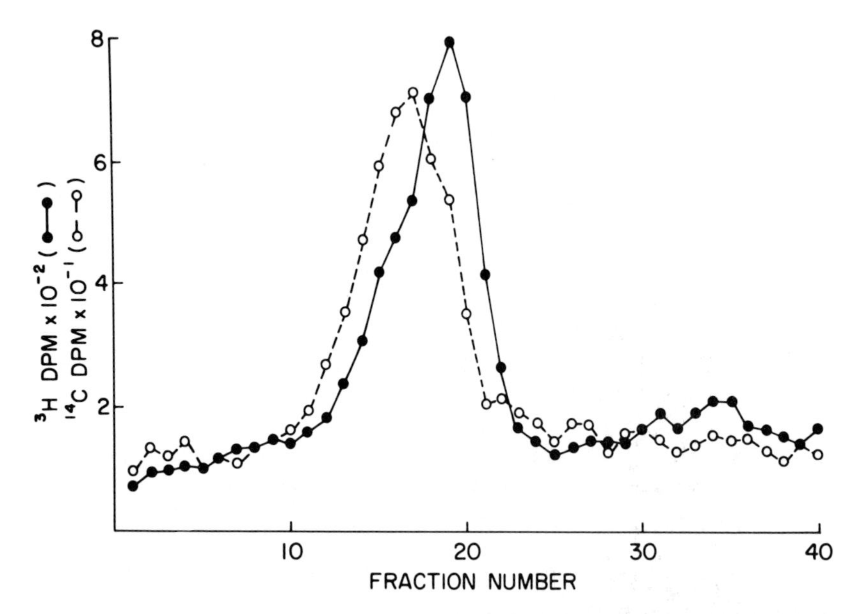

FIGURE 5. Developmental change in the apparent molecular weight of rat brain MAG. Myelin was purified from a mixed homogenate obtained by combining the brains of 14-day-old rats that had been injected with [^{14}C]fucose with those of 22-day-old rats that had been injected with [^{3}H]fucose. The doubly labeled myelin preparation was electrophoresed on a 5% polyacrylamide gel. The figure shows only that portion of the gel containing the MAG peak, which was cut into 1-mm slices for determination of radioactivity. Open circles, 14-day myelin labeled with ^{14}C; closed circles, 22-day myelin labeled with ^{3}H. (From Quarles et al., 1973*b*, with permission.)

An understanding of the functioning of MAG at the molecular level, and particularly the reason for the developmental change in apparent molecular weight, will require chemical analysis of the polypeptide and carbohydrate portions of the immature and mature glycoproteins. Work directed toward this goal is in progress. The recent development of a rapid and selective means of extracting MAG from myelin with lithium diiodosalicylate (LIS) and phenol (Quarles and Pasnak, 1977) is a significant advance toward characterizing this glycoprotein, which has proven very difficult to purify. This simple procedure results in a preparation greatly enriched in MAG containing only small amounts of contaminating proteins and can be applied to both the mature and immature glycoproteins (Figure 6). Removal of the minor contaminating proteins to yield very pure MAG can be achieved by preparative SDS polyacrylamide gel electrophoresis or gel filtration in the presence of neutral detergents. Chemical analyses of purified MAG are under way.

While methods for purifying MAG were being investigated, some indirect approaches were used to examine the chemical cause for the developmental change in molecular weight. When this molecular weight difference in MAG was first detected, there were a number of reports in the literature indicating that glycopeptide fractions prepared from the surfaces of rapidly dividing or virally transformed cells in tissue culture were larger than those from control cells (see Hughes, 1976, for review). The higher molecular weight of the glycopeptides was due in large part to a higher sialic acid content, and it seemed likely that the larger size of immature MAG could similarly be due to a greater degree of glycosylation. However, similar experiments involving gel filtration of glycopeptides prepared from partially purified MAG indicated that although there was a slight enrichment of larger sialic acid-rich oligosaccharide units in the immature glycoprotein, this was not the principal reason for its higher apparent molecular weight on SDS gels (Quarles, 1976). More recent experiments in which the mature and immature MAG were eluted from a variety of lectin affinity columns with differing sugar specificities also failed to give any indication that there was a substantial carbohydrate difference between the two forms of the glycoprotein (Quarles et al., 1977).

On the positive side, experiments in which fucose-labeled myelin was treated with trypsin showed that both the immature and mature forms of MAG were degraded to 80,000 dalton derivatives in which no developmental difference in molecular weight was apparent (McIntyre, Quarles, and Brady, 1978*a*). Elimination of the developmental difference in this manner did not liberate much radioactive fucose from the protein. The results suggest that the developmental difference in the intact MAG molecules may be due to an amino acid sequence on one end of the immature MAG molecule that is not present in mature MAG. This hypothesis still needs to be confirmed by peptide mapping and chemical analyses, which are in progress. However, there is currently a great deal of interest in posttranslational modification of proteins by limited proteolysis. Such a mechanism for modifying the biological activity of proteins has long been known in the case of the digestive proteolytic enzymes, for example. In the past few years, there has been considerable investigation of the "signal hypothesis" of Blobel and Dobberstein (1975), which postulates that an additional amino acid sequence is present in proteins to be secreted from the cell or incorporated into the

plasma membrane. This additional peptide on the amino terminus helps the larger precursor protein reach its proper association with membranes in the cell and is then proteolytically cleaved to give the authentic protein. Whether such a process is occurring with MAG and what significance it would have for the mechanism of myelination remains to be determined. However, it is of interest that a larger form of myelin basic protein with an additional amino acid sequence has recently been reported (Barbarese, Braun, and Carson, 1977) and is present in relatively larger amounts in immature myelin (Barbarese, Carson, and Braun, 1978).

LOCALIZATION OF MAG

The functions performed by glycoproteins associated with myelin and their possible role in glial-axonal interactions must be related to their precise locations within the myelin sheath. Biochemical studies in which CNS myelin and myelin-related membrane fractions have been subfractionated indicate that MAG is not uniformly distributed throughout multilamellar myelin, but rather has a more selective localization within the myelin-oligodendroglia units. The findings in these subfractionation experiments are summarized in Table 2. At one extreme is the light myelin subfraction consisting of large, multilamellar whorls of compact myelin (Matthieu, Quarles, Brady, and Webster, 1973). At the other extreme is a subfraction of the membranes released when crude myelin is osmotically shocked with water (McIntyre, Quarles, Webster, and Brady, 1978*b*). This fraction, termed "microsome-depleted $W_1 3$" for methodological reasons, consists primarily of single membranes and vesicles with a few small myelin fragments. The heavy myelin subfraction has a morphological makeup intermediate between these two extremes (Matthieu et al., 1973).

Table 2 shows the content of three representative chemical constituents of these subfractions. As would be expected, the light fraction has the highest content of basic protein, which is a major component of compact myelin. The level of basic protein is much lower in the

FIGURE 6. Electrophoresis of MAG preparations that had been partially purified by the LIS-phenol procedure. An adult rat MAG preparation was run in the left lane of a 12% polyacrylamide slab gel adjacent to a 14-day MAG preparation in the right lane. Coomassie brilliant blue staining reveals the slightly higher molecular weight of immature MAG. Arrows point to the MAG bands.

TABLE 2. *Composition of subfractions of myelin and myelin-related membranes*

	Basic protein (% of total)	CNP (μmol/h/mg protein)	MAG (units/mg protein)[a]
Myelin	35	1063	12
Light myelin[b]	47	550	5
Heavy myelin[b]	26	1350	17
Microsome-depleted $W_1 3$[c]	15	3909	32

[a] A unit is defined as the amount of MAG that stains with the same intensity as 1 μg of fetuin with periodic acid-Schiff reagents.

[b] Light and heavy subfractions of purified myelin were isolated as described by Matthieu et al. (1973). Data are from the same source, with permission.

[c] This fraction was isolated by density gradient centrifugation of the material released when myelin is osmotically shocked with water. Data are from McIntyre et al. (1978*b*), with permission.

fractions enriched in single membranes. By contrast, the so-called myelin marker enzyme, 2′:3′-cyclic nucleotide 3′-phosphodiesterase (EC 3.1.4.37; CNP), has an inverse distribution in the subfractions. The light, compact myelin has a relatively low level of activity, while the specific activity in microsome-depleted $W_1 3$ is three to four times higher than that of whole myelin. The distribution of MAG is similar to that of CNP, with the highest content in the single membranes and vesicles of microsome-depleted $W_1 3$. The same results were obtained for the similar SN-4 fraction (Waehneldt, Matthieu, and Neuhoff, 1977).

These findings clearly show that MAG and CNP are not most concentrated in compact myelin, as is basic protein. The origin of the single membranes and vesicles that are enriched in these components is not known, although one might speculate that they are derived from the membranous continuum between the oligodendrocyte surface and compact myelin. This could include membranes that peel off from the inner or outer surfaces of the myelin sheath during the isolation procedure or the paranodal lateral loops that form the glial-axonal junctions. However, interpretation of subfractionation studies is very difficult because there are no known morphological criteria for identifying such membranes after homogenization. Therefore, we have recently undertaken a more direct approach for localizing MAG.

This direct approach is the immunocytochemical demonstration of MAG by the peroxidase-antiperoxidase technique. The results of this study are reported in considerably more detail elsewhere (Sternberger,

Quarles, Itoyama, and Webster, 1979), but the most important findings are summarized here. MAG was selectively extracted from adult rat brain myelin by the LIS-phenol procedure and further purified by preparative polyacrylamide gel electrophoresis. Immunization of a rabbit with pure MAG resulted in the production of antibodies that could be demonstrated by Ouchterlony immunodiffusion or by a double antibody radioimmunoassay utilizing [^{3}H]fucose-labeled MAG as a test antigen. It is significant that this antiserum raised to MAG obtained from adult rat myelin also reacted well with the slightly larger MAG purified from immature rat myelin, further indicating the close structural relationship of these two forms of MAG.

The immunocytochemical procedure utilizing this antiserum stained oligodendrocytes and myelin sheaths in the developing CNS (Figure 7*A*). Furthermore, this technique showed that MAG is specifically found in the myelin-oligodendroglial units, since the MAG antibodies did not stain neurons, astrocytes, or other cells. Although the earlier biochemical studies had indicated that MAG was uniquely concentrated in myelin and oligodendrocytes, they did not rule out the presence of small amounts of this glycoprotein in other structures in the brain. It can now be said that this glycoprotein is specific for myelin and myelin-forming cells. With regard to the possible role of MAG in glial-axonal interactions, it is of particular interest that, as the brain develops and myelin sheaths thicken, the MAG staining remains confined to the periaxonal portion of the sheaths (Figure 7*B*). This is in contrast to the results with antiserum to myelin basic protein, where the thickening myelin sheaths were stained throughout by the same procedure (Sternberger, Itoyama, Kies, and Webster, 1978). Although the precise periaxonal localization of MAG cannot be defined at the light microscope level, a localization in the outer axoplasm or axolemma seems unlikely, since the glycoprotein first appears in the cytoplasm of oligodendrocytes and the surfaces of unmyelinated axons do not stain. Rather, MAG appears to be selectively localized in the inner layers of the myelin sheath adjacent to the axolemma, precisely the expected position if it is involved in axon-oligodendrocyte interactions.

MAG antiserum also stained Schwann cells and the periaxonal portion of PNS myelin sheaths in the trigeminal ganglion. This finding is particularly interesting in view of the results described by Spencer and by Aguayo in this volume, which show that the interactions between axons and myelin-forming cells that initiate myelination are in some respects similar in the central and peripheral nervous systems. How-

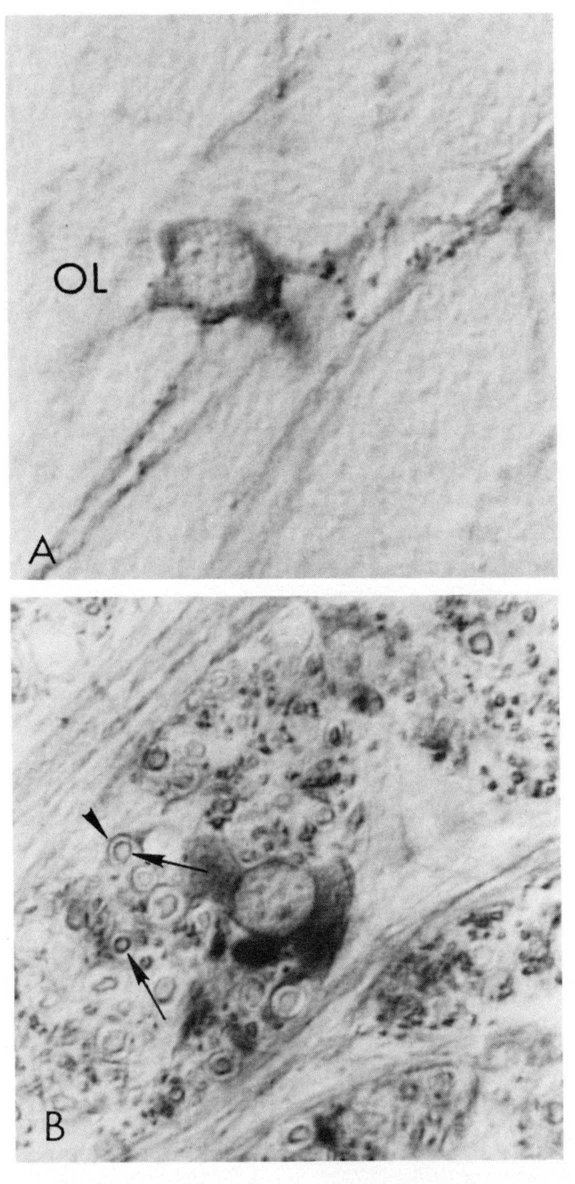

FIGURE 7. Micrographs of coronal sections of the developing rat pons. Sections were incubated with a 1:500 dilution of MAG antiserum and photographed with differential interference contrast optics. At 7 days (*A*), oligodendroglial (OL) cytoplasm and processes contain granular staining that extends along the developing

ever, it is also a surprising result, since previous biochemical studies had not revealed MAG in peripheral myelin (Everly et al., 1973; Quarles, 1979). The MAG antiserum does not form a precipitin line with the P_0 glycoprotein of PNS myelin under conditions in which an antiserum raised to P_0 does react. However, it must be remembered that the peroxidase-antiperoxidase immunocytochemical technique is much more sensitive than most biochemical procedures, as evidenced by its ability to demonstrate basic protein and MAG in newborn rat brain (Sternberger et al., 1978, 1979). Also, the large amount of P_0 glycoprotein in peripheral myelin may have masked the presence of a small amount of MAG in previous biochemical experiments. Furthermore, there have been recent reports of very small amounts of high-molecular-weight glycoproteins in purified peripheral myelin in addition to the major P_0 glycoprotein (Singh, Silberlicht, and Singh, 1978; Roomi, Ishaque, Kahan, and Eylar, 1978*b*). Very recent fucose-labeling experiments in our laboratory have confirmed the presence in peripheral myelin of small amounts of high-molecular-weight glycoproteins that electrophorese in the vicinity of MAG on SDS gels (unpublished results). These immunocytochemical and biochemical findings render it likely that MAG or a very similar glycoprotein is associated with myelin in the PNS, as it is in the CNS. Biochemical and immunochemical studies are in progress to further evaluate these possibilities.

CONCLUSIONS

The intent of this paper was to briefly review what is known about the glycoproteins in myelin and closely associated membranes and to relate this information to the rapidly evolving field concerning cell surface glycoproteins and their functions in recognition and cell-cell interactions. It is hoped that this will stimulate ideas, and perhaps experiments, about the molecular mechanisms underlying axon-glia interactions such as those described by the other speakers in this symposium. In order to achieve this objective, we have described our current hypotheses about MAG and tried to indicate how these ideas are influenced by findings about cell surface glycoproteins in other sys-

---

myelin sheaths. Cross sections of myelin sheaths at 14 days (*B*) are stained periaxonally (arrows). Axons and the compact portions of myelin sheaths are unstained. Thin, less dense rings are present at the outer margins of some sheaths (arrowhead) that are produced by the differential interference contrast illumination. (From Sternberger et al., 1979, with permission.)

tems. In some cases, these hypotheses are quite tentative and remain to be confirmed or disproved by direct chemical analyses of the purified glycoprotein. However, they serve to focus attention on certain aspects of the molecular structures of immature and mature MAG as their chemistry is worked out.

Certainly, the periaxonal localization of MAG is consistent with a role for this glycoprotein in establishing contacts between the myelin-forming cells and the axon. Recent evidence that periaxonal MAG is altered in the very early stages of multiple sclerosis (Itoyama, Sternberger, Quarles, Cohen, Richardson, Moser, and Webster, 1978) also suggests that it could have a critical function in the glial-axonal interactions involved in the formation, maintenance, and breakdown of myelin. Now that a specific antiserum to MAG is available, it may be possible to design experiments that directly probe the function of this glycoprotein in cell-cell interactions. It is hoped that this approach, along with the chemical characterization of MAG and other glycoproteins in glial and axonal membranes, will eventually lead to identification of the precise molecular determinants involved in glial-axonal recognition.

ACKNOWLEDGMENTS

The experimental work described here was supported in part by a grant from the National Multiple Sclerosis Society. The authors are grateful for the ideas and suggestions provided by Roscoe O. Brady and Henry deF. Webster during the course of these investigations.

REFERENCES

Aguayo, A. J., G. M. Bray, C. S. Perkins, and I. D. Duncan (1979). Axon-sheath cell interactions in peripheral and central nervous system transplants, *Soc. Neurosci. Symp.* **4**:361–383.

Autilio-Gambetti, L., P. Gambetti, and B. Shafer (1975). Glial and neuronal contribution to proteins and glycoproteins recovered in myelin fractions, *Brain Res.* **84**:336–340.

Barbarese, E., P. E. Braun, and J. H. Carson (1977). Identification of prelarge and presmall basic proteins in mouse myelin and their structural relationship to large and small basic protein, *Proc. Natl. Acad. Sci. USA* **74**:3360–3364.

Barbarese, E., J. H. Carson, and P. E. Braun (1978). Accumulation of four myelin basic proteins in mouse brain during development, *J. Neurochem.* **31**:779–783.

Blobel, G. and B. Dobberstein (1975). Transfer of proteins across membranes, *J. Cell Biol.* **67**:835–851.

Brady, R. O. and R. H. Quarles (1973). The enzymology of myelination, *Mol. Cell. Biochem.* **2**:23–29.

DeVries, G. H., J. M. Matthieu, M. Beny, R. Chicheportiche, M. Lazdunski, and M. Dolivo (1978). Isolation and partial characterization of rat CNS axolemma enriched fractions, *Brain Res.* **147**:339–352.

Everly, J. L., R. O. Brady, and R. H. Quarles (1973). Evidence that the major protein in rat sciatic nerve myelin is a glycoprotein, *J. Neurochem.* **21**:329–334.

Hughes, R. C. (1976). *Membrane Glycoproteins: A Review of Structure and Function.* Butterworth Publishers, Inc., Woburn, Mass.

Itoyama, Y., N. Sternberger, R. Quarles, S. Cohen, E. P. Richardson, Jr., H. W. Moser, and H. deF. Webster (1978). Successful immunocytochemical localization of myelin components in paraffin sections of human nervous tissue with preliminary observations on multiple sclerosis and metachromatic leukodystrophy, *Trans. Am. Neurol. Assoc.*, in press.

Kitamura, K., M. Suzuki, and K. Uyemura (1976). Purification and partial characterization of two glycoproteins in bovine peripheral nerve myelin membrane, *Biochim. Biophys. Acta* **455**:806–816.

Laemmli, U. K. and M. Favre (1973). Maturation of the head of bacteriophage T4. I. DNA packaging events, *J. Mol. Biol.* **80**:575–600.

Matthieu, J.-M., R. O. Brady, and R. H. Quarles (1974*a*). Anomalies of myelin-associated glycoproteins in quaking mice, *J. Neurochem.* **22**:291–296.

Matthieu, J.-M., R. O. Brady, and R. H. Quarles (1975). Change in a myelin associated glycoprotein in rat brain during development: metabolic aspects, *Brain Res.* **86**:55–65.

Matthieu, J.-M., A. Daniel, R. H. Quarles, and R. O. Brady (1974*b*). Interactions of concanavalin A and other lectins with CNS myelin, *Brain Res.* **81**:348–353.

Matthieu, J.-M., B. Koellreutter, and M. L. Joyet (1978*a*). Protein and glycoprotein composition of myelin and myelin subfractions from brains of quaking mice, *J. Neurochem.* **30**:783–790.

Matthieu, J.-M., R. H. Quarles, R. O. Brady, and H. deF. Webster (1973). Variation of proteins, enzyme markers, and gangliosides in myelin subfractions, *Biochim. Biophys. Acta* **329**:305–317.

Matthieu, J.-M., R. H. Quarles, H. deF. Webster, E. L. Hogan, and R. O. Brady (1974*c*). Characterization of the fraction obtained from the CNS of jimpy mice by a procedure for myelin isolation, *J. Neurochem.* **23**:517–523.

Matthieu, J.-M., H. deF. Webster, G. H. DeVries, S. Corthay, and B. Koellreutter (1978*b*). Glial vs. neuronal origin of myelin proteins studied by combined intraocular and intracranial labeling, *J. Neurochem.* **31**:93–102.

Matus, A., S. De Petris, and M. C. Raff (1973). Mobility of con A receptors in myelin and synaptic membranes, *Nat. New Biol.* **244**:278–279.

McIntyre, L. J., R. H. Quarles, and R. O. Brady (1978*a*). The effect of trypsin on myelin-associated glycoprotein, *Trans. Am. Soc. Neurochem.* **9**:106.

McIntyre, R. J., R. H. Quarles, H. deF. Webster, and R. O. Brady (1978*b*). Isolation and characterization of myelin-related membranes, *J. Neurochem.* **30**:991–1002.

Monticone, R. E. and J. S. Elam (1975). Isolation of axonally transported glycoproteins with goldfish visual system myelin, *Brain Res.* **100**:61–71.

Peterson, R. G. and D. C. Pease (1972). Myelin imbedded in polymerized glutaraldehyde-urea, *J. Ultrastruct. Res.* **41**:115–132.

Quarles, R. H. (1975). Glycoproteins in the nervous system, pp. 493–501 in *The Nervous System*, Vol. 1: *The Basic Neurosciences*, Tower, D. B., ed. Raven Press, New York.

Quarles, R. H. (1976). Effects of pronase and neuraminidase treatment on a myelin associated glycoprotein in developing brain, *Biochem. J.* **156**:143–150.

Quarles, R. H. (1979). Glycoproteins in myelin and myelin-related membranes, pp. 209–233 in *Complex Carbohydrates of the Nervous System*, Margolis, R. U. and R. K. Margolis, eds. Plenum Press, New York.

Quarles, R. H. and J. L. Everly (1977). Glycopeptide fractions prepared from purified central and peripheral rat myelin, *Biochim. Biophys. Acta* **466**:176–186.

Quarles, R. H., J. L. Everly, and R. O. Brady (1972). Demonstration of a glycoprotein which is associated with a purified myelin fraction from rat brain, *Biochem. Biophys. Res. Commun.* **47**:491–497.

Quarles, R. H., J. L. Everly, and R. O. Brady (1973*a*). Evidence for the close association of a glycoprotein with myelin in rat brain, *J. Neurochem.* **21**:1177–1191.

Quarles, R. H., J. L. Everly, and R. O. Brady (1973*b*). Myelin-associated glycoprotein: a developmental change, *Brain Res.* **58**:506–509.

Quarles, R. H., C. F. Foreman, J. F. Poduslo, and L. J. McIntyre (1977). Interactions of myelin-associated glycoproteins with immobilized lectins, *Trans. Am. Soc. Neurochem.* **8**:201.

Quarles, R. H. and C. F. Pasnak (1977). A rapid procedure for selectively isolating the major glycoprotein from purified rat brain myelin, *Biochem. J.* **163**:635–637.

Roomi, M. W., A. Ishaque, N. R. Kahn, and E. H. Eylar (1978*a*). The P_0 protein, the major glycoprotein of peripheral nerve myelin, *Biochim. Biophys. Acta* **536**:112–121.

Roomi, M. W., A. Ishaque, N. R. Kahn, and E. H. Eylar (1978*b*). Glycoproteins and albumin in peripheral nerve myelin, *J. Neurochem.* **31**:375–379.

Salzer, J. L., L. Glaser, and R. P. Bunge (1977). Stimulation of Schwann cell proliferation by a neurite membrane fraction, *J. Cell Biol.* **75**:118a.

Simpson, D. L., D. L. Thorne, and H. H. Loh (1977). Developmentally regulated lectin in neonatal rat brain, *Nature (London)* **266**:367–369.

Singh, H., I. Silberlicht, and J. Singh (1978). A comparative study of the polypeptides of mammalian peripheral myelin, *Brain Res.* **144**:303–311.

Spencer, P. S. (1979). Neuronal regulation of myelinating cell function, *Soc. Neurosci. Symp.* **4**:275–321.

Spencer, P. S. and H. J. Weinberg (1978). Axonal specification of Schwann cell expression and myelination, pp. 389–405 in *Physiology and Pathology of Axons*, Waxman, S. G., ed. Raven Press, New York.

Sternberger, N. H., Y. Itoyama, M. W. Kies, and H. deF. Webster (1978).

Immunocytochemical method to identify basic protein in myelin-forming oligodendrocytes of the newborn rat CNS, *J. Neurocytol.* **7**:251–263.
Sternberger, N. H., R. H. Quarles, Y. Itoyama, and H. deF. Webster (1979). Myelin-associated glycoprotein demonstrated immunocytochemically in myelin and myelin forming cells of developing rat, *Proc. Natl. Acad. Sci. USA* **76**:1510–1514.
Waehneldt, T. V., J.-M. Matthieu, and V. Neuhoff (1977). Characterization of myelin-related fraction (SN-4) isolated from rat forebrain at two developmental stages, *Brain Res.* **138**:29–43.
Wood, J. G. and R. H. Dawson (1973). A major myelin glycoprotein of sciatic nerve, *J. Neurochem.* **21**:717–719.
Wood, J. G. and B. J. McLaughlin (1975). The visualization of concanavalin A binding sites in the intraperiod line of rat sciatic nerve myelin, *J. Neurochem.* **24**:233–235.
Wood, P. and R. Bunge (1975). Evidence the sensory axons are mitogenic for Schwann cells, *Nature (London)* **256**:662–664.
Yamada, K. M. and K. Olden (1978). Fibronectins: adhesive glycoproteins of cell surface and blood, *Nature (London)* **275**:179–184.
Zanetta, J. P., L. L. Sarlieve, P. Mandel, G. Vincendon, and G. Gombos (1977). Fractionation of glycoproteins associated to adult rat brain myelin, *J. Neurochem.* **29**:827–838.

TRANSFER OF PHOSPHOLIPID CONSTITUENTS TO GLIA DURING AXONAL TRANSPORT

Bernard Droz, Marina Brunetti, Luigi Di Giamberardino,
Herbert L. Koenig, and Giuseppe Porcellati

Commissariat à l'Energie Atomique, Saclay, France
Universita di Perugia, Italy
Université Pierre et Marie Curie, Paris, France

The influence exerted by the axon on the maintenance of its myelin sheath derives mainly from the observation of Wallerian degeneration: the disintegration of the distal part of a severed axon is accompanied by the alteration and disappearance of its myelin sheath. In spite of the progress made in neuroscience since Waller's discovery in 1852, Suzuki (1978) noted that "we cannot yet talk intelligently, in biochemical terms, about the loss of the axonal 'signal' essential for normal maintenance of the myelin sheath." Critical reviews of the two main mechanisms proposed to account for the cooperation of the axon with myelin-forming cells have appeared recently (Aguayo, Charron, and Bray, 1976; Weinberg and Spencer, 1976; Spencer and Weinberg, 1978). First, a specific molecular interaction between the cell surface of axon and glia would initiate the metabolic activity of myelin-forming cells to manufacture the myelin sheath. Second, myelin-forming cells would be continuously supplied with a neuronal "trophic" factor formed in the nerve cell body, transported along the axon, and delivered to encompassing glia.

However, whether glial cells are supplied with neuronal molecules transported by axonal flow is a matter of controversy, except for nucleosides or nucleotides (Autilio-Gambetti, Gambetti, and Shafer, 1973; Schubert and Kreutzberg, 1974). When labeled proteins and glycoproteins, axonally transported along the optic nerve, have been recovered in myelin fractions (Elam, 1971; Giorgi, Karlsson, Sjöstrand, and Field, 1973), this finding has been interpreted as an axolemmal contamination rather than an axon-myelin transfer of macromolecules (Autilio-Gambetti, Gambetti, and Shafer, 1975; Matthieu, Webster, De Vries, Corthay, and Koellreutter, 1978). In peripheral nerves such as preganglionic fibers of the chicken ciliary ganglion, cell fractions (Di Giamberardino, Bennett, Koenig, and Droz, 1973) and autoradiographs (Bennett, Di Giamberardino, Koenig, and Droz, 1973; Droz, Koenig, and Di Giamberardino, 1973) failed to show a significant transfer of label to myelin in the course of the axonal transport of labeled proteins and glycoproteins. In contrast, when the axonal flow of phospholipids was recently investigated in the same material, the appearance of label in the myelin sheath (Droz, Di Giamberardino, Koenig, Boyenval, and Hässig, 1978) raised the question whether or not "the exchange of lipid molecules from axon to myelin seems feasible" (Norton, 1975).

EXPERIMENTAL APPROACH

In analyzing the mode of transfer of phospholipid constituents, the chicken ciliary ganglion offers several advantages. (1) The nerve cell bodies of the preganglionic neurons are grouped in the midbrain immediately below the cerebral aqueduct, into which labeled precursors of phospholipids are injected (Figure 1). (2) The myelinated axons, which terminate by forming a giant cholinergic synapse with the large ciliary ganglion cells, constitute a homogeneous population of the same diameter and length. (3) The characteristics of axonal transport are well defined in this system (Koenig, Di Giamberardino, and Bennett, 1973).

The choice of the labeled precursor was guided by the following considerations. Methyl-[^{3}H]choline has been used to study the renewal of phosphatidylcholine in myelin (Hendelman and Bunge, 1969; Gould and Dawson, 1976) and its axonal transport along peripheral nerves (Abe, Haga, and Kurakowa, 1973). This precursor is incorporated into newly formed phosphatidylcholine through net synthesis, but can also be substituted for another base by the freely reversible base-exchange reaction (Arienti, Brunetti, Gaiti, Orlando, and Porcellati, 1976), which also takes place in axons (Brunetti, Giuditta, and Porcellati, 1979). In other words, phosphatidylcholine tagged with radioactive choline can result not only from a net synthesis of phospholipid but also from an exchange of base. To palliate the versatility of this precursor, 2-[^{3}H]-glycerol, which was used by Grafstein, Miller, Ledeen, Haley, and

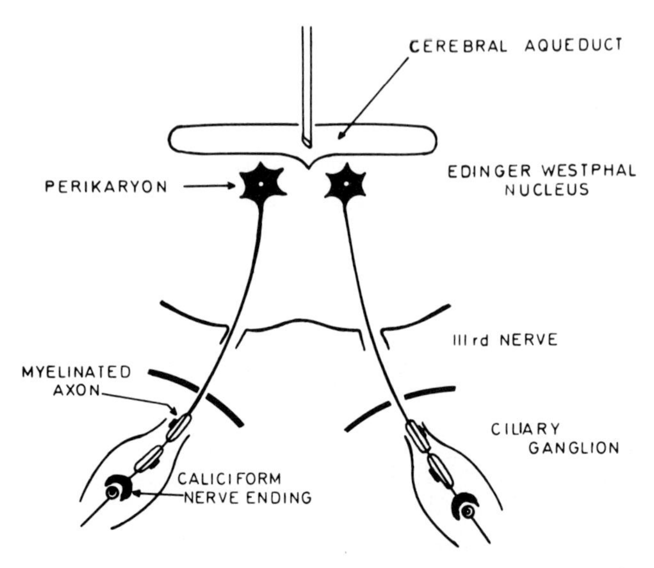

FIGURE 1. Use of the chicken ciliary ganglion for studies of axonal transport of phospholipids. Labeled precursors of phospholipids, e.g., radioactive choline or glycerol, are injected into the cerebral aqueduct. After incorporation, labeled phospholipids, made in perikarya of the preganglionic neurons in the Edinger-Westphal nucleus, move along the myelinated axons. They eventually reach the ciliary ganglia and accumulate in the large caliciform nerve endings. One ganglion is processed for autoradiography, the other for radioactivity counts.

Specht (1975) in their study on the axonal transport of phospholipids, was selected. 2-[^{3}H]Glycerol is indeed incorporated only into newly synthesized phospholipids (Benjamins and McKhann, 1973; Porcellati and Binaglia, 1976); by forming the backbone of labeled glycerophospholipids, 2-[^{3}H]glycerol is in fact neither exchanged nor recycled (Benjamins and McKhann, 1973). Thus [^{3}H]glycerol-labeled phospholipids may be considered excellent markers to study the transfer of intact phospholipid molecules from one cell to another.

In a first series of experiments, methyl-[^{3}H]choline or 2-[^{3}H]glycerol was injected into chicken cerebral aqueduct (Figure 1). Radioactive phospholipids, made in the cell bodies of the preganglionic nerves (Edinger-Westphal nucleus), were transported down their axons to the ciliary ganglia. One ganglion was processed to locate labeled phospholipids by autoradiography, and the other was homogenized to measure the radioactivity contained in the trichloracetic acid-soluble fraction and the chloroform-methanol lipid extract (Droz et al., 1978). In a second series of experiments, a double labeling analysis was undertaken by injecting methyl-[^{14}C]choline and 2-[^{3}H]glycerol simultaneously into the cerebral aqueduct to identify the phospholipid classes found to be labeled in the ciliary ganglion, to study the kinetics of their renewal, and to follow the evolution of the isotopic ratio $^{14}C/^{3}H$ in phosphatidylcholine (Table 1) (Brunetti, Droz, Di Giamberardino, and Porcellati, 1978).

AXONAL TRANSPORT OF LABELED PHOSPHOLIPIDS

When radioactive choline and glycerol were injected into the cerebral aqueduct of chickens, the radioactivity recovered in phospholipids of ciliary ganglia rose after a 1-hour delay (Figures 2 and 3). With both precursors, phosphatidylcholine constituted the major labeled component of phospholipids, accounting for 94% with [^{14}C]choline and 66% with 2-[^{3}H]glycerol at 20 hours after injection.

Light microscope autoradiographs showed that labeled phospholipids first invade the preganglionic axons and accumulate in the caliciform nerve endings (Figure 4) (Droz et al., 1978). The kinetics of labeled phospholipids in axons and nerve endings displayed a striking similarity to that of proteins and glycoproteins conveyed with fast axonal flow in the same material (Koenig et al., 1973). In electron microscope autoradiographs, the silver grains, which signal the presence of labeled phospholipids, were mainly found over or around profiles of the axonal smooth endoplasmic reticulum at early time intervals (Figure 7). When nerves were transected, both labeled phospholipids and tubules of smooth endoplasmic reticulum piled up within 1 hour in the proximal stump (Droz, 1975). These data indicate that phospholipids conveyed with fast axonal transport are preferentially associated with the smooth endoplasmic reticulum; at late time intervals, they are frequently incorporated into axonal mitochondria (Figures 8 and 11).

TABLE 1. *Changes of the isotopic ratio $^{14}C/^{3}H$ in phosphatidylcholine counted in chicken ciliary ganglia after a simultaneous injection of methyl-[^{14}C]choline and 2-[^{3}H]glycerol into the cerebral aqueduct*

Time after injection	1 h	6 h	10 h	20 h	40 h	3 d	10 d
$^{14}C/^{3}H$	0.21	0.44	0.77	1.09	1.23	1.59	3.62

The increase of the isotopic ratio $^{14}C/^{3}H$ indicates that [^{14}C]choline is reutilized for the synthesis of new phosphatidylcholine either by base-exchange or net synthesis in ganglion cells and Schwann cells (Brunetti et al., 1978).

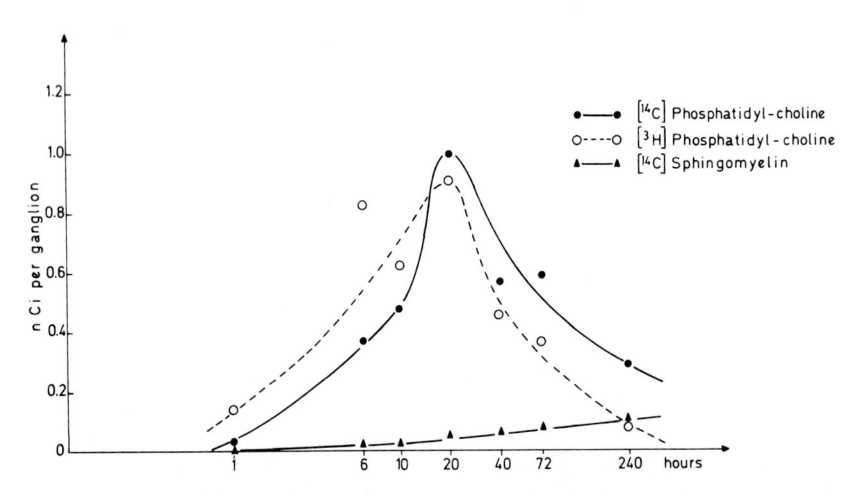

FIGURE 2. Radioactivity time-curves of [^{3}H]glycerol-labeled phosphatidylcholine, [^{14}C]choline-labeled phosphatidylcholine, and [^{14}C]choline-labeled sphingomyelin expressed in nCi per ganglion after a simultaneous injection of methyl-[^{14}C]choline and 2-[^{3}H]glycerol into the cerebral aqueduct (mean of 5 chickens). Note the slow and lasting increase of [^{14}C]choline-labeled sphingomyelin that coincides with the decrease of labeled phosphatidylcholine.

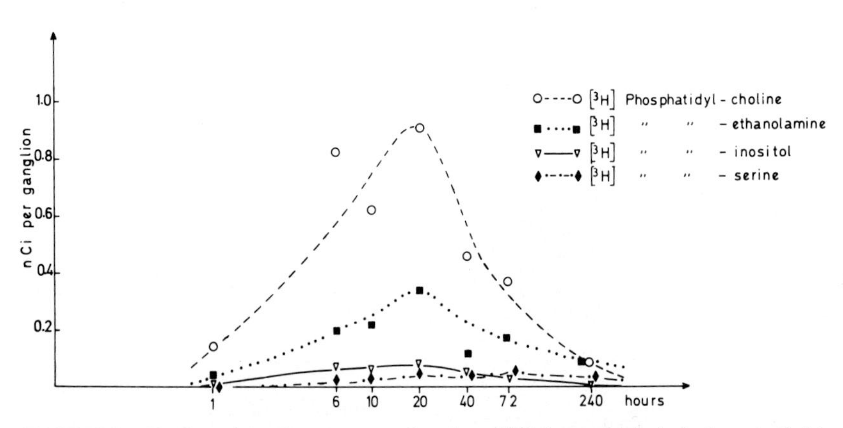

FIGURE 3. Radioactivity time-curves of various [^{3}H]glycerol-labeled phospholipids expressed in nCi per ganglion after a simultaneous injection of methyl-[^{14}C]choline and 2-[^{3}H]glycerol into the cerebral aqueduct (mean of 5 chickens). Most of the [^{3}H]glycerol-labeled phospholipids axonally transported and collected in the ciliary ganglion consist of phosphatidylcholine.

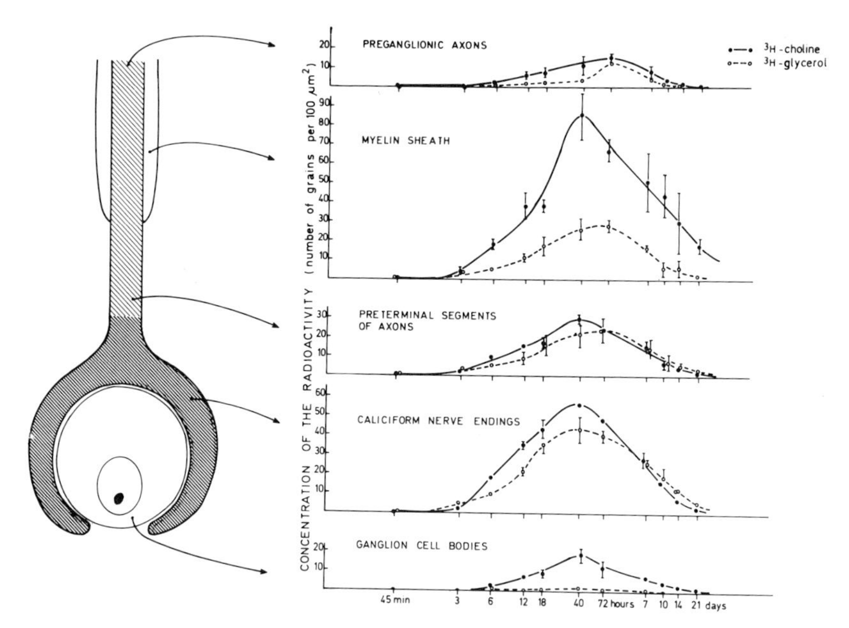

FIGURE 4. Time-curves of the concentration of label measured from light microscope autoradiographs of ciliary ganglion after intracerebral injection of methyl-[^{3}H]choline or 2-[^{3}H]glycerol. After a delay, labeled phospholipids appear in the preganglionic axons. Note the dramatic accumulation of label in myelin sheath at 40 hours after the injection of [^{3}H]choline or [^{3}H]glycerol. The high peak observed in myelin after injection of [^{3}H]choline and the moderate one recorded after the injection of [^{3}H]glycerol contrast with the similar concentration of label counted with both tracers in preterminal segments of axons and in caliciform nerve endings. The increase of radioactivity in the postsynaptic ganglion cells after the injection of [^{3}H]choline, but not of [^{3}H]glycerol, reflects a transsynaptic passage of choline or metabolite rather than intact phospholipids.

Besides the fast axonal flow of phospholipids synthesized in perikarya (Miani, 1963; Abe et al., 1973; Grafstein et al., 1975), intraaxonal incorporation of radioactive precursors of phospholipids could contribute in part to the labeling of axons (Gould, 1976; Brunetti et al., 1979). To determine to what extent a local incorporation occurs in preganglionic axons, ciliary ganglia were incubated in vitro with methyl-[^{3}H]choline or 2-[^{3}H]glycerol. The light labeling of axons and terminals observed with [^{3}H]choline, but not with [^{3}H]glycerol, suggests that such a process is limited. It is not known whether this result is due to limited entry of precursors and whether choline incorporation is due only to labeling by base-exchange.

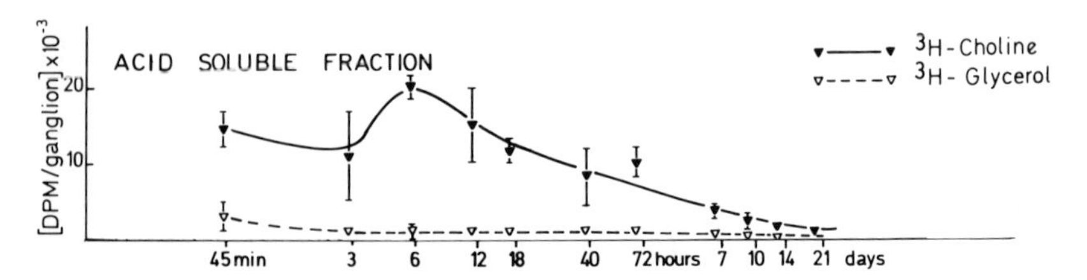

FIGURE 5. Radioactivity time-curves of the acid-soluble fraction of ciliary ganglion after an intracerebral injection of methyl-[^{3}H]choline or 2-[^{3}H]glycerol. The exceedingly low level of water-soluble radioactivity counted after the injection of [^{3}H]-glycerol indicates that local incorporation of the tracer is unlikely. After the injection of [^{3}H]choline, the peak of water-soluble radioactivity at 6 hours and the slow decline lasting for several days suggest a possible local reutilization of the tracer.

APPEARANCE OF LABELED PHOSPHOLIPIDS IN MYELIN

Autoradiographs of preganglionic nerve fibers conveying radioactive phospholipids show a progressive accumulation of label in the myelin sheath (Droz et al., 1978). Radioactive phospholipids were detected in myelin sheaths as early as 3 hours after the intracerebral injection of methyl-[^{3}H]choline or 2-[^{3}H]glycerol (Figure 4). However, the kinetics of the renewal of the label in myelin phospholipids was different if the radioactive atom was borne by choline or glycerol. The rate of accumulation and the total amount of label collected in myelin were much higher after an injection of radioactive choline than of radioactive glycerol (Figure 4). [^{3}H]Choline-labeled phospholipids were estimated to

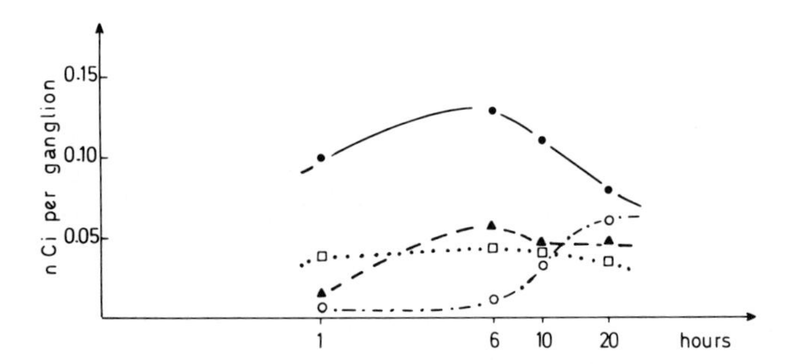

FIGURE 6. Time-curve of radioactive water-soluble components in the ciliary ganglion after the intracerebral injection of methyl-[^{14}C]choline (mean of 5 ganglia). ● —— ●, Choline; ▲ - - - ▲, phosphorylcholine; □ · · · · □, CDP-choline; ○ ·–·–· ○, glycerophosphorylcholine. In the absence of acetylcholinesterase inhibitor, the radioactivity due to acetylcholine could not be estimated.

stay in myelin for a longer period (mean time of about 13 days) than [^{3}H]glycerol-labeled phospholipids (mean time of about 5 days). In addition to these peculiarities, another striking difference of behavior between the two tracers was observed in Schwann cells (Table 2). After the intracerebral injection of [^{3}H]choline, a fair amount of labeled phospholipids was located in the outer Schwann cell cytoplasm (Figures 9 and 10), whereas after the injection of [^{3}H]glycerol the radioactivity counted in the Schwann cell cytoplasm remained at an exceedingly low level (Table 2) (Droz et al., 1978). The different behavior observed with the two tracers suggests that two mechanisms underlie the axon-myelin transfer of label. The appearance of labeled phospholipids in the myelin sheath may indeed result either from the passage of intact phospholipids from the axon to its myelin sheath or from a local incorporation of tracer into myelin phospholipids synthesized by Schwann cells. Both mechanisms are probably responsible for entry of label into peripheral and central nervous system myelin (Haley and Ledeen, 1979).

TRANSFER OF INTACT PHOSPHOLIPIDS FROM AXON TO MYELIN SHEATH

When 2-[^{3}H]glycerol is injected into the cerebral aqueduct, the very low level of acid-soluble radioactivity detected in the ciliary ganglion (Figure 5) and the low number of silver grains counted over Schwann cells (Table 2) and ganglion cells in autoradiographs (Figure 4) indicate that the local incorporation of [^{3}H]glycerol is negligible at any time. Since [^{3}H]glycerol incorporated into axonally transported phospholipids cannot be exchanged (Benjamins and McKhann, 1973), the [^{3}H]-glycerol-labeled phospholipids found in the myelin sheath could only derive from axonally transported phospholipids that have been transferred as intact macromolecules or as lysoderivatives from the neuron into the myelin leaflets.

The question may be raised whether the different classes or subclasses of lipids have the same probability of being transferred to myelin. Phosphatidylcholine, which accounts for 66% of the [^{3}H]glycerol-labeled phospholipids axonally transported to the ciliary ganglion (Figure 3), seems to be preferentially delivered to myelin. Preliminary results obtained by autoradiography after the intracerebral injection of [^{3}H]ethanolamine or [^{3}H]myo-inositol indicate that labeled phosphatidylethanolamine or phosphatidylinositol probably passes in minute amounts from the axon into the myelin. Furthermore, the rapid labeling

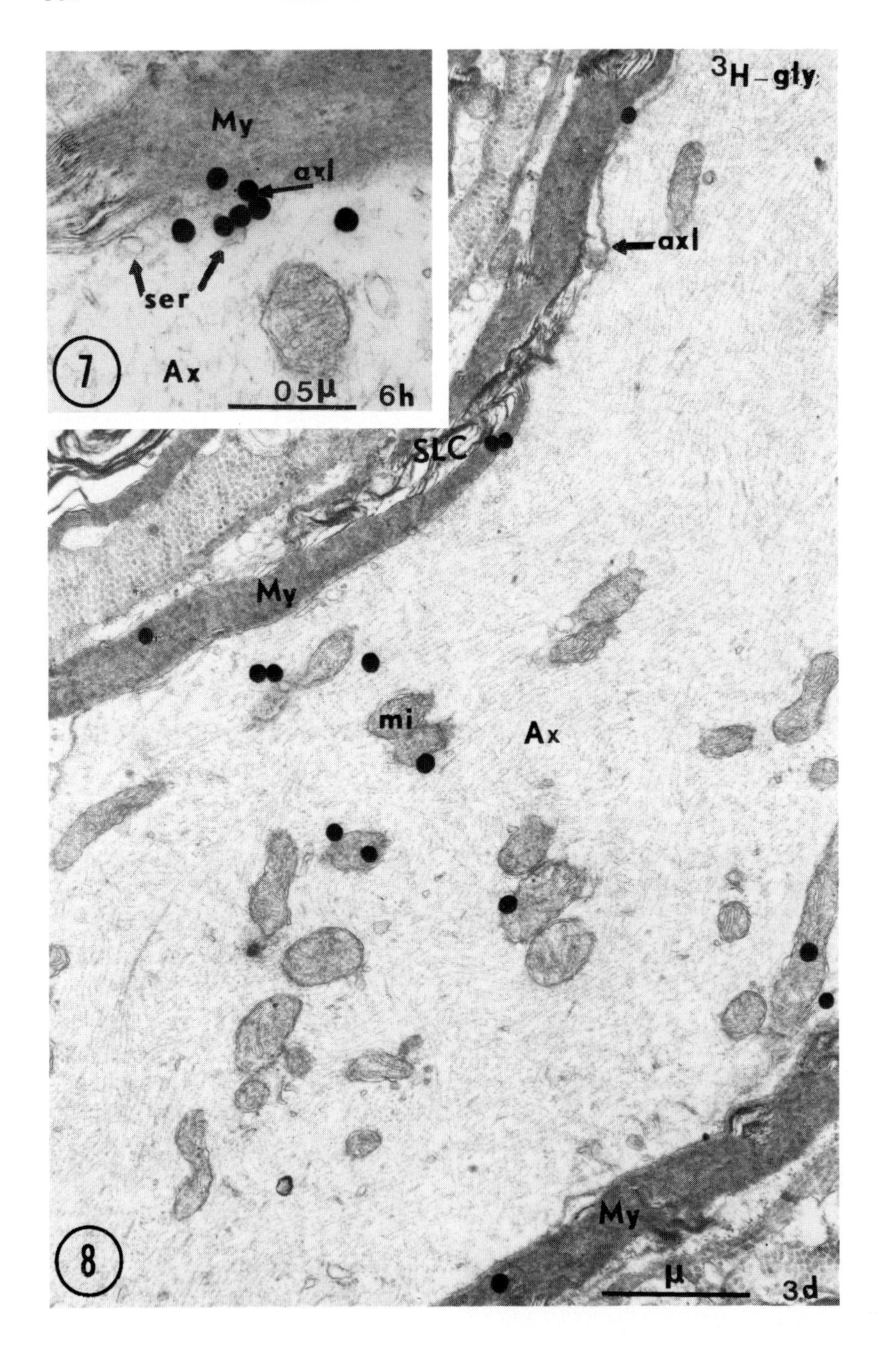
My
axl
ser
7
Ax
0.5μ
6h
³H-gly
axl
SLC
My
mi
Ax
My
8
μ
3d

TABLE 2. *Concentration of label in outer Schwann cell cytoplasm as measured from electron microscope autoradiographs after intracerebral injection of methyl-[^{3}H]choline or 2-[^{3}H]glycerol (expressed as number of silver grains per 100 μm^2)*

Time after injection	6 h	40 h	3 d	10 d
[^{3}H]choline	9.0 ± 2.1	32.5 ± 8.2	18.1 ± 6.8	4.5 ± 2.4
[^{3}H]glycerol	0.8 ± 0.2	1.4 ± 0.5	1.3 ± 0.3	0.3 ± 0.2

observed in the myelin sheath at early time intervals, such as 3 or 6 hours after the intracerebral injection of [^{3}H]choline (Figures 4 and 9), reflects the predominance of the axon-myelin transfer of axonally transported phosphatidylcholine.

AXONAL DELIVERY OF LABELED CHOLINE TO MYELIN PHOSPHOLIPIDS SYNTHESIZED IN SCHWANN CELLS

When radioactive choline was injected intracerebrally, axon-myelin transfer of labeled phosphatidylcholine alone could hardly account for the intense labeling of the Schwann cell cytoplasm and myelin sheath observed at later time intervals (Table 2 and Figure 4). A local incorporation of the tracer into myelin phospholipids made by Schwann cells should indeed contribute in large part to the accumulation of radioactive phospholipids in the myelin sheath, since free labeled choline and metabolites, after reaching a peak at 6 hours, remained at a relatively high level in water-soluble extracts of ciliary ganglia (Figures 5 and 6). Among the possible sources of labeled choline in the ciliary ganglion, the part taken by blood-borne tracer leaking out of the brain

FIGURES 7 and 8. Electron microscope autoradiographs visualizing the transfer of axonally transported [^{3}H]glycerol-labeled phospholipids from axons (Ax) to myelin (My).

Figure 7. At 6 hours after the intracerebral injection of 2-[^{3}H]glycerol, silver grains point to profiles of the smooth endoplasmic reticulum (ser) corresponding to subaxolemmal cisternae loaded with labeled phospholipids, to the adjacent axolemma (axl), and to inner leaflets of the myelin sheath (My).

Figure 8. By 3 days, mitochondria (mi) contain most of the axonally transported phospholipids labeled with 2-[^{3}H]glycerol. Silver grains are present over the myelin sheath; they extend from the inner to the outer layers of myelin, including Schmidt-Lanterman clefts (SLC).

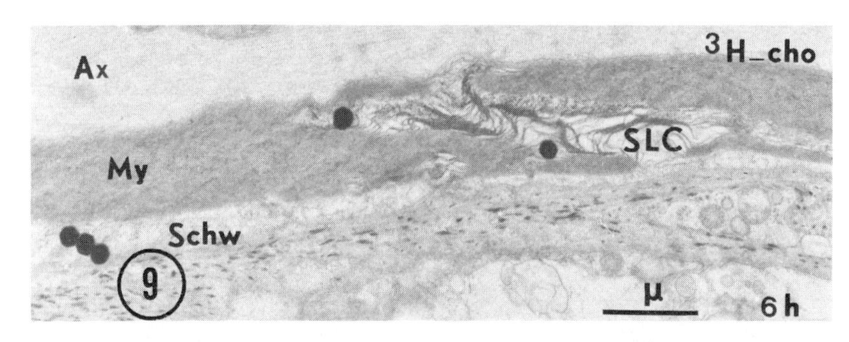

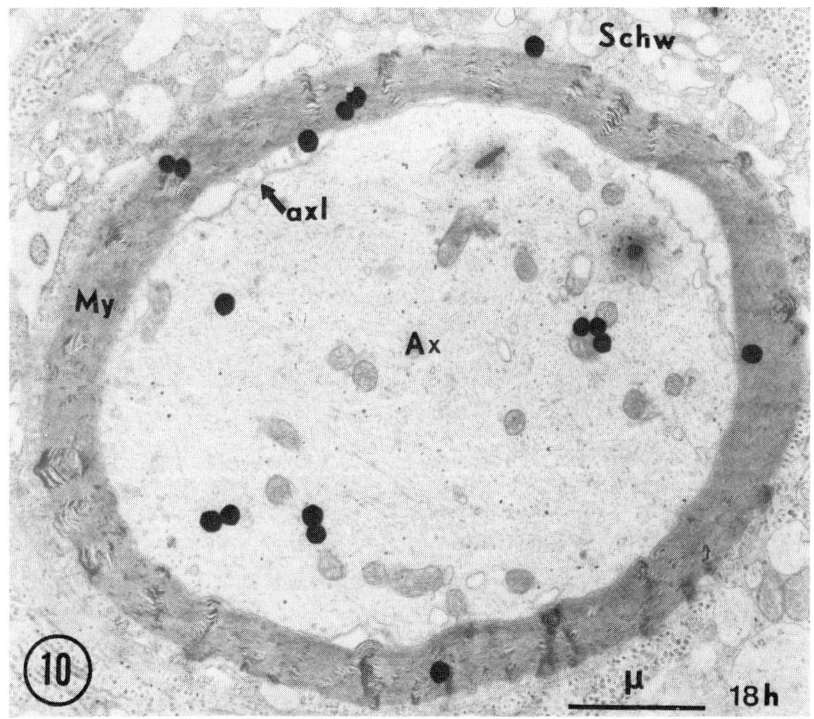

FIGURES 9, 10, and 11. Electron microscope autoradiographs showing the transfer of label to myelin sheath (My) and Schwann cell cytoplasm (Schw) in the course of the axonal transport of [^{3}H]choline phospholipids.

Figure 9. At 6 hours after the intracerebral injection of methyl-[^{3}H]choline, silver grains are seen over the Schmidt-Lanterman clefts (SLC) of the myelin sheath and over the outer Schwann cell cytoplasm.

Figure 10. By 18 hours, silver grains found over the axon (Ax) signal the presence of radioactive phospholipids transported along the axon. Many silver grains are located over the myelin sheath (My); the label is associated with the inner leaflets (as distinct from the axolemma, axl) and with the outermost layers, including the outer Schwann cell cytoplasm.

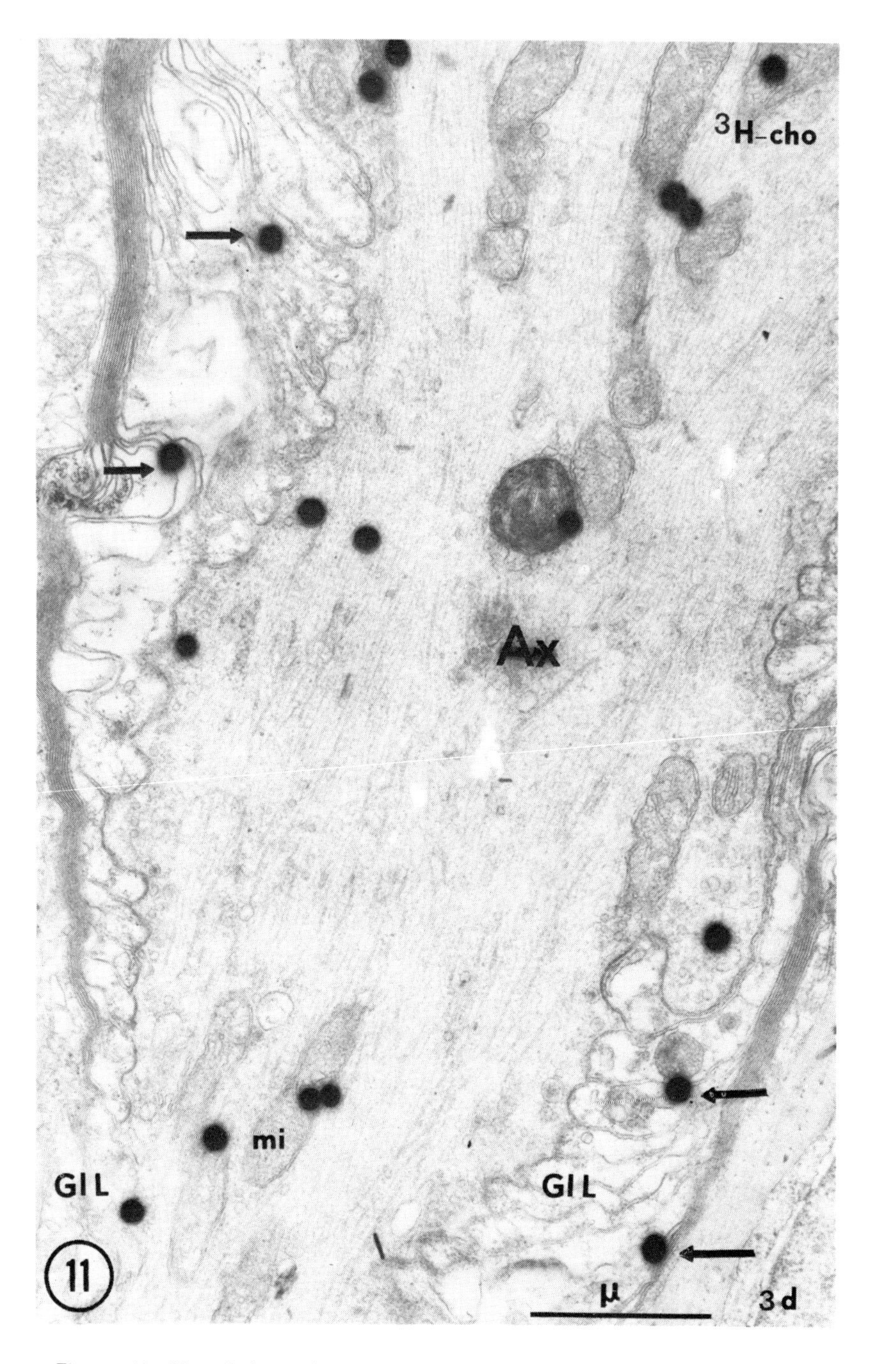

Figure 11. After 3 days, the paranodal regions, adjacent to nodes of Ranvier, display an increased number of silver grains over the myelin leaflets (arrow) giving rise to the glial loops (GIL).

seems to be negligible, because an equivalent amount of [^{3}H]choline administered intraperitoneally failed to accumulate in ganglia; as noted by Dawson and Gould (1976), the penetration of choline from plasma into peripheral nerves seems to be impeded. The rise of radioactivity in water-soluble compounds could be produced by the arrival of free labeled choline or metabolites flowing down the preganglionic nerves toward the ciliary ganglia, but the existence of such a flow remains to be proved. Radioactive choline incorporated into axonally transported phosphatidylcholine can constitute an appreciable store of base that can be freed by a phospholipase D-like activity or by base-exchange reaction, released from the axon and taken up by the adaxonal glial cytoplasm. Whatever the origin of axonal choline, it has been shown that [^{3}H]choline introduced into an axon by iontophoresis is reincorporated into myelin (Alvarez and Chen, 1972). The recycling of radioactive choline in the ciliary ganglion is supported by two facts: first, after a double labeling experiment, the isotopic ratio of methyl-[^{14}C]-choline-labeled versus 2-[^{3}H]glycerol-labeled phosphatidylcholine was not constant (Table 1); its increase with time reflects a large reutilization of the choline moiety (Brunetti et al., 1978). Second, the temporal relationship of ^{14}C-labeled phosphatidylcholine and sphingomyelin (Figure 3) suggests that radioactive phosphatidylcholine could be a donor of choline for most of the sphingomyelin synthesized, as has been proposed for the central nervous system (Jungalwala, 1974).

POSSIBLE PATHWAYS OF AXON-MYELIN TRANSFER OF MOLECULES

Axonally transported phospholipids labeled either with [^{3}H]choline or [^{3}H]glycerol were found to accumulate first in focal regions of the axon occupied frequently by subaxolemmal cisternae (Figure 7); simultaneously, the label started to appear in inner layers of myelin (Figure 7) as well as in Schmidt-Lanterman clefts (Figure 9). The initial labeling of the myelin suggests that phospholipids axonally transported with the smooth endoplasmic reticulum are translocated to the axolemmal surface by means of subaxolemmal cisternae, as has been proposed for membrane glycoproteins by Markov, Rambourg, and Droz (1976). Phospholipids could pass from the axons to the Schwann cells by means of specific transfer proteins (Wirtz, 1974) and integrate the inner myelin membranes. The Schmidt-Lanterman clefts, by offering a breach through the compact myelin (Figures 9 and 12), could facilitate the insertion of transferred phospholipids into outer myelin leaflets (Figures 8 and 10). The Schmidt-Lanterman incisures could also play the role of cytoplasmic channels (Singer and Bryant, 1969; Mugnaini, 1978) along which free choline or metabolites released from the axon could easily diffuse from the inner to the outer cytoplasm of Schwann cells (Figure 12). There, according to the remarkable studies of Gould and Dawson (1976),

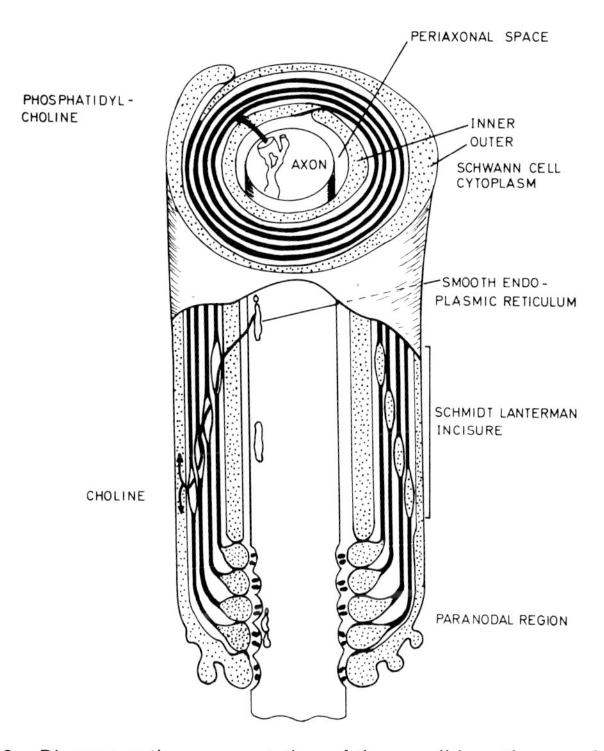

FIGURE 12. Diagrammatic representation of the possible pathways of transfer of label from the axon to the myelin sheath and the outer Schwann cell cytoplasm after an intracerebral injection of labeled precursor of phospholipid.

After the administration of radioactive choline or glycerol in the vicinity of neuronal perikarya, labeled phospholipids consisting mainly of phosphatidylcholine are transported along the axon and transferred as a whole to the adjacent myelin sheath. Electron microscope autoradiographs suggest an accumulation of labeled phospholipids in subaxolemmal cisternae prior to their passage into Schwann cells. After the intercellular transfer, phospholipids could integrate the innermost leaflets of myelin or more external layers by running along the cytoplasmic channels of the Schmidt-Lanterman incisures.

When radioactive choline has been injected, free choline or metabolites released from the axon are taken up by the encompassing Schwann cell. These small molecules probably diffuse along the Schmidt-Lanterman incisures and reach the outer region of the Schwann cell cytoplasm; there they are incorporated into newly formed phosphatidylcholine and sphingomyelin and integrated into myelin leaflets.

[^{3}H]choline could be reincorporated into newly formed phospholipids, which eventually move inwardly along the spirals of myelin. Simultaneously, a longitudinal redistribution of labeled phospholipids was visualized by their progressive accumulation in the paranodal loops (Figure 11), where they are probably stopped by the specialized membrane junctions (Akert, Sandri, Livingston, and Moor, 1974; Mugnaini, 1978).

In conclusion, myelinated axons provide their Schwann cells with phospholipids by delivering axonally transported phosphatidylcholine to the myelin sheath. After accumulation in subaxolemmal cisternae, phosphatidylcholine molecules are transferred from the axolemma to the adaxonal Schwann cell membrane; then they integrate inner or more external myelin leaflets through the Schmidt-Lanterman clefts. In addition to the transfer of intact phosphatidylcholine molecules, choline or other derivatives are released from the axon, taken up by Schwann cells, and reincorporated into newly formed phosphatidylcholine or sphingomyelin. With time, phospholipids integrated within myelin membranes move inwardly along the leaflet spirals and longitudinally toward the paranodal regions. The intercellular movement of phospholipids and choline from the axon to its Schwann cells may be considered a contribution of the neuron to the maintenance of the myelin sheath.

ACKNOWLEDGMENTS

M. Brunetti was a Fellow of the European Training Program in Brain and Behaviour Research. L. Di Giamberardino is Chargé de Recherches, INSERM. The authors express their gratitude to J. Boyenval and R. Hässig, who carried out electron microscope autoradiography and contributed to the illustrations. Funds were provided by the Commissariat à l'Energie Atomique , UER 61 of the Université Pierre et Marie Curie, and the Fondation pour la Recherche Médicale Française.

REFERENCES

Abe, T., T. Haga, and M. Kurakowa (1973). Rapid transport of phosphatidylcholine occurring simultaneously with protein transport in the frog sciatic nerve, *Biochem. J.* **136**:731–740.

Aguayo, A. J., L. Charron, and G. M. Bray (1976). Potential of Schwann cells from unmyelinated nerves to produce myelin: a quantitative ultrastructural and autoradiographic study, *J. Neurocytol.* **5**:565–573.

Akert, K., C. Sandri, R. B. Livingston, and H. Moor (1974). Extracellular spaces and junctional complexes at the node of Ranvier, pp. 9–22 in *Actualités Neurophysiologiques,* Vol. 10, Monnier, A. M., ed. Masson, Paris.

Alvarez, J. and W. Y. Chen (1972). Local incorporation of choline by a myelinated fibre, *Acta Physiol. Lat. Am.* **22:**270–273.

Arienti, G., M. Brunetti, A. Gaiti, P. Orlando, and G. Porcellati (1976). The contribution of net synthesis and base-exchange reaction in phospholipid biosynthesis, *Adv. Exp. Med. Biol.* **72:**63–78.

Autilio-Gambetti, L., P. Gambetti, and B. Shafer (1973). RNA and axonal flow. Biochemical and autoradiographic study in the rabbit optic system, *Brain Res.* **53:**389–398.

Autilio-Gambetti, L., P. Gambetti, and B. Shafer (1975). Glial and neuronal contribution to proteins and glycoproteins recovered in myelin fractions, *Brain Res.* **84:**336–340.

Benjamins, J. A. and G. M. McKhann (1973). 2-[^{3}H]Glycerol as a precursor of phospholipids in rat brain: evidence for lack of recycling, *J. Neurochem.* **20:**1111–1120.

Bennett, G., L. Di Giamberardino, H. L. Koenig, and B. Droz (1973). Axonal migration of protein and glycoprotein to nerve endings. II. Radioautographic analysis of the renewal of glycoproteins in nerve endings of chicken ciliary ganglion after intracerebral injection of [^{3}H]fucose and [^{3}H]glucosamine, *Brain Res.* **60:**129–146.

Brunetti, M., B. Droz, L. Di Giamberardino, and G. Porcellati (1978). A study on the axonal flow of phospholipids in the ciliary ganglion of the chicken, p. 484 in *Proceedings of the European Society for Neurochemistry*, Vol. 1, Neuhoff, V., ed. Verlag Chemie, Weinheim.

Brunetti, M., A. Giuditta, and G. Porcellati (1979). The synthesis of choline phosphoglycerides in the giant fibre system of the squid, *J. Neurochem.* **32:**319–324.

Dawson, R. M. C. and R. M. Gould (1976). Renewal of phospholipids in the myelin sheath, *Adv. Exp. Med. Biol.* **72:**95–113.

Di Giamberardino, L., G. Bennett, H. L. Koenig, and B. Droz (1973). Axonal migration of protein and glycoprotein to nerve endings. III. Cell fraction analysis of chicken ciliary ganglion after intracerebral injection of labeled precursors of proteins and glycoproteins, *Brain Res.* **60:**147–159.

Droz, B. (1975). Synthetic machinery and axoplasmic transport: maintenance of neuronal connectivity, pp. 111–127 in *The Nervous System,* Vol. 1, Tower, D. B., ed. Raven Press, New York.

Droz, B., L. Di Giamberardino, H. L. Koenig, J. Boyenval, and R. Hässig (1978). Axon-myelin transfer of phospholipid components in the course of their axonal transport as visualized by radioautography, *Brain Res.* **155:** 347–353.

Droz, B., H. L. Koenig, and L. Di Giamberardino (1973). Axonal migration of protein and glycoprotein to nerve endings. I. Radioautographic analysis of the renewal of protein in nerve endings of chicken ciliary ganglion after intracerebral injection of [^{3}H]lysine, *Brain Res.* **60:**93–127.

Elam, J. S. (1971). Association of axonally transported proteins with goldfish brain myelin fractions, *J. Neurochem.* **28:**345–354.

Giorgi, P. P., J. O. Karlsson, J. Sjöstrand, and E. J. Field (1973). Axonal flow and myelin protein in the optic pathway, *Nature New Biol.* **244:**121–124.

Gould, R. M. (1976). Inositol lipid synthesis localized in axons and unmyelinated fibers of peripheral nerves, *Brain Res.* **117**:169–174.

Gould, R. M. and R. M. C. Dawson (1976). Incorporation of newly formed lecithin into peripheral nerve myelin, *J. Cell Biol.* **68**:480–496.

Grafstein, B., J. A. Miller, R. W. Ledeen, J. Haley, and S. C. Specht (1975). Axonal transport of phospholipids in goldfish optic system, *Exp. Neurol.* **46**:261–281.

Haley, J. E. and R. W. Ledeen (1979). Incorporation of axonally transported substances into myelin lipids, *J. Neurochem.* **32**:735–742.

Hendelman, W. J. and R. P. Bunge (1969). Radioautographic studies of choline incorporation into peripheral nerve myelin, *J. Cell Biol.* **40**:190–208.

Jungalwala, F. B. (1974). The turnover of myelin phosphatidylcholine and sphingomyelin in the adult rat brain, *Brain Res.* **78**:88–108.

Koenig, H. L., L. Di Giamberardino, and G. Bennett (1973). Renewal of proteins and glycoproteins of synaptic constituents by means of axonal transport, *Brain Res.* **62**:413–417.

Markov, D., A. Rambourg, and B. Droz (1976). Smooth endoplasmic reticulum and fast axonal transport of glycoproteins: an electron microscope radioautographic study of thick sections after heavy metals impregnation, *J. Microsc. Biol. Cell.* **25**:57–60.

Matthieu, J. M., H. deF. Webster, G. H. De Vries, S. Corthay, and B. Koellreutter (1978). Glial versus neuronal origin of myelin proteins and glycoproteins studied by combined intraocular and intracranial labelling, *J. Neurochem.* **31**:93–102.

Miani, N. (1963). Analysis of the somato-axonal movement of phospholipids in the vagus and hypoglossal nerves, *J. Neurochem.* **10**:859–874.

Mugnaini, E. (1978). Fine structure of myelin sheaths, pp. 3–31 in *Proceedings of the European Society for Neurochemistry,* Vol. 1, Neuhoff, V., ed. Verlag Chemie, Weinheim.

Norton, W. T. (1975). Myelin: structure and biochemistry, pp. 467–481 in *The Nervous System,* Vol. 1, Tower, D. B., ed. Raven Press, New York.

Porcellati, G. and L. Binaglia (1976). Metabolism of phosphoglycerides and their molecular species, pp. 75–88 in *Lipids,* Vol. 1, Paoletti, R., G. Porcellati, and G. Jacini, eds. Raven Press, New York.

Schubert, P. and G. W. Kreutzberg (1974). Axonal transport of adenosine and uridine derivatives and transfer to postsynaptic neurons, *Brain Res.* **76**:526–530.

Singer, M. and S. V. Bryant (1969). Movement in the myelin Schwann sheath of the vertebrate axon, *Nature (London)* **221**:1148–1150.

Spencer, P. S. and H. J. Weinberg (1978). Axonal specification of Schwann cell expression and myelination, pp. 389–405 in *Physiology and Pathobiology of Axons,* Waxman, S. G., ed. Raven Press, New York.

Suzuki, K. (1978). Biochemistry of myelin disorders, pp. 337–347 in *Physiology and Pathobiology of Axons,* Waxman, S. G., ed. Raven Press, New York.

Weinberg, H. J. and P. S. Spencer (1976). Studies on the control of myelinogenesis. II. Evidence for neuronal regulation of myelination, *Brain Res.* **113**:363–378.

Wirtz, K. W. A. (1974). Transfer of phospholipids between membranes, *Biochim. Biophys. Acta* **344**:95–117.

AXON-SHEATH CELL INTERACTIONS IN PERIPHERAL AND CENTRAL NERVOUS SYSTEM TRANSPLANTS

Albert J. Aguayo, Garth M. Bray, C. Suzanne Perkins, and Ian D. Duncan

The Montreal General Hospital and McGill University, Montreal, Canada

Orderly function in neural tissues, both simple and complex, depends on interactions of neurons with each other (see Black, this volume), with their sheath cells (Varon and R. P. Bunge, 1978), with the physicochemical milieu of their immediate environment, and with the peripheral fields they innervate (Smith and Kreutzberg, 1976). The connective tissue that surrounds peripheral nerves is important for the mechanical support of nerve fibers as well as the regulation of the intraneural environment (Krnjevic, 1954; Olsson and Reese, 1971; Low, Marchand, Know, and Dyck, 1977). In addition, recent in vitro experiments have indicated that collagen and other elements of the extracellular matrix of nerves may have an inductive effect on the ensheathment of axons by Schwann cells (Bunge and Bunge, 1978).

The study of cell interactions in the peripheral and central nervous systems has been aided by the availability of new in vitro and in vivo techniques. In tissue culture preparations, it has been possible to separate and recombine neurons and sheath cells and to investigate mechanisms controlling myelin production and Schwann cell proliferation (Wood and R. P. Bunge, 1975; McCarthy and Partlow, 1976; Bunge, Bunge, Okada, and Wood, 1978). In vivo combinations of axons and Schwann cells, each originating from different animals or from animal and human nerves, have been used to investigate specific aspects of Schwann cell function (Aguayo, Epps, Charron, and Bray, 1976*a*;

Aguayo, Charron, and Bray, 1976*b*) and the pathogenesis of certain hereditary neuropathies in man and experimental animals (Aguayo, Attiwell, Trecarten, Perkins, and Bray, 1977*a*; Aguayo, Kasarjian, Skamene, Kongshavn, and Bray, 1977*b*; Aguayo, Perkins, Duncan, and Bray, 1978*a*). It is also feasible to combine in vitro and in vivo techniques by transplanting cultured Schwann cells into peripheral nerves (Aguayo, R. P. Bunge, Duncan, Wood, and Bray, 1979*a*). Finally, peripheral and central nervous tissue transplants have been used to investigate regeneration and remyelination in the central nervous system (Blakemore, 1976; Kao, Chang, and Bloodworth, 1977; Aguayo, Dickson, Trecarten, Attiwell, Bray, and Richardson, 1978*b*; Katz and Lasek, 1978; Rosenstein and Brightman, 1978; Stenevi and Bjorklund, 1978; Richardson, McGuinness, and Aguayo, 1979; Weinberg and Spencer, 1979).

This review will describe the results of our experience using experimental nerve grafts to investigate axon-sheath cell interactions, particularly between axons and Schwann cells in the peripheral nervous system, but also between axons and glia in grafts using central nervous system tissues.

METHODS OF EXPERIMENTAL NERVE TRANSPLANTATION

This experimental approach is based on the observation that, when a segment of one nerve is grafted between the cut ends of another, Schwann cells in the donor segment survive, multiply, and eventually ensheath axons that regenerate from the proximal stump of the recipient nerve (Figure 1). By selecting host and donor nerves, it is possible to study interactions between axons and Schwann cells in a variety of combinations: (1) different nerves in the same species of animal, (2) the same nerve in different animal species, and (3) animal and human nerves.

In the experimental grafts described in this report, a segment of donor nerve measuring approximately 5 mm in length was placed between the stumps of a transected sciatic nerve in a recipient animal. Care was taken to match the caliber of host and grafted nerves and to avoid misalignment and traction. Under a dissecting microscope, the grafts were secured to the stumps of the recipient nerve with 10/0 nylon sutures. Nerves were allowed to regenerate for variable periods of time. In the case of allotransplants, which involve genetically dissimilar animals of the same species, segments of sciatic nerves were obtained

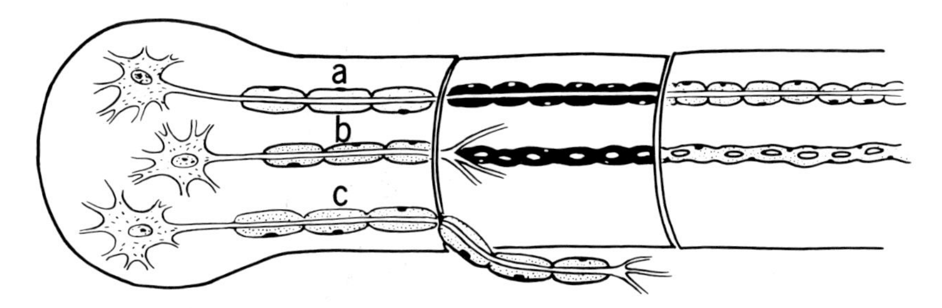

FIGURE 1. Schematic representation of axon-Schwann cell relationships during the degenerative and regenerative phases that follow nerve grafting. (*a*): In early stages the axons of fibers in the graft and distal (host) segments undergo Wallerian degeneration, leaving columns of "denervated" Schwann cells. (*b*): Axons regenerating from the proximal stump are ensheathed by two populations of Schwann cells: those in the graft (black) originate from the donor nerve, while those in the distal stump (white) are host Schwann cells. (*c*): Aberrant axons that grow outside the perineurium are ensheathed and myelinated by Schwann cells migrating from the proximal stump. (From Aguayo, A. J., G. M. Bray, and C. S. Perkins [1979*b*]. Axon-Schwann cell relationships in neuropathies of mutant mice, *Ann. N.Y. Acad. Sci.* **317**:512–533. With permission of the New York Academy of Sciences.)

from mice of different strains. In xenogenic transplants of human nerves, single fascicles of sural nerves obtained from diagnostic biopsies or limb amputations were grafted into adult mice.

Schwann cells grown in tissue culture preparations (rather than those derived from whole nerve grafts) were also used for transplantation studies. For this purpose, Holtzman rat Schwann cells, cultured and separated from neurites (Wood, 1976), were transplanted into a reservoir created by suturing a 5-mm segment of Holtzman rat blood vessel between the stumps of a mouse sciatic nerve (Aguayo et al., 1979*a*). Finally, to study the possible interactions between peripheral nerve axons and CNS glia, optic nerves from mice were grafted between the stumps of a transected sciatic nerve or one of its branches in recipient mice (Aguayo et al., 1978*b*; Weinberg and Spencer, 1979). For these experiments, optic nerve segments approximately 3 mm in length were obtained by sectioning the donor optic nerve between the retina and the optic foramen. Thus, these grafts contained glia but no neurons.

To prevent rejection of the allo- or xenografts in all these experiments, antilymphocytic serum (ALS) was administered subcutaneously to host animals in doses of 0.5 ml twice weekly beginning on the day of transplantation. The ALS was prepared by injecting rabbits with a suspension of 10^9 living thymocytes (Levey and Medawar, 1966). Thymus-deficient mouse mutants known as "nude" (Wortis, 1971) have also been used to study human nerve xenografts (Dyck, Lais, and Low, 1978; Dyck, Lais, Sparks, Oviatt, Hexum, and Low, 1979).

Identification of the Origin of Schwann Cells in Regenerated Grafted Nerves

Because Schwann cells from normal host and donor mammalian nerves are morphologically identical, it has been necessary to prove that, in fully regenerated grafted nerves, the Schwann cells within the regenerated graft segment originate from the donor nerve and not by migration from the stumps of the recipient nerve. Proof of the origin of these cells within the graft was obtained by autoradiographic and immunologic methods as well as by the use of cytopathologic and other cell markers.

Autoradiographic Methods

In one group of experiments, recipient nerves were transected and labeled by injecting host animals with tritiated thymidine at a time when intense Schwann cell proliferation is known to occur (Bradley and Asbury, 1970). Subsequently, unlabeled nerve segments were grafted into these nerves to determine if there was migration from host to graft. In another group of animals, only the donor nerve was labeled, in order to establish if transplanted cells persisted in the graft after regeneration. Experiments in the first group of animals showed that the regeneration of axons across the grafted segment was not accompanied by an increase of labeled cells in the graft; conversely, in the second group of experiments, large numbers of labeled cells remained in the grafted segment, confirming that Schwann cells indigenous to the donor nerve, rather than cells that had migrated from the host, were the supporting cells for regenerating axons (Aguayo et al., 1976*a*,*b*).

Because tritiated thymidine labeling of the cell nucleus may persist after cell division, it was necessary to determine if Schwann cells in the regenerated graft were the transplanted cells themselves or daughter cells resulting from mitotic division. Repeated daily injections of label given to the host animal established that more than 70% of the cells within a nerve graft divided once or more during the first 10 days after nerve transplantation (Figure 2).

Immunologic Methods

In the immune-suppressed animals with allogenic or xenogenic nerve transplants and in those with Schwann cells from tissue culture, the origin of the cells in the graft was determined by discontinuing the immunosuppression of the host animal at different intervals after transplantation. In addition, 2 weeks after the discontinuation of ALS, host mice received a dose of 10^8 lymphoid cells obtained from the spleens and lymph nodes of syngeneic mice hyperimmunized against tissues of the donor (Steinmuller, 1970). This immune cell transfer was used to enhance allo- or xenograft rejection by the recipient animal. Although discontinuation of ALS is followed by xenograft rejection, the additional transfer of immune cells permits a more precise timing of the rejection response and is therefore advantageous for the sequential study of the cellular events associated with regenerated grafts (see below). The selective rejection of most of the Schwann cells and fibro-

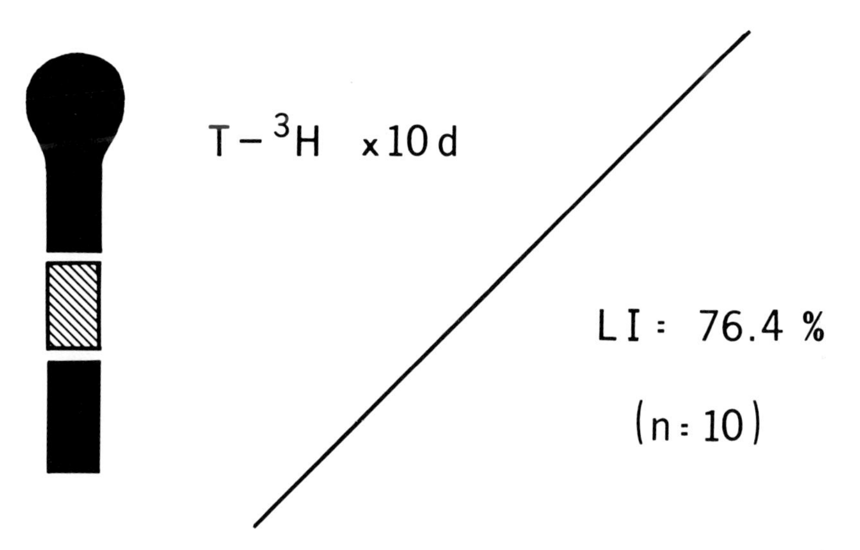

FIGURE 2. In 10 adult mice, a 5-mm long segment was severed from the sciatic nerve but left in its original place. These animals were injected with tritiated thymidine [^{3}H-T] in a dose of 4 μCi/g body weight daily for 10 days and then sacrificed. Labeling indices (LI) in these nerves averaged 76.4%.

blasts within the grafted segments of regenerated nerves and the sparing of the sheath cells in the proximal and distal stumps of these same nerves confirmed that the donor cells had persisted in the graft after regeneration (Aguayo et al., 1977*b*).

Cytopathologic and Other Cell Markers

Because there is a generalized deficiency of myelin in the Schwann cells of certain mouse mutants such as Trembler (Ayers and Anderson, 1976), it is possible to identify these cells after they are transplanted into normal nerves or to use the nerves of these mutants as hosts for Schwann cells originating from other sources. It is anticipated that other biologic markers such as the clearly distinguishable nucleolus of quail cells (Le Douarin, 1973), the sex chromatin (Barr bodies), or specific Schwann cell surface antigens (Brockes, Fields, and Raff, 1977) could also be used to identify the origin of cells in regenerated experimental grafts. Unlike autoradiographic nuclear labels, which are diluted by successive cell division, these genetically determined cell markers have the advantage of persisting after mitosis.

In the regenerated grafted nerves studied by these methods, there is usually a distinct boundary between the Schwann cell population derived from the host proximal and distal stumps and those originating in the graft. In the grafted nerves, where there is little evidence of Schwann cell migration, it appears that interactions between the contiguous populations of host and transplanted Schwann cells may, possibly through contact inhibition, check the progression of cells into or out of the graft. However, migration of Schwann cells, presumably accomplished by repeated cell division during the ensheathment of regenerating axons, must be important for the regrowth of nerves deprived of a distal stump, for bridging of short nerve gaps (Aguayo et al., 1976*a*; Aguayo, Bray and Perkins, 1979*b*), and for the ensheathment of axons that follow an aberrant, extraneural course (Spencer, 1971).

With the documentation of the origin of Schwann cells, it has been possible to demonstrate that axons and Schwann cells from a variety of nerves in widely different species can be associated in regenerated nerve grafts. Thus, Schwann cells from different donors (auto-, iso-, allo-, or xenogenic) and from nerves with different structure (myelinated or unmyelinated) and function (sensory, motor, or autonomic) will ensheath and myelinate regenerating host axons (Aguayo et al., 1976*a, b*, 1978*a*; Weinberg and Spencer, 1975, 1976; Dyck et al., 1978, 1979).

Myelination by normal, allogenic Schwann cells transplanted into immune-suppressed mice has been found to be similar to that produced by host Schwann cells. However, when the thickness of the myelin sheath produced by transplanted mouse and human Schwann cells was compared, myelination of mouse axons by allogenic mouse Schwann cells approached normal values by 2 months after transplantation, but equivalent myelination of the mouse axons by human Schwann cells was only accomplished by 6 months after grafting (Aguayo et al., 1978*a*). These differences in myelination of mouse axons by human and mouse Schwann cells may be due to peculiar axon-Schwann cell interactions in xenografted nerves, true species-related differences in the speed of myelination, different experimental conditions in the graft, or the age of the donor nerves (Schroder, Bohl, and Brodda, 1978).

IMMUNOLOGIC ASPECTS OF SCHWANN CELL TRANSPLANTATION

Methods used to graft and reject allogenic and xenogenic cells have helped to document morphologic features of cell-mediated immune responses directed against Schwann cells and fibroblasts. The earliest signs of rejection of allo- and xenografts in regenerated grafted nerves were observed approximately 8 days after immune cell transfer (nearly 3 weeks after stopping ALS). They consisted of a mononuclear cell infiltration of the graft that tended to involve Schwann cells and fibroblasts but spared most axons (Aguayo et al., 1977*b*). Schwann cells of myelinated and unmyelinated fibers were surrounded, penetrated, and eventually replaced by mononuclear cells. In later stages of rejection, it was observed that many axons had been deprived of their sheath cells and were surrounded only by a basal lamina. The basal lamina of Schwann cells, presumably because of weak antigenicity, generally escaped the immune reaction. Schwann cells and fibroblasts in the proximal and distal stumps of the grafted nerves were not rejected. Damage to axons within the grafts varied from animal to animal and could represent a secondary, bystander effect of the immune response (King, Thomas, and Pollard, 1977; Madrid and Wisniewski, 1977). A similar rejection response is seen in nerve grafts containing Schwann cells transplanted from tissue culture (Aguayo et al., 1979*a*).

Thus, in these experiments, the immune response was directed against Schwann cells rather than myelin, as occurs in experimental allergic neuropathy (Arnason, 1975). This type of immune response may hence be related to histocompatibility antigens on the surface of the sheath cells, because such antigens appear not to be present in myelin (Lennon, Lesley, and Thompson, 1978), even though it is derived from the sheath cell surface membrane.

In animals that had recovered their immunologic competence and rejected allo- or xenografts, the host axons were rapidly re-ensheathed by host Schwann cells that must have migrated from the recipient nerve stumps. This Schwann cell migration may have been facilitated and guided by the axons and basal lamina that survived graft rejection. The removal of the transplanted cells by the immune response presumably reactivated Schwann cell division and allowed these cells to migrate into the graft.

AXON-SCHWANN CELL INTERACTIONS IN EXPERIMENTAL NERVE GRAFTS

The development, regeneration, and function of normal peripheral nerves depend on interactions between axons and Schwann cells (Varon and R. P. Bunge, 1978). These interactions also influence the morphological and functional manifestations of peripheral nerve disease. Because experimental nerve grafts permit new combinations of different nerves, they have been used to investigate interactions of axons and Schwann cells from normal nerves as well as nerves from patients and animals with hereditary neuropathies. The results of these investigations are summarized in the following paragraphs.

Differentiation of Normal Schwann Cells

During development, Schwann cells proliferate intensely (Asbury, 1967; Terry, Bray, and Aguayo, 1974), become associated with axons, and differentiate in one of two directions: in unmyelinated fibers, individual Schwann cells encompass several axons, while in myelinated fibers, each Schwann cell ensheaths only a single axon. In both instances this differentiation is accomplished by a radial and longitudinal extension of the Schwann cell cytoplasm and plasma membrane, although this growth is quantitatively more extensive and qualitatively more complex in the myelin-forming Schwann cells. By using nerve transplantation techniques (Aguayo et al., 1976*a*,*b*) and cross-anastomosis between myelinated and unmyelinated nerves (Weinberg and Spencer, 1975, 1976; Aguayo et al., 1976*a*), it has been demonstrated that each Schwann cell is potentially able to form myelin (Webster,

1971) or the complex spatial arrangement observed in Remak fibers (Aguayo, Bray, Terry, and Sweezey, 1976*c*). These findings also suggest that Schwann cell differentiation is modulated by the axons they ensheath (for full discussion of this subject see Spencer, this volume).

Myelination in Hereditary Neuropathies

Axons and Schwann cells from mutant mice with disorders of myelination in their peripheral nerves have been combined in experimental nerve grafts to determine if axons, Schwann cells, or general systemic factors are responsible for the neuropathy. Trembler mice (Ayers and Anderson, 1973; Low, 1977) have a dominantly inherited neuropathy with severe abnormalities in peripheral nerves (Figure 3). All axons are surrounded by Schwann cells, but myelin sheaths are either abnormally thin, poorly compacted, or totally absent. Schwann cells are frequently surrounded by multiple layers of basal lamina and occasionally by elongated, concentrically arranged Schwann cell processes resembling the "onion bulbs" observed in human hypertrophic neuropathy (Dyck,

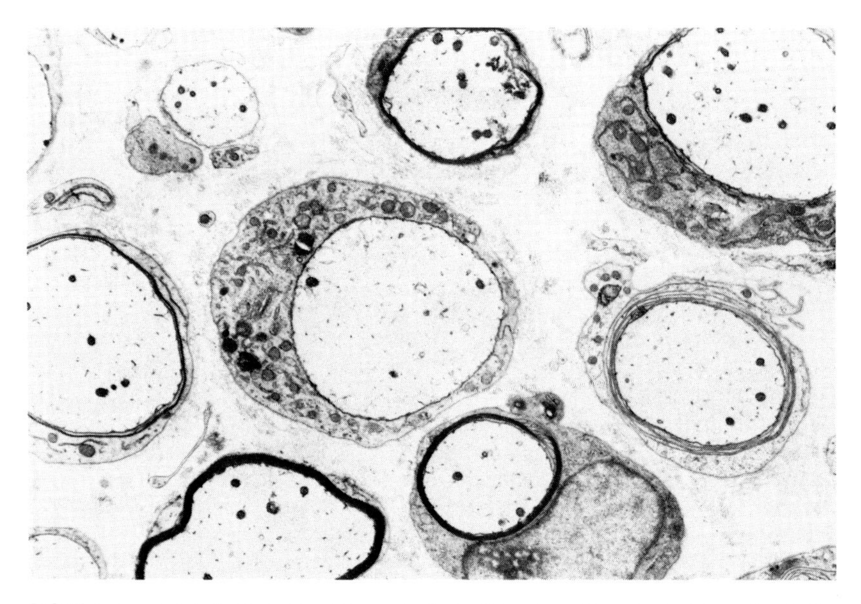

FIGURE 3. In adult Trembler mouse peripheral nerves there is a generalized disorder of myelination. Most fibers lack myelin, while others are hypomyelinated (ventral root). Electron micrograph, ×6,900.

1975). When segments of Trembler nerves were grafted into the sciatic nerves of normal mice, regenerated fibers within the graft were deficient in myelin, but in regenerated grafts of normal Schwann cells in Trembler nerves, the abnormality was corrected (Figure 4) (Aguayo et al., 1977*a*). The electrophysiologic consequences of transplanting normal Schwann cells into Trembler nerves have also been demonstrated. In intact Trembler nerves, conduction velocities are reduced to less than 10 m/sec, while in controls they are approximately 50 m/sec (Low and McLeod, 1975). When segments of normal nerve were grafted into Trembler

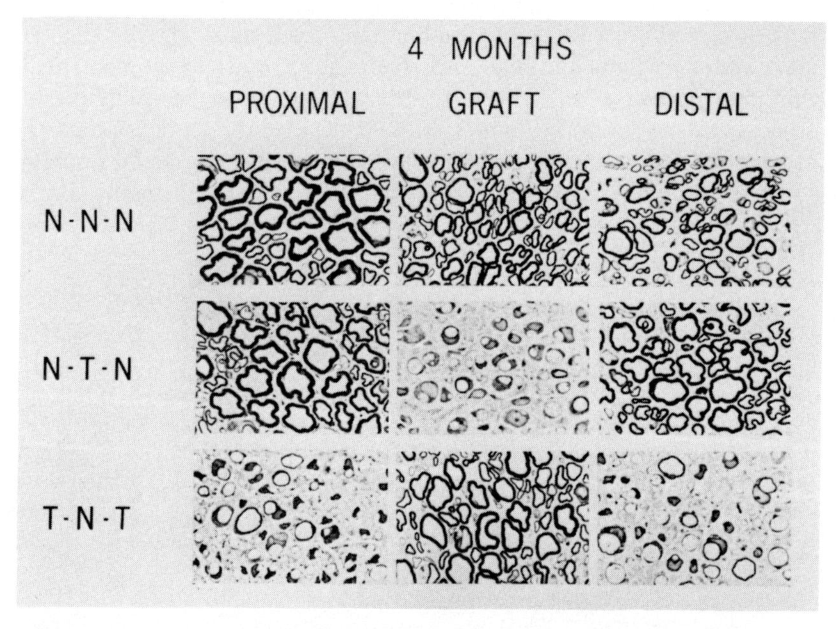

FIGURE 4. Proximal stumps, grafted segments, and distal stumps of regenerated nerves 4 months after grafting. In normal-normal combinations (N-N-N), the graft and distal stumps contain many regenerated myelinated fibers that resemble those in the proximal stump. When Trembler nerves are grafted into normal nerves (N-T-N), the proximal and distal stumps appear normal, but fibers in the graft lack myelin or are hypomyelinated. Normal nerves grafted into Trembler nerves (T-N-T) show deficient myelination in the proximal and distal stumps, whereas the grafted portion of the same nerve is normally myelinated. Phase micrographs, ×733. (From Aguayo, A. J., G. M. Bray, and C. S. Perkins [1979*b*]. Axon-Schwann cell relationships in neuropathies of mutant mice, *Ann. N.Y. Acad. Sci.* **317**:512–533. With permission of the New York Academy of Sciences.)

sciatic nerves, however, the conduction velocity across the grafted segment increased to 25–36 m/sec after 6 months (Pollard, 1978).

Mice homozygous for the recessive quaking gene show hypomyelination in both the central and peripheral nervous systems (Sidman, Dickie, and Appel, 1964; Samorajski, Friede, and Reimer, 1970). In grafts of quaking and normal nerves arranged in combinations similar to those used for Trembler nerves, it was demonstrated that grafted quaking Schwann cells failed to produce sufficient myelin when ensheathing normal axons. However, grafts originating from normal nerves resulted in normal myelination of the segments grafted into quaking nerves (Aguayo et al., 1979*b*).

These findings indicate that the manifestations of the genes responsible for the Trembler and quaking neuropathies are fully expressed by the transplanted mutant Schwann cells and their daughter cells. Furthermore, the abnormalities that characterize these neuropathies are corrected by transplants of normal Schwann cells. Taken together, these findings indicate that Schwann cells rather than axons or systemic factors are responsible for the disorder of myelination in these mutants.

Myelin deficits are also present in peripheral nerves in a dominantly inherited human disease, the hypertrophic neuropathy of Charcot-Marie-Tooth (CMT) (Dyck, 1975). In this disease, it is believed that recurrent demyelination and remyelination cause proliferation of Schwann cell processes (onion bulb formation) and hypertrophy of nerves. When fascicles from a sural nerve biopsy from a patient with this disease were grafted into immune-suppressed mice (Aguayo et al., 1978*a*), the CMT Schwann cells failed to myelinate host axons as effectively as Schwann cells from control human nerves, and the appearance of the regenerated graft resembled that of the original biopsy (Figure 5). In addition, by injecting host animals with tritiated thymidine before sacrifice, it was shown that Schwann cell labeling indices in the CMT graft were abnormally high 6 months after transplantation (Perkins, Aguayo, and Bray, 1978*a*). This application of the nerve graft technique demonstrates that principles applicable to nerve disorders in experimental animals can also be used to study biologic properties of Schwann cells in human neuropathies. However, although the CMT nerve graft studies indicate that Schwann cell abnormalities play a major role in this disease, additional axonal or systemic influences cannot be excluded as definitely as in the neuropathies of mutant mice, where normal Schwann cells can be grafted into nerves of the affected animals.

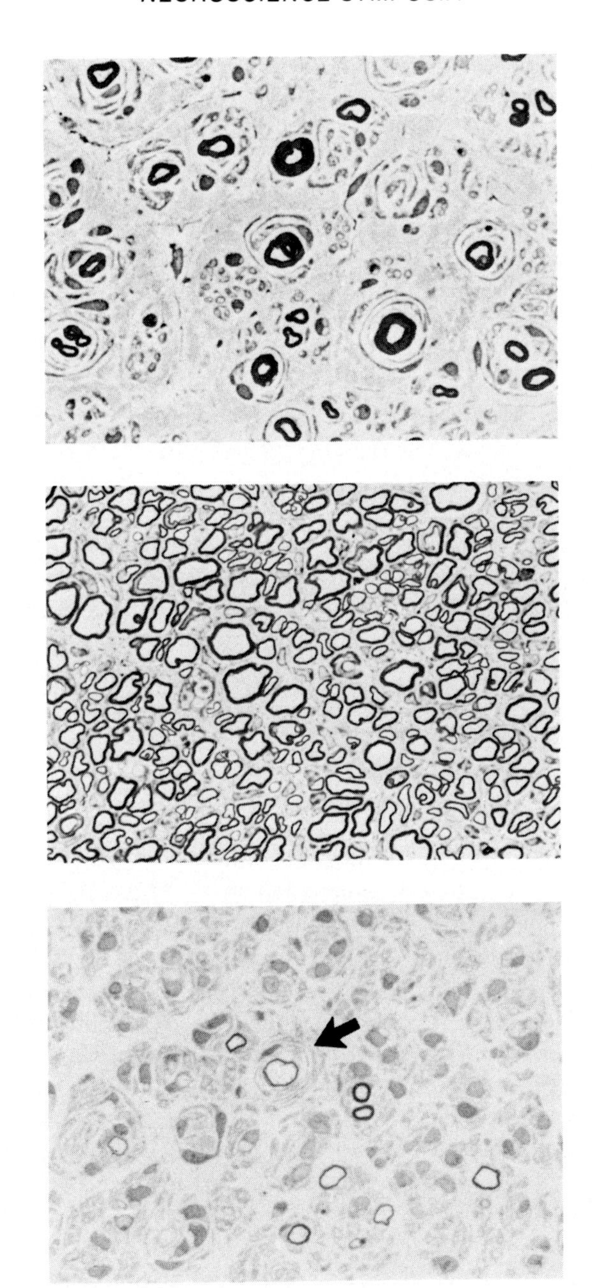

Schwann Cell Influences on Axons

Schwann cells may supply nutrients and trophic substances to axons (Lasek, Gainer, and Barker, 1977; Varon and Bunge, 1978). Schwann cells also influence axonal caliber. Axon diameters are decreased in demyelinating neuropathies (Lubinska, 1958; Cragg and Thomas, 1964; Raine, Wisniewski, and Prineas, 1969; Raine, Wisniewski, Dowling, and Cook, 1971; Prineas and McLeod, 1976), in Trembler mice (Low, 1976), and in the naked axons seen in murine muscular dystrophy (Bradley and Jenkison, 1973). Axon size increases toward normal with the remyelination of demyelinated axons (Raine et al., 1969, 1971) and with the ensheathment of naked axons in dystrophic mice (Aguayo et al., 1979*b*). When segments of normal nerves were transplanted into Trembler nerves, axon diameters became normal in the regenerated grafts but remained less than normal in the proximal and distal recipient nerves. Conversely, in regenerated grafts of Trembler Schwann cells in normal nerves, axon diameters were less than normal in the grafted segment, although diameters of the same axons in the proximal and distal recipient nerves remained normal (Aguayo et al., 1979*b*). These observations raise the possibility that full maturation of axonal size depends not only on factors arising from the cell body and the periphery (Aitken, 1949; Aguayo, Peyronnard, and Bray, 1973), but also on the presence of normal Schwann cells and myelin sheaths. Although the mechanism for this local effect is unknown, it could be due to a segmental reconstitution of a barrier to fluid loss from the axon (Prineas and McLeod, 1976) or to a trophic effect on axons by Schwann cells (Varon and Bunge, 1978).

Stimuli for Schwann Cell Multiplication

Schwann cells multiply intensely during neonatal development (Asbury, 1967; Terry et al., 1974). Although their multiplication vir-

---

FIGURE 5. *Top*: Sural nerve biopsy from patient with Charcot-Marie-Tooth (CMT) disease demonstrates myelin deficit and "onion bulbs." *Middle*: A cross section of the grafted nerve segment from a control nerve 6 months after transplantation into a mouse sciatic nerve shows many myelinated fibers. *Bottom*: The graft originating from the CMT nerve contains only a few myelinated fibers. Concentrically arranged Schwann cell processes are seen around one of these fibers (arrow). Phase micrograph, ×720. (From Aguayo, A. J., S. Perkins, I. Duncan, and G. Bray [1978*a*]. Human and animal neuropathies studied in experimental nerve transplants, pp. 37–48 in *Peripheral Neuropathies,* Canal, N. and G. Pozza, eds. With permission of Elsevier/North Holland Biomedical Press.)

tually ceases in normal adult animals, Schwann cells remain able to proliferate in response to injury (Bradley and Asbury, 1970; Romine, Bray, and Aguayo, 1976). In Trembler nerves, Schwann cell numbers increase with age, and autoradiographic studies have demonstrated that Schwann cell multiplication persists in adult animals (Perkins, Mizuno, Aguayo, and Bray, 1977). Persistent Schwann cell multiplication is also present in regenerated grafts of these abnormal nerves but not in the stumps of the normal recipients (Perkins, Bray, and Aguayo, 1978*b*; Aguayo et al., 1979*b*).

Axons have an important effect on the initiation of Schwann cell multiplication in vitro (Wood and Bunge, 1975; McCarthy and Partlow, 1976). If the putative axonal mitogenic signal is exposed in the inadequately ensheathed Trembler nerves, the persistent multiplication of Schwann cells in these nerves could also be axonally triggered. For this effect, however, it is not necessary to postulate an abnormality of the Trembler axons themselves, because Trembler Schwann cells continue to divide even when they ensheath normal axons (Perkins et al., 1978*b*). An alternative explanation is that active demyelination with the release of myelin breakdown products may be an important stimulus for Schwann cell multiplication in these nerves (Hall and Gregson, 1975).

The increased number of Schwann cells in Trembler animals is confined to nerves that, in normal animals, are myelinated. In the cervical sympathetic trunk (CST), which in normal mice is composed almost entirely of unmyelinated Remak fibers (Romine et al., 1976), the structure of individual Schwann cells and total population of Schwann cell nuclei are similar in mutants and controls (Aguayo et al., 1979*b*). Thus, there must not be a generalized and uncontrolled disorder of Schwann cell proliferation in Trembler mice. Additional information about Schwann cell behavior in Trembler mice was obtained by grafting their unmyelinated CSTs into normal sural nerves that contained many myelinated fibers. When this experiment was done with normal animals, Schwann cells in the CST graft were induced to form myelin (Aguayo et al., 1976*b*). Although Schwann cells from the Trembler CST ensheathed individual axons regenerating from the normal sural nerves, they failed to produce myelin normally, and the regenerating graft resembled a Trembler nerve (Perkins, Aguayo, and Bray, unpublished observations). It would appear that although the abnormal gene is present in all Trembler Schwann cells, the abnormal phenotype is only expressed by those Trembler Schwann cells required to differentiate into myelin-forming cells. The basic abnormality of the Trembler mutant may therefore be regarded as an impairment of the radial and longitudinal extension of Schwann cell plasma membranes necessary to form myelin. The morphology of unmyelinated Remak fibers is normal in Trembler nerves, presumably because the plasma membrane extensions of Schwann cells required for the ensheathment of unmyelinated axons is less than that needed for myelination.

TRANSPLANTATION OF CNS TISSUE INTO PERIPHERAL NERVES

In adult mice and rats, it is possible to investigate interactions between regenerating peripheral axons and transplanted CNS tissue by grafting short segments of optic nerves into the main trunk of the sciatic nerve or one of its branches (Aguayo et al., 1978*b*; Weinberg and Spencer, 1979). Optic nerves (ON) were selected for these experiments because: (1) the strong connective tissue sheath that surrounds them permits the secure attachment of the graft to the host nerve and directs the entry of regenerating axons to the ends of the graft; (2) the cylindrical shape of the optic nerve facilitates the study of the penetration and progression of axons through the graft; (3) they contain astrocytes and oligodendroglial cells, but no neurons; (4) in the ON grafts, both Schwann cells (in the distal stump of the host nerve) and glial cells (in the transplant) are available to the regenerating axons; (5) under normal circumstances ON glia are not associated with peripheral nerve axons.

Following transplantation, the fibers within the ON graft degenerated, but in contrast to the rapidity with which myelin and axon remnants were cleared from degenerating peripheral nerves, such debris was observed in ON segments for nearly 3 months after grafting. The majority of axons arising from the proximal stump of the recipient nerve bypassed the transplant and re-entered the distal stump (Figure 6). These axons were ensheathed by Schwann cells that presumably had migrated from the host nerve. The ON transplants, examined 4–11 months after grafting, contained some axons that were surrounded by astrocytic processes and occasionally by myelin of the CNS type (see Spencer, this volume). In serial cross sections, it was shown that although a considerable number of peripheral axons succeeded in penetrating the grafts (Figure 7), the longitudinal growth of most was limited to less than 1 mm, and only occasional axons reached the most distal end of the graft. Thus, ON transplants are less receptive to the re-

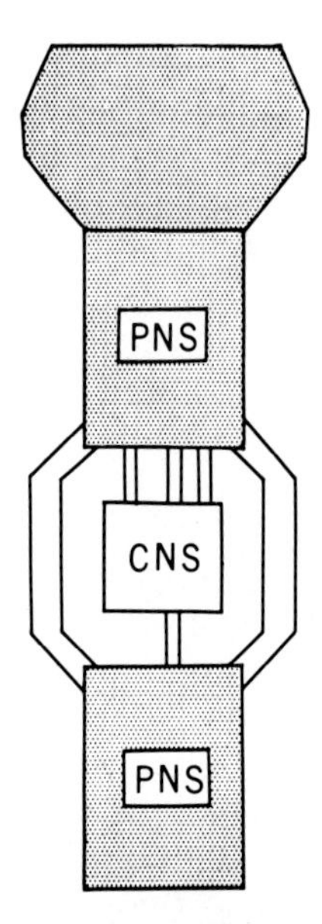

FIGURE 6. Schematic representation of changes in grafts of optic nerve segments (CNS) into sciatic nerves (PNS). After transplantation, most PNS fibers bypass the CNS graft and re-enter the distal nerve stump. Although many axons enter the glial transplant, only few succeed in growing across the graft.

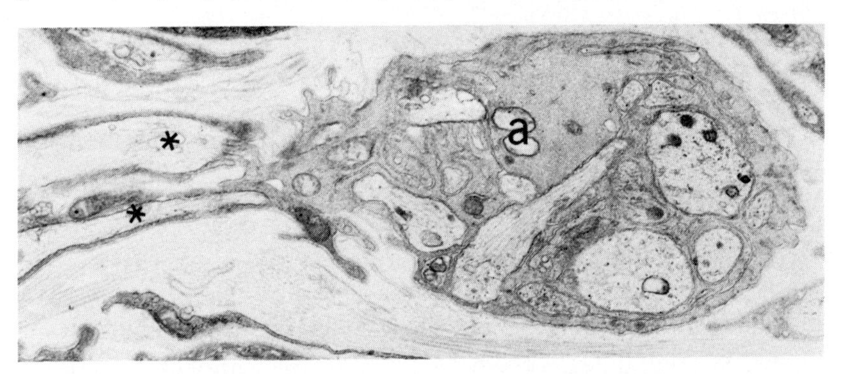

FIGURE 7. Cross section of a dome-like protuberance at the junctional zone between the host peripheral nerve and the optic nerve graft 6 months after transplantation. Astrocytic processes in the graft surround several axons (a). Two unmyelinated peripheral nerve fibers (*) are seen entering the glial transplant. Electron micrograph, ×13,000. (From Aguayo, A. J., R. Dickson, J. Trecarten, M. Attiwell, G. M. Bray, and P. Richardson [1978*b*]. Ensheathment and myelination of regenerating PNS fibers by transplanted optic nerve glia, *Neurosci. Lett.* **9**:97–104. With permission of Elsevier/North Holland Biomedical Press.)

generation of peripheral axons than are peripheral nerve grafts or the distal stumps of transected peripheral nerves. Factors responsible for diverting most axons away from the ON graft and for impairing axonal growth through the graft may include the relatively slow removal of myelin and axon debris from degenerating ON segments, but are more likely due to the formation of a "glial scar" made of dense intertwining astrocytic processes and/or the gradual disappearance of oligodendroglial cells in denervated optic nerves (Fulcrand and Privat, 1977). Although only a small proportion of the regenerating axons were able to penetrate the CNS grafts, astrocytes and oligodendroglial cells in the graft were capable of imitating their previous patterns of axonal ensheathment. Similar findings were also described in *Xenopus* optic nerve transplants (Reier, 1979) and in in vitro preparations that combine rat Schwann cell-free sensory neurites and glia from optic nerves (R. P. Bunge et al., 1978).

The limited reinnervation of the ON transplants by peripheral axons contrasts strikingly with the response in PNS grafts and resembles the ineffective regeneration that occurs after CNS injury. Thus, in addition to being relevant to the mechanisms involved in the signaling of myelin formation by oligodendrocytes (see Spencer, this volume), the ON transplants may contribute to the understanding of factors influencing CNS regeneration in vertebrates (Clemente, 1964; Guth, 1975; Sidman and Wessells, 1975). It is an advantage of the present studies that the ON grafts represent a less complex model for the study of regeneration because they are innervated by axons known to be capable of vigorous regrowth. Furthermore, the possibility that regenerating axons might be diverted by early synaptogenesis with neurons (Bernstein and Bernstein, 1971) can be excluded because there are no nerve cells in these grafts. Thus, the success or failure of axons to reinnervate the ON grafts must be largely dependent on glial responses. An impairment of regenerating PNS axons to grow into glia has also been observed after crushing dorsal roots in adult mammals. In these experiments, axons grew within the Schwann cell-ensheathed portion of the root but failed to cross the CNS-PNS junction near the spinal cord (McCouch, 1955; Carlsson and Thulin, 1967; Aguayo, Perkins, Mizuno, and Bray, unpublished observations). Thus, while glia in amphibians (Nordlander and Singer, 1978) and fish (Murray, 1976) may provide the substrate and guidance for regenerating axons, glial responses in adult mammals seem to constitute an important barrier to the regrowth of axons. The ON transplantation experiments provide additional evidence for such an impairment but also indicate that the denervated glial tissues remain responsive to the effects of axonal reinnervation.

CONCLUSIONS

Morphologic studies have been described in which experimental nerve grafts were used to investigate axon-sheath cell interactions involved in Schwann cell multiplication, differentiation, and myelin formation, as well as axon growth and regeneration in normal and abnormal peripheral nerves. In addition, it has been shown that grafts of CNS tissue into peripheral nerves may be useful in studying certain problems related to CNS regeneration. Most of the interactions discussed in this review are undoubtedly mediated biochemically, probably at cell surfaces. If the understanding of axon-sheath cell interactions is to advance, future investigations must address the molecular basis of these phenomena. Thus, it is anticipated that biochemical, histochemical, and special ultrastructural studies of experimental nerve grafts may provide useful new information about the interrelationships of axons and their sheath cells.

ACKNOWLEDGMENTS

The Medical Research Council of Canada, the Muscular Dystrophy Association of Canada, the Multiple Sclerosis Society of Canada, and the Dysautonomia Foundation of America supported the research conducted in our laboratories. The authors wish to acknowledge the assistance of Margaret Attiwell, Jane Trecarten, and Wendy Wilcox.

REFERENCES

Aguayo, A. J., M. Attiwell, J. Trecarten, S. Perkins, and G. M. Bray (1977*a*). Abnormal myelination in transplanted Trembler mouse Schwann cells, *Nature (London)* **265**:73–75.

Aguayo, A. J., G. M. Bray, and C. S. Perkins (1979*b*). Axon-Schwann cell relationships in neuropathies of mutant mice, *Ann. N.Y. Acad. Sci.* **317**: 512–533.

Aguayo, A. J., G. M. Bray, L. C. Terry, and E. Sweezey (1976*c*). Three-dimensional analysis of unmyelinated fibres in normal and pathologic autonomic nerves, *J. Neuropathol. Exp. Neurol.* **35**:136–151.

Aguayo, A. J., R. P. Bunge, I. D. Duncan, P. M. Wood, and G. M. Bray (1979*a*). Rat Schwann cells, cultured *in vitro*, can ensheath axons regenerating in mouse nerves, *Neurology* **29**:589.

Aguayo, A. J., L. Charron, and G. M. Bray (1976*b*). Potential of Schwann cells from unmyelinated nerves to produce myelin: a quantitative ultrastructural and autoradiographic study, *J. Neurocytol.* **5**:565–573.

Aguayo, A. J., R. Dickson, J. Trecarten, M. Attiwell, G. M. Bray, and P. Richardson (1978*b*). Ensheathment and myelination of regenerating PNS fibers by transplanted optic nerve glia, *Neurosci. Lett.* **9**:97–104.

Aguayo, A. J., J. Epps, L. Charron, and G. M. Bray (1976*a*). Multipotentiality of Schwann cells in cross-anastomosed and grafted myelinated and unmyelinated nerves: quantitative microscopy and radioautography, *Brain Res.* **104**:1–20.

Aguayo, A. J., J. Kasarjian, E. Skamene, P. Kongshavn, and G. M. Bray (1977*b*). Myelination of mouse axons by Schwann cells transplanted from normal and abnormal human nerves, *Nature (London)* **268**:753–755.

Aguayo, A. J., S. Perkins, I. Duncan, and G. Bray (1978*a*). Human and animal neuropathies studied in experimental nerve transplants, pp. 37–48 in *Peripheral Neuropathies*, Canal, N. and G. Pozza, eds. Elsevier/North Holland, New York.

Aguayo, A. J., J. M. Peyronnard, and G. M. Bray (1973). A quantitative ultrastructural study of regeneration from isolated proximal stumps of transected unmyelinated nerves, *J. Neuropathol. Exp. Neurol.* **32**:256–269.

Aitken, J. T. (1949). The effects of peripheral connections on the maturation of regenerating nerve fibers, *J. Anat.* **83**:32–43.

Arnason, B. G. (1975). Inflammatory polyradiculoneuropathies, pp. 1110–1148 in *Peripheral Neuropathy*, Dyck, P. J., P. K. Thomas, and E. H. Lambert, eds. W. B. Saunders, Philadelphia.

Asbury, A. K. (1967). Schwann cell proliferation in developing mouse sciatic nerve, *J. Cell Biol.* **34**:735–743.

Ayers, M. M. and R. Mc. D. Anderson (1973). Onion bulb neuropathy in the Trembler mouse: a model of hypertrophic interstitial neuropathy (Dejerine-Sottas) in man, *Acta Neuropathol.* **25**:54–70.

Ayers, M. M. and R. Mc. D. Anderson (1976). Development of onion bulb neuropathy in the Trembler mouse, *Acta Neuropathol.* **36**:137–152.

Bernstein, J. J. and M. E. Bernstein (1971). Axonal regeneration of formation of synapses proximal to the site of the lesion following transection of the spinal cord, *Exp. Neurol.* **24**:336–351.

Black, I. B., M. D. Coughlin, and P. Cochard (1979). Factors regulating neuronal differentiation, *Soc. Neurosci. Symp.* **4**:184–207.

Blakemore, W. F. (1976). Invasion of Schwann cells into the spinal cord of the rat following local injections of lysolecithin, *Neuropathol. Appl. Neurobiol.* **2**:21–39.

Bradley, W. G. and A. K. Asbury (1970). Duration of synthesis phase in neurilemma cells in mouse sciatic nerve during degeneration, *Exp. Neurol.* **26**:275–282.

Bradley, W. G. and M. Jenkison (1973). Abnormalities of peripheral nerves in murine muscular dystrophy, *J. Neurol. Sci.* **18**:227–247.

Brockes, J. P., K. L. Fields, and M. C. Raff (1977). A surface antigenic marker for rat Schwann cells, *Nature (London)* **266**:364–366.

Bunge, R. P. and M. B. Bunge (1978). Evidence that contact with connective tissue matrix is required for normal interaction between Schwann cells and nerve fibers, *J. Cell Biol.* **78**:943–950.

Bunge, R. P., M. B. Bunge, E. Okada, and P. Wood (1978). Myelination of sensory ganglion neurites by oligodendrocytes, *J. Neuropathol. Exp. Neurol.* **37**:596.

Carlsson, C. A. and C. A. Thulin (1967). Regeneration of feline dorsal roots, *Experientia* **23**:125–126.

Clemente, C. D. (1964). Regeneration in the vertebrate central nervous system, pp. 257–301 in *International Review of Neurobiology*, Vol. 6, Pfeiffer, C. C. and J. R. Smythies, eds. Academic Press, New York.

Cragg, B. G. and P. K. Thomas (1964). The conduction velocity of regenerated peripheral nerve fibres, *J. Physiol. (London)* **171**:164–175.

Dyck, P. J. (1975). Inherited degeneration and atrophy affecting peripheral motor sensory and autonomic neurons, pp. 825–867 in *Peripheral Neuropathy*, Dyck, P. J., P. K. Thomas, and E. H. Lambert, eds. W. B. Saunders, Philadelphia.

Dyck, P. J., A. C. Lais, and P. A. Low (1978). Nerve xenografts to assess cellular expression of the abnormality of myelination in inherited neuropathy and Friedreich ataxia, *Neurology* **28**:261–265.

Dyck, P. J., A. C. Lais, M. F. Sparks, K. F. Oviatt, L. A. Hexum, and P. Low (1979). Apportioning the role of Schwann cell and axon in the hypomyelination of HMSN-III using human sural nerve xenografts to nude mice, *Neurology* **29**:588.

Fulcrand, J. and A. Privat (1977). Neuroglial reactions secondary to Wallerian degeneration in the optic nerve of the postnatal rat: ultrastructural and quantitative study, *J. Comp. Neurol.* **176**:189–224.

Guth, L. (1975). History of central nervous system regeneration research, *Exp. Neurol.* **48**:3–15.

Hall, S. M. and N. A. Gregson (1975). The effects of mitomycin C on the process of remyelination in the mammalian peripheral nervous system, *Neuropathol. Appl. Neurobiol.* **1**:149–170.

Kao, C. C., L. W. Chang, and J. M. B. Bloodworth (1977). Axonal regeneration across transected mammalian spinal cords: an electron microscopic study of delayed microsurgical nerve grafting, *Exp. Neurol.* **54**:591–615.

Katz, M. J. and R. J. Lasek (1978). Eyes transplanted to tadpole tails send axons rostrally in two spinal cord tracts, *Science* **199**:202–204.

King, R. H. M., P. K. Thomas, and J. D. Pollard (1977). Axonal and dorsal root ganglion cell changes in experimental allergic neuritis, *Neuropathol. Appl. Neurobiol.* **3**:471–486.

Krnjevic, K. (1954). The connective tissue of the frog sciatic nerve, *Q. J. Exp. Physiol. Cogn. Med. Sci.* **39**:55–72.

Lasek, R. J., H. Gainer, and J. L. Barker (1977). Cell-to-cell transfer of glial proteins to the squid giant axon. The glia-neuron protein transfer hypothesis, *J. Cell Biol.* **74**:501–523.

Le Douarin, N. (1973). A biological cell labelling technique and its use in experimental embryology, *Dev. Biol.* **30**:217–222.

Lennon, V. A., J. F. Lesley, and M. Thompson (1978). CNS myelin lacks histocompatibility antigens. Relevance to potential modes of cell-mediated cytotoxicity, *Neurology* **28**:394.

Levey, R. H. and P. B. Medawar (1966). Nature and mode of action of antilymphocytic antiserum, *Proc. Natl. Acad. Sci. USA* **56**:1130–1137.

Low, P. A. (1976). Hereditary hypertrophic neuropathy in the Trembler mouse. II. Histopathological studies: electron microscopy, *J. Neurol. Sci.* **30**:343–368.

Low, P. A. (1977). The evolution of "onion bulbs" in hereditary hypertrophic neuropathy of the Trembler mouse, *Neuropathol. Appl. Neurobiol.* **3**:81–92.

Low, P., G. Marchand, F. Know, and P. J. Dyck (1977). Measurement of endoneurial fluid pressure with polyethylene matrix capsules, *Brain Res.* **122**:373–377.

Low, P. A. and J. G. McLeod (1975). Hereditary demyelinating neuropathy in the Trembler mouse, *J. Neurol. Sci.* **26**:565–574.

Lubinska, L. (1958). Short internodes "intercalated" in nerve fibers, *Acta Biol. Exp.* **18**:117–136.

Madrid, R. C. and H. M. Wisniewski (1977). Axonal degeneration in demyelinating disorders, *J. Neurocytol.* **6**:103–117.

McCarthy, K. D. and L. M. Partlow (1976). Neuronal stimulation of [^{3}H] thymidine incorporation by primary cultures of highly purified non-neuronal cells, *Brain Res.* **114**:415–426.

McCouch, G. P. (1955). Comments on regeneration of functional connections, pp. 171–180 in *Regeneration in the Central Nervous System*, Windee, W. F., ed. Charles C. Thomas, Springfield, Ill.

Murray, M. (1976). Regeneration of retinal axons into the goldfish optic tectum, *J. Comp. Neurol.* **168**:175–196.

Nordlander, R. H. and M. Singer (1978). The role of ependyma in regeneration of the spinal cord in the urodele amphibian tail, *J. Comp. Neurol.* **180**:349–373.

Olsson, Y. and T. S. Reese (1971). Permeability of vasa nervorum and perineurium in mouse sciatic nerve studied by fluorescence and electron microscopy, *J. Neuropathol. Exp. Neurol.* **30**:105–119.

Perkins, S., A. J. Aguayo, and G. M. Bray (1978*a*). Increased multiplication rates of Schwann cells transplanted from Charcot-Marie-Tooth (CMT) neuropathy, *Clin. Res.* **26**:873A.

Perkins, S., G. M. Bray, and A. J. Aguayo (1978*b*). Persistence of abnormal proliferation in Schwann cells from Trembler mice transplanted into normal nerves—a radioautographic study, *Neurology* **28**:381.

Perkins, S., K. Mizuno, A. J. Aguayo, and G. M. Bray (1977). Increased Schwann cell proliferation in Trembler mouse nerves—quantitative microscopy and radioautography, *Neurology* **27**:377.

Pollard, J. D. (1978). Nerve allografts in Trembler mice, *4th International Congress on Neuromuscular Diseases*, abstr. no. 8.

Prineas, J. W. and J. G. McLeod (1976). Chronic relapsing polyneuritis, *J. Neurol. Sci.* **27**:427–458.

Raine, C. S., H. Wisniewski, P. C. Dowling, and S. D. Cook. (1971). An ultrastructural study of experimental demyelination and remyelination. IV. Recurrent episodes and peripheral nervous system plaque formation in experimental allergic encephalomyelitis, *Lab. Invest.* **25**:28–34.

Raine, C. S., H. Wisniewski, and J. Prineas (1969). An ultrastructural study of experimental demyelination and remyelination. II. Chronic experimental allergic encephalomyelitis in the peripheral nervous system, *Lab. Invest.* **21**:316–327.

Reier, P. J. (1979). Penetration of grafted astrocytic scars by regenerating optic nerve axons in *Xenopus* tadpoles, *Brain Res.* **164:**61–68.

Richardson, P. M., U. M. McGuinness, and A. J. Aguayo (1979). Regeneration following sciatic nerve grafting to the rat spinal cord, *Can. J. Neurol. Sci.*, in press.

Romine, J. S., G. M. Bray, and A. J. Aguayo (1976). Multiplication of unmyelinated Schwann cells after crush injury, *Arch. Neurol.* **33:**49–54.

Rosenstein, J. M. and M. W. Brightman (1978). Intact cerebral ventricle as a site for tissue transplantation, *Nature (London)* **276:**83–85.

Samorajski, T., R. L. Friede, and P. R. Reimer (1970). Hypomyelination in the quaking mouse. A model for the analysis of disturbed myelin formation, *J. Neuropathol. Exp. Neurol.* **29:**507–523.

Schroder, J. M., J. Bohl, and K. Brodda (1978). Changes of the ratio between myelin thickness and axon diameter in the human developing sural nerve, *Acta Neuropathol.* **43:**169–178.

Sidman, R. L., M. M. Dickie, and S. H. Appel (1964). Mutant mice (quaking and jimpy) with deficient myelination in the central nervous system, *Science* **144:**309–311.

Sidman, R. L. and N. J. Wessells (1975). Control of direction of growth during elongation of neurites, *Exp. Neurol.* **48:**237–251.

Smith, B. H. and G. W. Kreutzberg, eds. (1976). Neuron-target cell interactions, *Neurosci. Res. Program Bull.* **14.**

Spencer, P. S. (1971). *Light and electronmicroscopic observations on localized peripheral nerve injuries.* Ph.D. thesis, University of London.

Spencer, P. S. (1979). Neuronal regulation of myelinating cell function, *Soc. Neurosci. Symp.* **4:**275–321.

Steinmuller, D. (1970). Cross-species transplantation in embryonic and neonatal animals, *Transplant. Proc.* **2:**438–446.

Stenevi, U. and A. Bjorklund (1978). Transplantation techniques for study of regeneration in the central nervous system, pp. 101–112 in *Maturation of the Nervous System*, Corner, M. A., R. E. Baker, N. E. Vandepoll, D. F. Swanband, and H. B. M. Uylings, eds. Elsevier Scientific Publishing Co., New York.

Terry, L. C., G. M. Bray, and A. J. Aguayo (1974). Schwann cell multiplication in developing rat unmyelinated nerves—a radioautographic study, *Brain Res.* **69:**144–148.

Varon, S. S. and R. P. Bunge (1978). Trophic mechanisms in the peripheral nervous system, *Annu. Rev. Neurosci.* **1:**327–362.

Webster, H. deF. (1971). The geometry of peripheral myelin sheaths during their formation and growth in rat sciatic nerves, *J. Cell Biol.* **48:**348–367.

Weinberg, E. L. and P. S. Spencer (1979). Studies on the control of myelinogenesis. III. Signalling of oligodendrocyte myelination by regenerating peripheral axons, *Brain Res.* **162:**273–279.

Weinberg, H. J. and P. S. Spencer (1975). Studies on the control of myelinogenesis. I. Myelination of regenerating axons after entry into a foreign unmyelinated nerve, *J. Neurocytol.* **4:**395–418.

Weinberg, H. J. and P. S. Spencer (1976). Studies on the control of myelinogenesis. II. Evidence for neuronal regulation of myelin production, *Brain Res.* **113**:363–378.

Wood, P. M. (1976). Separation of functional Schwann cells and neurons from normal peripheral nerve tissue, *Brain Res.* **115**:361–375.

Wood, P. M. and R. P. Bunge (1975). Evidence that sensory axons are mitogenic for Schwann cells, *Nature (London)* **256**:662–664.

Wortis, H. H. (1971). Immunological responses of "nude" mice, *Clin. Exp. Immunol.* **8**:305–317.

KEY WORD INDEX

NOTES

NOTES

NOTES

NOTES